BRAIN TALK

Also by David Schnarch

Passionate Marriage
Keeping Love and Intimacy Alive
in Emotionally Committed Relationships

Constructing the Sexual Crucible
An Integration of Sexual and Marital Therapy

Resurrecting Sex
Resolving Sexual Problems &
Rejuvenating Your Relationship

Intimacy & Desire
Awaken the Passion In Your Relationship

BRAIN TALK

HOW MIND MAPPING BRAIN SCIENCE CAN CHANGE YOUR LIFE & EVERYONE IN IT

DR. DAVID SCHNARCH

STERLING PUBLISHERS

Copyright © 2018 by David Schnarch, PhD. All rights reserved. Printed in the United States of America. No part of this book may be used or reproduced in any manner whatsoever without written permission except in the case of brief quotations embodied in critical articles and reviews. For more information, contact Sterling Publishers, 2922 Evergreen Parkway, Suite 310, Evergreen, Colorado, 80439, USA.

Limited liability/disclaimer of warranty: While the author and publisher used their best efforts in preparing this book, they make no representations or warranties with respect to the accuracy or completeness of the contents of this book, and specifically disclaim any implied warranties of merchantability or fitness for a particular purpose. Neither the author nor publisher shall be liable for any loss of profit or any other commercial damages, including but not limited to special, incidental, consequential, or other damages.

The content contained herein is provided for informational purposes regarding the subject matter covered and is not intended as a substitute for advice or treatment that may or should be prescribed by a physician or therapist. The methods and strategies contained herein may not be suitable for your situation. Before adhering to any information or recommendations given here you should consult your physician or therapist. This book is sold with the understanding that the author and publisher are not rendering therapy services. If medical, psychological, legal, or other expert assistance is required, the services of a competent local professional person should be sought.

Case examples are composites of cases from clinical practice.

Tables 1 and 2 and Figures 5 and 6 are reprinted with permission from "Neuroanatomical and neurochemical bases of theory of mind" by Ahmad Abu-Akel and Simone Shamay-Tsoory, 2011, *Neuropsychologia*, 49(11), 2971–2984, by Elsevier.

Table 3 is reprinted with permission from "Two systems for empathy: A double dissociation between emotional and cognitive empathy in inferior frontal gyrus versus ventromedial prefrontal lesions" by Simone Shamay-Tsoory, Judith Aharon-Peretz, and D. Perry, 2009, *Brain*, 132(3), 617–627, by Oxford University Press.

"Crucible®," "Sexual Crucible®," "Sexual Crucible Approach™," "Crucible Approach™," "Passionate Marriage®," "Passionate Couples®," "Four Points of Balance™," "Solid Flexible Self™," "Quiet Mind–Calm Heart™," "Grounded Responding™," and "Meaningful Endurance™," are trademarks owned and pending by David Schnarch, PhD. Programs, trainings, services, and materials using these trademarks can only be provided by the Crucible Institute of Evergreen, Colorado and its authorized designates. Website: www.Crucible4Points.com

Zephyr™ BioModule™ is a trademark of Medtronic Corporation.

FIRST EDITION

Cover design by Vikiana
Printed by CreateSpace

ISBN: 9781548371531
Sterling Publishers
Publisher number: 1492934

In memory of

Stanley and Rose Schnarch

beloved parents
who passed on after 70 years together at age 92.
I had no idea how good I had it growing up,
until I learned about my clients' parents.

In honor of

His Holiness Pope Francis I
(Jorge Mario Bergoglio)

a truly good man, Father to the world,
who recognizes the worst in us, but believes in the best in us.

If there is rightness in the heart,
there will be beauty in the character.

If there is beauty in the character,
there will be harmony in the home.

If there is harmony in the home,
there will be order in the nation.

If there is order in the nation,
there will be peace on earth.

Lao Tzu
5th century BCE

Brain Talk

Table of Contents

Acknowledgments xiii
Introduction 1
Chapter 1. What Is Mind Mapping? 11
    The Driving Wheel of Relationships 12
    We're Native-Born Psychologists 13
    If You Want Great Sex… 14
    Everywhere, Everyone, All the Time................................ 17
Chapter 2. The Brain Science Behind Mind Mapping 19
    Theory of Mind 19
    Mind Mapping Is a Survival Skill 21
    The Nuts and Bolts of Mind Mapping 24
    Mind Mapping Shapes Your Life Story 28
    Mind Mapping Impacts Your Brain 31
Chapter 3. Mind Mapping in Children 33
    Implicit Mind Mapping.. 33
    Explicit Mind Mapping.. 34
    Enjoy Your Brief Moment of Omniscience 36
    Four-Year-Olds: Budding Lie Detectors......................... 37
    Little Lie, Big Moment 38
    End of Your Privacy 39
    See the World Through Your Children's Eyes 41
Chapter 4. Does Everyone Have Mind-Mapping Ability? ..45
    Dogs Have Mind-Mapping Ability 45
        If children and dogs have mind-mapping ability,
        does everyone?... 48
    Parents Impact on Children's Mind-Mapping Abilities.......50
    What if Your Mother Wasn't Mind-Minded? 52
    What if You Come From a Troubled Home? 57

Where Do Schizophrenia, Autism, and
Asperger's Syndrome Fit In? 60
    Mind mapping in children with autism 63
Wait! My Husband Has Mind-Mapping Ability?! 64
Chapter 5. Mind Mapping in Adult Life 67
    Mind Mapping in Marriage 67
    Mind Mapping and Sexual Desire 70
    Mind Mapping and Sexual Dysfunctions 73
    Mind Mapping in Affairs 75
    Singles and Mind Mapping 77
    There's a Whole Lot More Going on Than You Thought ... 78
Chapter 6. Mind Masking: Defeating Mind Mapping 81
    Second Level of Mind-Mapping Ability: Mind Masking 81
        Why do people mask their minds? 82
    People From Bad Homes Develop Terrific
    Mind-Masking Ability 85
    What Does It Take To Mask Your Mind? 86
        Mind masking in adult love relationships 87
    Third Level of Mind-Mapping Ability: Implanting
    False Beliefs 88
        Good therapists need to be good liars 89
    Fourth Level of Mind-Mapping Ability: Mind Twisting 90
Chapter 7. Do You Know Your Own Mind? 93
    Our Selves Are Connected 94
        Sources of self-knowledge 95
        Reflected sense of self 95
    Perceived Specialness of Introspection 97
    Fallibility of Introspection 98
        Distortions in mapping your own mind 100
        Ways of fooling yourself 100
    Other People Know You Better Than You Know Yourself .. 102
    You Can Be Right About Your Partner but
    Wrong About Yourself 103
        Did you marry a complete lunatic? 104
Chapter 8. Traumatic Mind Mapping 109
    Antisocial Applications of Mind Mapping 110

Traumatic Mind Mapping.. 112
Traumatic Mind Mapping Impairs Your Response to
Stress...115
    The closer the relationship, the bigger the impact............ 121
Traumatic Mind Mapping in Troubled Homes..................... 122
    Sexual abuse most often occurs through traumatic
    mind mapping..124
PTSD from Traumatic Mind Mapping.....................................127
    Mind mapping distortions from traumatic mind mapping 129
    Example of holes in mind-mapping radar......................... 131
Anticipatory Traumatic Mind Mapping................................. 132

Chapter 9. Antisocial Empathy...135
Wrong Notions About Empathy.. 136
What Is Antisocial Empathy... 138
Schadenfreude: Harm-Joy 140
    Examples of antisocial empathy.. 141
    Mirror neurons and embodied knowledge......................... 144
How You Can Tell Someone Has Antisocial Empathy...... 146
Studying Who Fails in Therapy 149
Disgusting Parenting... 154
    Therapists, teachers, and doctors overlook
    disgusting parenting 156

Chapter 10. Impacts of Traumatic Mind Mapping.........159
Short-Term Impacts of Traumatic Mind Mapping............. 160
    Cognitive impairments... 160
    Mind mapping shuts down 161
    Repeatedly triggered primary emotions 163
    Impaired emotional functioning.. 164
    Mind mapping fails to collapse... 166
Long-Term Impacts of Repeated Traumatic Mind Mapping.167
    Steady state regressions.. 169
    Impaired disgust reaction 170
    Cruel mental "voice"... 174
    Hard-wired thought patterns 175
    Holes in your mind-mapping radar...................................... 177
    Autobiographical memory gaps and distortions 179

Antisocial empathy 182
Being taken hostage 183
The Benefits of Seeing Dark Things 185
Recovering From "Trump Trauma" 186
Crucible® Neurobiological Therapy 189
Chapter 11. Detecting Mind Mapping in Others 193
Identify Other People's Mind-Mapping Abilities 194
Detecting Mind-Masking Ability 197
    Discover deception and lying 201
    Improving deception detection 203
Spotting Someone Playing Three Moves Ahead 204
    Rule out mind blindness 206
    Hearing is a large part of mind mapping 208
Mind Mapping Isn't Perfect 209
    When mind-mapping ability is revealed 212
Chapter 12. Reversing Traumatic Mind Mapping 215
Visualization and Targeted Mind Mapping 218
    See the setting 220
    Watch the movie unfold 222
    Use a first-person view 226
    Eye contact isn't always necessary 228
Solutions for Resistant Problems 231
    Shift to a third-person view 232
Bang Your Head Against the Wall Until the Wall Moves ... 235
Use Your Left Brain To Corner Your Right Brain 237
Chapter 13. Repairing Your Autobiographical Memory ... 243
Keep People's Mind-Mapping Abilities in Mind 244
    Autobiographical memory gaps make this difficult 245
Detecting Holes in Your Mind-Mapping Radar 249
Repair Autobiographical Gaps Through Revisualization .. 251
    Mealtime memories are rich resources 255
Cross-Reference Memories of Traumatic Events 258
    Analyze correspondence 260
    Analyze audio and video recording 264
Have Mental Dialogs With Your Antagonist 266

Chapter 14. Dealing with Destructive People................273
   Dealing with People Who Do Disgusting Things 273
      Background 274
      Parents messing with couple's decision to have a child ... 276
      Disclosure of dead brother 277
      The wedding 278
      The bachelor party 278
      Mother 279
      Father 280
   Situationally Accessible Memory (SAM) 283
   Handling People Who Make Moves on You 284
      Written mental dialogs 285
      Decipher your antagonists' moves 286
      Concrete steps for dealing with your antagonists 288
      Show your antagonists you can see them 290
   Hold On To Yourself (Differentiation) 293
      Dialogue with father 294
   Make "Gold-standard" Responses 295
      Don't let your antagonist get around you 297
      Another comment brings things to a head 301
   Email Interactions 304
   Face-to-Face Meetings 308
      Keep up with your antagonist in real time 308
      Look for "news of a difference" 311
      Did Mother and Father know what they were doing? 312
   Impacts Ripple Through Dysfunctional Families 314
      "False memories" 315
   Post-Traumatic Growth 316
      The moral of the story 317

Chapter 15. Create Positive Moments of Meeting 319
   Positive Moments in Love Relationships 320
      Give "the gift of mind" to those you love 321
      Stop masking your mind 322
      Resolving extramarital affairs 323
      Deeper intimacy in and out of bed 324
      Hugging 'till Relaxed 324

Heads on Pillows 326
Eyes-Open Sex and Orgasms 327
Positive Interactions with Your Children 327
    Give your children permission to see you 327
    Freedom to unmask their minds 328
    Enter young children's mental worlds 329
    Brain-oriented sex education 331
    Three-step repair strategy 332
Positive Moments with Friends 335
Creating a Healthier Workplace 337
Reconnecting with Siblings 339
Cherishing Aging Parents 341
Let's Do This! 344

Appendix A: Mind Mapping 349
Parts of the Brain Involved in Mind Mapping 350
How the Brain Tracks Whose Mind You're Mapping 351
    Ventral and dorsal attentional systems 352
    Differential routing through the brain 352
    Mapping emotions and feelings 353
Mapping Out Thoughts Versus Feelings 353
Neurochemistry of Mind Mapping 354
    Mapping thoughts and knowledge 354
Disorders Impacting Mind-Mapping Ability 357
Psychopaths' Mind-Mapping Ability 358

Appendix B: Traumatic Mind Mapping 361
How Traumatic Mind Mapping Fits with PTSD 364
DESNOS Symptoms 365
Traumatic Mind Mapping and PTSD Diagnostic Criteria .. 367
Beyond DSM and DESNOS 368
Varieties of Interpersonal Neurobiological Problems 370
    Mind-mapping impairments 370
    Non-psychotic thinking disorders 370
    Non-psychotic emotional regulation disorders 371
    Disgust reaction impairments 372
    Additional impairments 372
Impacts of Traumatic Mind Mapping on the Body 375

Trauma and the central nervous system 376
Trauma and the autonomic nervous system 379
Appendix C: Antisocial Empathy 383
How the Brain Creates Empathy 384
Neuroanatomical Basis of Antisocial Empathy 387
Schadenfreude! 388
Cognitive and Emotional Empathy:
A Brain-Based Perspective 392
Machiavellianism, Narcissism, Sociopathy, and Sadism. 395
"Gaslighting" 401
Incredible Things We Can Learn From Disgust 404
    Anterior insula and disgust 405
    Spindle neurons and antisocial empathy 408
Appendix D: Creating Neuroplasticity 413
Typical PTSD Treatment 414
Considerations for More Effective Therapy 416
    Insight is not enough 417
    Focus on inner experience 418
    Synchronize implicit and explicit memory 418
    Exploit the IFO (anterior insula) 422
Methods of Crucible® Neurobiological Therapy 425
    Visualization and revisualization 425
    Written dialogs 427
    Conjoint neuroplastic activities 430
    Modify current relationship with antagonist 435
    Increase clients' differentiation 436
    Role of the therapist in CNT 436
    In-session psychophysiological monitoring 440
    Heart-rate variability biofeedback 441
    High arousal and "safe emergencies" 443
    Window of tolerance 444
    Post-traumatic growth 446
About the Author 542

LIST OF TABLES

Table 1. Brain Regions Involved in Mind Mapping....350

Table 2. Disorders Impacting Mind-Mapping Ability..356

Table 3. Differences Between Emotional and Cognitive Empathy..............394

Table 4. Aspects of Visualization Activities in Crucible® Neurobiological Therapy..............428

Table 5. Aspects of Written Dialogs in Crucible® Neurobiological Therapy..............431

Table 6. Aspects of Hugging 'till Relaxed, Heads on Pillows and Eyes-Open Sex..............434

LIST OF FIGURES

Figure 1. Display of in-session psychophysiology readings of clients and therapist..............116

Figure 2. In-session psychophysiological monitoring during family therapy..............118

Figure 3. Zephyr™ BioModule™ system used by the author..............118

Figure 4. Heart-rate variability readings during psychotherapy session..............119

Figure 5. Brain systems differentiating mental maps of self and other people..............351

Figure 6. Serotonin–dopamine (DS) system in cognitive and affective mind mapping..............355

Figure 7. Right anterior and posterior insula..............384

Acknowledgments

Looking back on my life, I'm grateful for what I've observed and learned about people and relationships. I consider these my riches, my rewards from a life well spent in service to others. Every time I discover something new I feel like The Great Oneness has smiled on me. This frequently involves discovering subtle brain difficulties I've never seen before and, eventually, being able to solve them.

My epiphanies usually coincide with the most difficult times in my clients' lives. Imagine if your new therapist tells you he thinks you have a brain problem. He thinks you have gaps in your autobiographical memory (life story), and you don't accurately understand your own history. He proposes you're blind to important people in your life. He also says he's never seen a problem quite like yours. He's hopeful he can help you–but no guarantees. This is not the way anyone wants to embark on therapy.

Despite (or because of) this, my clients let me study and learn from them. I encouraged my "epiphany" clients to get better because their improvement proved such problems could be treated. Their successes outlined treatment paths for others to follow. Over time, my insights evolved into treatment protocols.

Clients who came later had a somewhat easier time of it. At least I could say I'd seen other people with similar problems, and some basic methods helped these clients get remarkably better. Still no guarantees of success, but this put my later clients in a different position. They had a scientific contribution to make as well: I knew for a fact their problem could be treated. They could prove how quickly it could be done.

None of my clients had an easy time in therapy. Treatment involved difficult epiphanies about themselves and the people they love. When dealing with destructive parents who threatened to disown them, my clients manifested the awesome power of the human spirit. They demonstrated how sheer stubbornness can change one's brain for the better when it's driven by the best in a person. Their successes evolved into Crucible Neurobiological Therapy. This underscores my indebtedness to those who have sought my help. A mere "thank you" does not suffice.

I also want to acknowledge some people I've never met: hard-core scientists and researchers who devote their lives to revealing the secrets of the human brain. Their realizations, derived from long careers of painstaking research, appear as brain science in this book. Their protracted efforts make it possible for clinicians like me to create new brain-based psychotherapies.

This book may read somewhat differently compared to my prior books because the writing process differs. I language things differently when I'm teaching versus when I'm writing on a keyboard. In this case, I spent a number of days teaching Storms Reback, an accomplished writer himself, the contents of this book. He developed the initial draft from transcripts of our sessions, aided by his understanding of the material. From this, I wrote a second draft, which Storms edited. I then wrote the significantly expanded final draft. Together we shaved years off my normal book-writing process. I deeply appreciate his help. Ultimately, however, I am responsible for everything you read here.

Linda Kirkpatrick, former editor and publisher of our local "good news" newspaper, polished the final manuscript. Her urging led me to completely revise Chapter 14, which now has greater depth and detail. Linda has been my friend for over 20 years and my neighbor for much of it because Linda suggested we buy the wonderful house next door. Editing this book is one more way my life is better because of Linda.

I am also grateful to the people in my life who made writing this book possible. First and foremost, I give thanks to my incredibly patient wife, Dr. Ruth Morehouse, who makes my life richer, deeper, and vastly more meaningful. Besides being a wonderful wife and mother, Ruth is a terrific therapist in her own right. Thankfully, the meanness I describe in this book is absent from our marriage. Living with Ruth makes me want to be a better person.

My brother, Steve, enriches my life through his good heart. He supports our professional work as our operations manager and handles our monthly webinar. His easy-going nature wins people over. He's a really good person and loyal to the core. He has our parents' impact (and genetics) written all over him.

When an author writes about dark aspects of relationships, there's a tendency to wonder *Boy, what kind of terrible childhood did he have?!* So I'll tell you: Our parents, Stan and Rose Schnarch, gave us more than we could possibly appreciate as children. They treated each other (and us) with kindness. They gave us parents we could respect, which is the greatest gift of a parent can give. I learned virtually everything in this book from working with my clients, rather than from personal experience. Thank God.

Our daughter, Sarah, has become the sweet, good-natured, claw-hammer-banjo-picking person she is because she grew up with Ruth, Steve, and my parents. Her friends nicknamed her "Sunny," which says it all. Besides being into biodynamic farming and medicinal herbs, Sarah diligently lets me know where I'm less than perfect.

I am fortunate to have people who really love me, and I couldn't ask for a better wife, daughter, or brother. I just wish they could say the same about me, especially when I'm engrossed in writing and lost to the world.

Along the way, this manuscript benefited from Ruth's valuable comments and suggestions, as well as from those of Dr. John Thoburn, Grace Whitman, and Barbara Fairfield.

Finally, thanks to my book agents, Michael Wright, and Leslie Garson (Garson and Wright Public Relations) for their professional expertise and friendship. Aside from being consummate professionals, I admire what good parents they are.

Introduction

People who seek therapy frequently want a magic bullet that wipes away all their ills after just one visit. Psychotherapy doesn't work that way, of course. But I've discovered some things that can vastly improve your functioning with a single exposure. Over the past 12 years I've developed Crucible® Neurobiological Therapy, which I've observed to be truly life-changing and transformative. I don't think it's hyperbole to say this book has the power to greatly improve your life.

This revolutionary therapy emerged from troubled origins. During my four-decade-long career helping couples with sex, intimacy, and relationship problems, many went on to have happier marriages, healthier families, and more functional lives. Of particular interest to me, however, were those who didn't thrive. What you'll read here comes from studying my clients who did the worst.

It's not like I was dealing with large sample sizes. From time to time I'd have a couple that didn't do well. Sometimes I would even have two. But on one occasion I got incredibly lucky. I had *three* couples that weren't making progress! I say "lucky" because, as many scientists attest, groundbreaking discoveries often emerge from disappointing failures.

But to tell the truth, I wasn't feeling very scientific or fortunate at the time. Having three fail-to-thrive couples at once bothered me. I was frustrated. However, this also allowed me to do something having just one failing couple didn't. *What do they have in common?* I asked myself. *Do couples who don't do well in therapy share a common factor? How can I help these people?*

Around the same time I started studying my treatment failures, the nascent field of interpersonal neurobiology emerged. Modern brain science was in full swing, especially Theory of Mind, the study of humankind's ability to recognize other people's mental states as well as our own. I found this language too abstract and unwieldy, so I devised my own take on the subject: *Mind mapping*, the brain's ability to make a mental map of another person's mind, was born. You'll get a nuanced and elaborate explanation of mind mapping in the first part of this book. Pretty quickly I think you'll see why mind mapping completely captivated me.

To make a long story short, it turned out that the missing link among my fail-to-thrive couples had everything to do with mind mapping. (I'll tell you what I discovered in Part Two.) Applying what I learned about mind mapping proved to be a game changer. Mind mapping exponentially advanced my prior decades of work. This helped me identify root problems facing my fail-to-thrive couples and allowed me to develop new treatments to help them. This eventually evolved into Crucible Neurobiological Therapy. (We'll cover this in Part Three.)

By looking at their lives through the lens of mind mapping, my clients began to see themselves, their partners, and their relationships in entirely new ways. Long-avoided difficult truths not only came to the surface, but my hard-core couples were also more willing and able to deal with them. Even highly troubled couples from very difficult backgrounds revitalized their marriages and developed better relationships, healthier families, and more productive lives.

Two things really caught my attention: The first was how learning about mind mapping really helped couples *thrive*. Partners weren't just happier with themselves and each other. They reported being better able to handle *all* the important relationships in their lives. Many said their typical thoughts and feelings–their general state of mind–changed for the better. I also noticed definite changes in their appearance and demeanor. They looked better; their eyes were more alive, and their faces were more appealing. Some women reported girlfriends asking if they'd had Botox treatments!

The other astonishing thing was how *quickly* this happened. For the last three decades, my practice has consisted of people from around the world who fly to Evergreen, Colorado for four days of three-hour intensive therapy sessions. My no-nonsense, direct-talking therapy is well known for being fast-paced. So trust me, I'm used to fast and intense. But applying mind mapping to Crucible Therapy was like igniting a booster rocket. It's been an integral component of my therapy ever since.

Although I've been developing my methodology over the last 12 years, I haven't shared it with the general public until now. Only recently have I started teaching this to other therapists. I'm a natural skeptic, especially of my own work. It's taken me this long to study mind mapping's effects on the brain, package my methods into a teachable protocol, and convince myself of its effectiveness.

After I gave a lecture in Germany in 2015, I knew it was time to share this with a broader audience. It was scheduled for two hours (I was working with a translator), but it actually lasted three. There were 700 people in attendance; and, although I talked an hour longer than I was supposed to, virtually no one got up and walked out. You would think after two hours of fairly heavy material most would leave, but everyone stayed. People were riveted.

Afterwards I was blown away by the audience's response. One person told me, "My life has opened up. Mind mapping creates an entirely new reality for me. It makes me view my past and present completely differently, and I see a whole new future for me." That's the sort of impact mind mapping can have on you. It radically changes your view of your life and everyone in it.

The same thing happens when therapists learn about mind mapping. I discovered this when I conducted the initial professional training workshops in Europe and America. The jaws of therapists dropped in shock. Because many had spent lots of time and money in their own personal therapies, they were stunned by the new personal insights and revised life histories they developed during the workshops. They also understood their own patients so very differentlyt hat their prior therapeutic methods looked naive. This showed me how one exposure to this information can have tremendous impact.

I think one reason for this powerful effect is that it enters your brain so quickly and easily. Your brain just eats up this information. Then it starts generating new understandings of the important people and past and current experiences in your life. This happens so fast, it's almost like mind mapping speaks the brain's language. This prompted me to coin the phrase "brain talk." It's the best way to describe mind mapping's incredible accessibility and potency.

In *Brain Talk*, you'll read a lot about the brain. As you learn about mind mapping, your brain gains knowledge of itself. For instance, there can be undetected gaps in your mind-mapping radar. Did you know your mind mapping shuts down under certain conditions? Once your brain knows this, it starts to take this into account. Without this knowledge, you probably underestimate other people's mind-mapping ability.

Did you know that you can also be blind to what's going on in your own mind at the very moment you're insisting you're the master of it? Once you learn about mind mapping, your brain reorganizes the autobiographical story of your life, and the meaning of events that populate it. Your picture of who you are and how you got this way changes, as does your view of all the people in your life.

This transformation brings forth moments of profound insight. Sometimes it's an "aha!" moment. Other times it's more like "Oh no!" Pleasing or not, most people find this new awareness helps them function better in daily life. (Parts Two and Three show you how to trigger this process.) That's why learning about mind mapping is so rewarding. It applies to every aspect of your life: sex, intimacy, money, fractious children, elderly parents, touchy in-laws, difficult bosses, petulant co-workers, and on and on.

Pretty quickly, you'll see mind mapping everywhere you look. Trips to the supermarket or department store will never be the same. As you walk down the aisles, other people's interactions turn into dramatic vignettes. You recognize a husband trying to tell his wife she doesn't look good in a dress she's tried on without explicitly saying so. You watch a teenage daughter trying to manipulate her mother into buying her a provocative outfit by implanting the false belief all her girlfriends dress this way. Elsewhere a mother tells her screaming three-year-old she's going to leave him in the store if he doesn't take her hand and walk out with her. You realize the cashier at the checkout counter is being overly solicitous to the attractive shopper standing in front of you. Then you go home and use mind mapping to convince your husband you really need something you bought! All the dramas surrounding you are suddenly illuminated.

Mind mapping isn't some abstract theory. It's a tangible process that's clearly observable. It's been analyzed with sophisticated brain scanners in hundreds of research studies. Mind mapping occurs constantly in interpersonal relations. It never stops. You can't help but be profoundly influenced by it. Mind mapping lies at the core of the best interpersonal experiences you can have (including sex). Unfortunately, however, it's also part of the worst ones (sex too).

For better and worse, humans are natural-born mind mappers. It's an important innate ability that can be further developed and has many practical applications. Learning about the negative brain impacts of *traumatic* mind mapping takes away your complacency about dealing with highly troubled relationships. It lights a fire under you. It makes you get off your butt and do what's necessary to improve your life and the lives of everyone around you.

A large part of Crucible Neurobiological Therapy can be self-administered. You don't need a therapist to guide you through it. (I recommend discussing this with your therapist if you have one. Moreover, this book is not a substitute for therapy you may be needing.) However, therapists are unavailable in many places in the world. *Brain Talk* is my response to this obstacle. The pragmatic step-by-step things I'll describe can help people who might otherwise not have access to a therapist.

In order to give you something that might really help you, what I'll describe may be troubling at times. This can also trigger disturbing memories. Part Two focuses on *traumatic mind mapping*, the damage done when mapping out someone's mind negatively impacts your brain and mind. There I'll describe troubling things that often occur in marriage and parenting, including dark secrets the general public refuses to acknowledge. This no-nonsense view of human nature might disturb folks who'd rather see the world through rose-colored glasses.

When I describe unfortunate life experiences in Part Two, take comfort that Part Three provides effective solutions for problems caused by these experiences. You'll find detailed information about how to resolve traumatic mind mapping. I'll also teach you how to handle difficult people who are capable of frying your brain. To conclude, I'll show how to use mind mapping to create really lovely experiences—like increasing intimacy in your marriage, enhancing relationships with your children and aging parents, improving the atmosphere at your workplace, and deepening your friendships.

Doing therapy for over 30 years has made me a realistic optimist. Make no mistake, *Brain Talk* appeals to the best of human nature, rather than the worst. I've certainly seen the worst in people, yet I've always been more impressed by just how good the best in people can be. I constantly work with people from truly deplorable backgrounds, harmed by the most severe traumatic mind mapping. But I haven't developed a negative worldview. That's because I've learned how to help these folks turn their lives around. Most times I get to watch people from highly dysfunctional families dramatically improve their lives. I hope reading *Brain Talk* infuses you with a similar positive attitude.

Suggestions for reading. Based on my experiences doing therapy, training therapists, and lecturing about mind mapping, here are some suggestions on how best to read this book. Take these suggestions to heart and apply them from start to finish. I can't stress enough how important they are to getting what this book has to offer.

Here's the first thing: Read *Brain Talk* differently than any of my prior books. *Resist* your urge to underline text or make notes. (Boy, is this a change! I previously suggested underlining.) *Don't* read this as a collection of ideas. *Don't* try to memorize passages of text. *Don't* try to convert this to familiar terminology. *Forgo* your impulse to use deductive logic. *Stop* yourself from developing mental counterarguments.

All of these use your left brain, which is where your language ability and analytical thinking comes from. As much as I like using my left brain to figure things out, the left brain frequently interferes with learning about mind mapping and resolving traumatic mind mapping, which are more right-brain processes.

Your right brain operates through pictures. Pictures are its "language." It constructs vivid dynamic images that simultaneously contain critical details and holistic meaning. Reading *Brain Talk* is designed to be a right-brain activity. (Granted, reading is inherently rooted in text and language.)

You'll get the most out of *Brain Talk* if you repeatedly develop pictures in your mind, envisioning what you're reading. You need to *watch* the vignettes as they develop. Map the minds of the people you meet in each chapter by *seeing* them. You'll learn and remember more than if you tried to memorize what you're reading.

Revisualizing personal memories through the lens of mind mapping often provides rapid shifts in perspective and meaning. Use this to envision noteworthy past and present events in your life. Study yourself and other people as they appear in the scenes. *Look* at these pictures, rather than intellectualizing or theorizing about them. You'll learn more by *watching* than by applying familiar concepts and theories. As intuitive as mind mapping is, this will take some effort on your part and willingness to see things you might otherwise want to avoid.

You can read *Brain Talk* at whatever depth you prefer. First and foremost, *Brain Talk* is designed for the general public. But from the outset, *Brain Talk* was also designed to be a "crossover book" for mental health professionals, including clinicians, teachers, researchers, and students. The scientific background and relevant documentation they require are contained in four Appendices at the conclusion. Think of this book as the neuroscience version of *Passionate Marriage* meets *Constructing the Sexual Crucible*. It's got the readability of the former and over 400 references like the later.

If you just want self-application or a lighter read, stick with the main text. Read the book from front to back as you normally would. You can jump back and forth between the main text and delving into the underlying, amazing, scientific research in the Appendices using the hyperlinked Table of Contents.

But if you're a mental health professional or academic, you

can read the book from back to front. The extensive Index is your table of contents (also hyperlinked in the ebook version). It offers a completely different topical organization and ready access to the brain science and therapy details reported within. When accessing information through the Index, *Brain Talk* looks like a completely different book organized around the technical aspects of Crucible® Neurobiological Therapy.

If you're a brain wonk, you'll be in Seventh Heaven if you read the Appendices straight through. You can also jump back and forth between the main text and delving into the underlying, truly amazing, scientific research.

Ultimately, I hope *Brain Talk* makes you appreciate how much we are all inextricably interconnected and how we impact each other's brains with our minds and not simply our behavior. You can make the world around you a better place, simply by cleaning up your inner mental world. Legislation and law enforcement will never, by themselves, end sexual and emotional abuse, racism, terrorism, and other forms of discrimination, persecution, and exploitation. For this, we must look into our own minds and hearts. That's where the world we all want to live in starts.

Dr. David Schnarch
Evergreen, Colorado
January 2, 2018

Part One

Understanding Mind Mapping

1

WHAT IS MIND MAPPING?

During my career as a clinical psychologist, I've learned many helpful things. But if you asked me to pick out what transformed my life and my work, the cutting-edge brain science of mind mapping would share top honors. For me, learning about mind mapping was like learning about orgasms. I knew about sex long before I learned orgasms existed. When I finally discovered them from personal experience, I couldn't believe they had been happening all around me but that nobody had let me in on the secret! *How could this be going on and no one told **me**?!*

That's how I felt when I stumbled onto mind mapping in neurobiology research journals at the turn of this century. I was shocked! The topic was unknown in the mental health field when I was in training, and for the most part this continues today. Mind mapping didn't even exist as a concept until the 1980s, when two primatologists wondered whether chimpanzees understood that other chimps might have their own thoughts about things or see things differently. They referred to this as "Theory of Mind," but I call it mind mapping.

Mind mapping is not a core part of training in any mental health discipline. It's not required for licensure as a psychologist, psychiatrist, social worker, or pastoral counselor.

Relatively few psychotherapists know anything about it. But humans have been mapping each other's minds for hundreds of thousands—if not millions—of years. We do this constantly in virtually every situation, and it's a fundamental function of the human brain. So you can appreciate my shock when I read my first journal article about this. Wouldn't you think this is something therapists should know about?

According to the Introduction's brief definition, mind mapping is your brain's ability to make a mental map of another person's mind. It also applies to mapping out your own mind too. This is concise but not particularly illuminating. Let me explain mind mapping more thoroughly so you'll understand why it's so important.

The Driving Wheel of Relationships

Mind mapping is an intuitive process. Whenever you interact with people, your brain automatically creates mental pictures of their minds. It then looks at these pictures and makes inferences about them. *What does he want? What is she like? Is he smart or cunning? Does she want to have sex with me? Why is he looking (or not looking) at me? Should I ask her for a date? Should I fear this guy?* Then your brain uses these attributions to predict what other people are going to do and adjusts your agendas and behaviors accordingly. First and foremost, mind mapping is about predicting other people's behaviors.

Like most people, you have pictures of your father's and mother's (or stepparents') minds in your head. You know how they think. You see how they perceive themselves and how they feel about each other. You also know where they lie to themselves, where they're blind to themselves and where they are self-aggrandizing. You recognize the areas where they don't really see you but think they do. You even know how they react to strangers and other family members. In short, you know what makes them tick. You glean all this information from the mental maps in your head. You then use this to predict what they're going to do in a given situation, which is how you decide to approach them.

For example, let's say you want your parents to give you some ice cream or, if you're a bit older, the keys to the car and money for gas. Your mind starts working off your maps of them and devises a plan: *Who should I go to first? What's the best strategy to use to get them to agree? What's the best time for me to ask? How can I convince them if they initially say no?*

Mind mapping is the brain-based psychology of interacting with other people. Once you recognize mind mapping in daily life, you'll be better able to detect other people's motivations and predict their behaviors, whether it's your spouse, your children, your parents and siblings, or your boss and co-workers. In interpersonal relationships, particularly ones loaded with tension, drama, conflict, high risk or great meaning, no tool is more valuable.

WE'RE NATIVE-BORN PSYCHOLOGISTS

Mind-mapping satisfies our inherent desire to figure out other people. Let's say you're walking down the street and you notice a man walking towards you. In a split second, you look at his clothing. You observe his gait. You determine what he's looking at. Is he looking at you or the person behind you, or is he gazing in a different direction? You study his facial expression, particularly his eyes and mouth. You try to assess his mood. You start attributing a personality to him. From information gleaned in a fraction of a second, you'll decide whether to put your head down and ignore him or meet his gaze as you pass on the street. This entire drama occurs in the blink of an eye, and you'll repeat it every time a stranger approaches. By the end of the block, you've constructed a half dozen mind maps of passersby without exchanging a single word with them.

Humans are natural-born psychologists. We're so good at mind mapping that trained therapists aren't necessarily any better at this than the average person. In fact, 30 years of doing therapy and training therapists has shown me that people seeking counseling are often better mind mappers than their therapists and often beat their therapists' mind-mapping radar.

Mind mapping is an innate survival skill we've inherited from reptiles. To ensure our own safety and well-being, we need to know what makes other people tick. We need to understand what's going on in their minds to be able to anticipate their behavior. It's all about being able to predict what they're going to do. This doesn't require extraordinary skill or effort. It's not related to your IQ level.

People are pretty predictable. If you know their beliefs, values, and intentions, and if you know what they know, you'll be able to predict their behavior with a high degree of certainty. Mind-mapping is about figuring out what they really want. The better you understand someone's motivations and agendas, the more accurately you can anticipate what he or she will do next.

IF YOU WANT GREAT SEX...

You use mind mapping every time you walk into a room full of strangers. The first thing you do is map everyone in the room, attributing personalities, agendas, and social status to those who are there. This helps you ascertain how you fit in: *Which girl is the prettiest? Who's the alpha male? Who should I talk to? What's my role? Has anyone noticed me? Is anyone interested in me?* Your mind wants to know where you stand in the social hierarchy. Mind-mapping helps you figure this out.

This is one of countless examples. In fact, I challenge you to imagine an interaction between two or more people where mind mapping doesn't play a role. For instance, if you want to teach people effectively, you need to figure out how they learn best. Some people are visual learners. Others are auditory learners. And some of us learn best when we're physically involved. Effective teachers map their students' minds and figure out how each one absorbs information most readily, what works and what doesn't. They see when to encourage plugging away at a difficult problem and when to back off and let them take a break. This requires mind-mapping ability.

If you're going to empathize with people, you have to intuit their feeling and why they feel the way they do. This involves more than "feeling what they feel." You have to map out what's going through their minds to understand them. The old idea of "putting yourself in other people's shoes" is more accurately stated as "working off your map of their minds." You have to figure out how they tick. To do this, you need mind-mapping ability.

If you're going to offer other people guidance and support, you need to know when to stop. You must recognize when they find you intrusive or meddlesome. For example, let's say your son's bicycle has a flat tire and you offer to help him fix it. However he says, "I can fix my bike by myself!" You can see he doesn't want your help, but you also know he doesn't know how to patch an inner tube. Now you need the artistry good parents develop, of figuring out how to offer him the help he needs without offending his autonomy. Mind mapping helps you know when to speak up or say nothing and let him struggle on his own.

If you're going to negotiate a business deal, you need to figure out what's most important to your customer, supplier, or business partner. You need to know what they are willing to give up and what they won't. You try to psych out their personalities to fathom how best to handle them. *Are they meek or combative? When they say, "No way!" Does that really mean no, or are they bluffing?* To figure this out, you try to map out how they perceive you. You'll be a better negotiator if you hone your mind mapping abilities.

If you're trying to navigate the thorny world of dating in search of a suitable partner, you have to accurately gauge the interest of potential mates. If a woman returns your gaze, is she looking at you because your fly is open or because she wants you to approach her? To answer this crucial question, you'll use your mind-mapping ability (and also check your fly). If it turns out she's looking at you for both reasons, and you're interested in her, this might be your lucky day!

If you're thinking about throwing your husband a surprise birthday party, you'd better be able to predict how he'll respond. Will he be pleased or will he'll be embarrassed? If you don't know him as well as you think you do, it could ruin what's supposed to be a happy day for him. Likewise, when buying sexy lingerie for your wife, you'd better know whether she identifies more with Shy Sally, Nurse Nancy, or Doris the Dominatrix. If you pick the wrong outfit, the gesture you intended to instill passion could cause an argument. Mind mapping helps you avoid a week of chilly silence.

If you underestimate the importance of mind mapping, you're going to be two steps behind everyone else. Here's what I'm talking about. Let's say you have a job that pays well; and you enjoy spending money on expensive meals, nice clothes, and whatever gadget catches your eye. Your husband is the opposite. You have him pegged as a tightwad, and you think he's very controlling. Your conflicting views on handling money often lead to arguments. To avoid another fight, you hide your new purchase from your husband, and to do that effectively you have to be able to map him. If you know he wouldn't wash a dish if his life depended on it, you can safely hide your shopping booty beneath the kitchen sink because there's no chance he'll ever look there. If he stumbles upon your stash, you tell him these are early-purchased Christmas presents.

Now let's look at this from the husband's perspective. You tend to be fiscally responsible. You delay immediate gratification for something better down the road. You prefer to save money. You don't include your wife when choosing investments because you know it usually leads to a fight. You don't trust her judgment, and including her in decisions makes it harder for you to do what you want. Getting away with this requires managing investments without your wife finding out. You know she never balances her checkbook or goes through your joint tax returns. You're pretty confident your wife won't find out. Besides, if she does, this will blow over if you let her spend some money. You decide, all things considered, simply not to tell her.

Like I said, you're going to be two steps behind if you underestimate the importance of mind mapping in daily life, You'll miss out on opportunities to conduct successful business negotiations, throw wonderful surprise parties, provide effective teaching, offer productive emotional support, and enjoy many other things that create a satisfying life. You'll also be handicapped if your partner is playing three moves ahead.

Still not convinced of the importance of mind mapping in daily life? One last example might finally persuade you. If you want mind-blowing sex, mind mapping will be involved. Bulging biceps and a gym-hardened butt can only get you so far. Mind mapping during sex gets you into each other's minds, hearts, souls, and brains. How about having an orgasm looking right into your partner's eyes? Mind mapping is a core part of the best sex you'll ever have, whether you're establishing deep emotional connection or playing out fantasies.

It's no coincidence that our mind is the part of ourselves we most often hide during sex. It's a lot easier to let someone see your body than it is to reveal what really turns you on. Turning humdrum sex into sizzling passion requires unmasking your mind and allowing your partner to map your eroticism.

Now do I have your attention?

Everywhere, Everyone, All the Time

Mind mapping is happening all around you all the time (just like people having orgasms–mostly with their eyes closed). Once you know the signs–and I'll show you how to spot them–you'll recognize mind mapping everywhere you look. It's going on in good relationships and, unfortunately, in bad ones too. (Did you expect me to say good relationships have mind mapping but bad relationships don't?) For better and worse, in good times and bad, mind mapping is happening in all your interactions with your spouse, kids, parents, and classmates, customers and co-workers, and everywhere else.

Given its out-sized role, it's important to remember mind mapping is neither inherently good nor bad. As you'll see, it's involved in the absolute best things human beings do and, unfortunately, also the worst ones. Negative uses of mind mapping are as numerous as positive ones. Mind mapping itself is value-neutral. This becomes more understandable when you know the brain science behind this incredible ability we all take for granted. I'll tell you all about this in the next chapter.

2

The Brain Science Behind Mind Mapping

Mind mapping is the scientific basis of what people used to refer to as "folk psychology." Folk psychology offered simple, direct explanations of our ability to predict other people's behavior using everyday reasoning and basic assumptions about human nature. "People seek pleasure and avoid pain" is a good example of folk psychology.

However, lots of research has been conducted on how your brain predicts what other people are going to do. In fact, it's one of the hottest areas of modern brain science. We still have much to learn. But after reading hundreds of research articles, I'm thrilled by how much we actually know about how the brain pulls off this amazing feat.

Theory of Mind

The scientific origins of mind mapping can be traced to the article "Does the Chimpanzee Have a Theory of Mind?" published by two primatologists, David Premack and Guy Woodruff, in the journal *Behavior Brain Science* in 1978. They coined the term "Theory of Mind," which refers to the realization that others possess beliefs and desires different from our own; and we can predict other people's behavior by ascertaining what's in their minds.

Premack and Woodruff wondered if chimpanzees had mind-mapping ability (my term for Theory of Mind). To figure this out, they let a 14-year-old chimpanzee named Sarah watch a series of videos showing a man struggling with various problems (trying to get out of a locked cage, for example) followed by several photographs suggesting potential solutions to those problems (one photos showed a key). After Sarah consistently chose the correct photograph, Premack and Woodruff assumed she recognized the problem presented in the video, understood the man's needs, and made a choice based on his mental state.

However, because chimpanzees can't talk, it's difficult to ascertain their intentions with certainty. In another study, when prompted to locate a peanut hidden among several closed cups, chimpanzees choose randomly even when the researcher pointed to the one with the peanut. Whether or not chimpanzees have mind-mapping ability is still debated today.

Research took a great leap forward in the 1980s when child psychologists began studying mind-mapping abilities in infants and small children. The results were astonishingly clear and sometimes unexpected. For instance, famed child expert Jean Piaget had proposed that young children are so completely self-absorbed that they don't show interest in other people's minds until around age eight. However, study after study convincingly demonstrated that most children develop remarkable mind-mapping ability around the age of four.

Thanks to technological and scientific advances in recent years, there is now a strong foundation of mind mapping research. But before I dive into the brain science behind mind mapping, here are three general facts you should know.

First, the role of mind mapping in a wide range of disorders has been studied using state-of-the-art MRI, fMRI, and PET brain scanners, from autism and schizophrenia to alcoholism and depression.

Second, the neurophysiology and neurochemistry of how the brain makes a mental map of another person's mind is surprisingly well known. (At least I was surprised.) So much so, this has fueled the emergence of an exciting new field of brain science called *interpersonal neurobiology*, which explains how other people's actions, thoughts, and feelings shape our brains—and vice versa—for better or worse.

Third, although mind-mapping ability is incredibly sophisticated, it is also extremely robust. It's not easy to stop it. Although some brain injuries and a small group of diseases can impair mind-mapping ability, the vast majority of people have it. If you have to guess whether or not a complete stranger has mind-mapping ability, you'll be right far more often than you'd be wrong if you presumed he did, at least until proven otherwise.

Mind Mapping Is a Survival Skill

Where does our mind-mapping ability come from? What are its biological origins and evolutionary purposes? These questions go right to the heart of human nature, and the answers might surprise you.

Given mind mapping's presence in humans—and dogs and to a lesser extent dolphins and elephants—you might assume its origins are mammalian. If you believe bonding is the primary drive wheel of human relationships (as proponents of attachment theory believe), this probably seems intuitive to you. You might assume the human brain developed the mind-mapping ability to facilitate positive emotional connections between people. After all, mind mapping plays an integral role in empathy and compassion.

However, if you came to these conclusions, you'd be wrong. Mind mapping's origins are actually *reptilian*. That's right. This highly evolved ability, which seems so quintessentially human, in large part springs from the part of our brain we share in common with snakes, lizards, and frogs. And when you realize reptiles don't bond with their sex partners or offspring like humans do, mind mapping's original purpose becomes clearer: *survival*. Although mind mapping *can* facilitate human bonding, it's primarily a survival skill.

Reptiles and amphibians have a primitive behavior-prediction system. Picture a frog sitting on a lily pad in a pond looking at all the other frogs around it. It tries to pick up information that might enhance its chances of survival. It wants to know: *What are the other frogs eating? Should I be eating that* too? *Does that big frog keep looking at me because it wants to eat me or mate with me? Do I have to kill it, or are we going to have sex?* (Sounds like some marriages, doesn't it?) To divine answers to these questions, the frog relies on highly specialized brain cells that monitor two parts of other frogs' bodies, their eyes and mouths. By tracking other frogs' eyes and mouths, it can more or less predict what other frogs are going to do.

Likewise, when a snake corners a mouse, it tracks the mouse's eyes to figure out which way the mouse is likely to run. Snakes who consistently guess wrong go hungry, and their chances of long-term survival go down. Your life probably doesn't literally depend on your mind-mapping ability; but the health of your relationships with your spouse, children, parents and co-workers do. But if you're in a physically abusive relationship with a human viper, your life could very well hinge on your mind-mapping ability at some point.

For years researchers generally accepted the idea that the human brain consists of three sub-brains: the "reptilian" portion in the back of your head, the "mammalian" part in the middle, and the uniquely human part, the prefrontal cortex, located in your forehead.[1] It turns out you have cells in the reptilian portion of your brain that track other humans' eyes and mouths. Whenever we look at other people's eyes or mouths, those cells start firing, providing important data that help us predict what that person is going to do.

Other parts of your reptilian brain participate in your mind-mapping system, as do entirely different portions as well. It's a complex modular system distributed throughout your brain. Some mind-mapping processes, such as understanding sarcasm, involve your prefrontal cortex, the most highly evolved part of your brain. However, mind-mapping functions you might think of as fairly sophisticated—like detecting lying, for instance—are actually detected by your reptilian brain.

Rebecca Saxe, associate professor of cognitive neuroscience in MIT's Brain and Cognitive Sciences Department, has devoted her entire career to studying a single part of the human reptilian brain called the temporoparietal junction (TPJ). As a graduate student, Saxe identified the TPJ as an important part of the brain's mind-mapping system that helps you understand what other people are thinking and feeling. Saxe likened it to a modular, integrated, computer chip that, among other things, detects lies. Detecting deception is a basic brain function that doesn't involve sophisticated cognitive processes such as deductive logic. If someone lies or does something deceptive, it triggers cells in your reptilian brain and *bam!* You're now on red alert!

The implications of this are important and far reaching. Mind mapping is very much (but not exclusively) a "bottom-up process," emerging from the most primitive part of your brain, which explains how it is so robust. The profound implications of this will become more apparent in subsequent chapters, but I'll point out several of them now. For one thing, this explains how people who grow up in highly destructive families still develop mind-mapping ability. (I'll tell you all about this in Chapter Four.) For another, this clarifies how children end up mapping out things in their parents' minds they really don't want to know. (We'll cover traumatic mind mapping in Chapter Eight.)

But the final noteworthy implication concerns human nature and adult love relationships. Humans are wired to detect deception, lying, misdirection, and manipulation in parallel with our ability to deceive and manipulate. Does this facilitate stable, long-term bonding? On face value it doesn't because who wants to be in a relationship with a deceptive manipulative partner? Such relationships don't offer much safety or security.

However, most married people acknowledge that "emergency lies"–your response, for example, to your wife asking you if her new dress makes her look fat–are a necessary component of long-term relationships. Same goes for "helping your partner see the wisdom of your position" when the two of you disagree, which is called social intelligence or manipulation, depending on whether you or your partner are doing it.

The sad truth is many people are only able to keep their relationships going by lying, deceiving, and manipulating. As I'll explain next chapter, successful lying, manipulation and deception requires mind-mapping ability. Some of us need this to keep our partners in the dark or confused because they'd be out the door if they ever got an accurate picture of who we really are. If your marriage is supposedly monogamous but you're having an extramarital affair, for instance, the only way you can keep your marriage together is through lies, deception, and manipulation. Almost half of all married men and women report extramarital affairs, and for them this is standard operating procedure to "preserve" their marriages.

This highlights the centrality of mind mapping in emotionally committed relationships, and provides a stunning observation about human nature. I hope the irony isn't lost on you. I started by saying humans didn't evolve mind mapping to facilitate human bonding, in the "safety and security" sense so popular today. Mind mapping is a survival skill. It is value-neutral. You or your partner couldn't hide an affair if you didn't have mind-mapping ability. The same goes for being able to detect your partner is messing around.

I frequently work with couples recovering from affairs, and I'm all too aware of how affairs erode relationships and frequently lead to divorce. So here's what this comes down to: Mind mapping aids and abets affairs (and a panoply of other bad behaviors) which destabilize marriages. And mind mapping facilitates the lies and deceptions many people need to keep their relationships going. In his movie "The Invention of Lying" Comedian Ricky Gervais offers a sad and hilarious commentary on what relationships would be like if people couldn't mask their minds from each other.[2]

The Nuts and Bolts of Mind Mapping

I love studying mind mapping because it reveals the incredibly sophisticated processes your brain goes through to accomplish things we all take for granted. For instance, when making a mental map of another person's mind, your brain has to keep track of whose thoughts are whose in your own mind. Are you picturing your own thoughts and feelings or someone else's? It turns out one part of your reptilian brain is responsible for representing the contents of your own mind[3] while another part tracks the minds of others.[4]

Your brain also relies on two separate but related "attention networks" to alternately focus your concentration. Your "dorsal" system tracks other people while your "ventral" system tracks yourself. By switching back and forth between your ventral and dorsal attention networks, your brain forms a cohesive picture in your mind of other people's minds as well as your own. When I first learned this, I was amazed! I'd never even considered that distinguishing between your map of your own mind from your map of another person's mind could be a problem, and yet scientists already figured out how your brain does it! I could hardly believe brain science had advanced so far! I began reading everything I could find on this subject. I had to know how the brain actually pulls off this whole mind-mapping thing!

My knowledge advanced greatly in 2011, when I encountered a sophisticated model of mind mapping published by Drs. Ahmed Abu-Akel and Simone Shamay-Tsoory. I'd followed both scientists' extensive mind-mapping research careers ever since they'd collaborated on a simpler model in 2003. This time around, they not only integrated the latest neurophysiology research, they added crucial new information about mind-mapping's neurochemistry that was missing from their previous model. That's how much brain research had been conducted in the intervening eight years. I was literally jumping out of my chair as I read it!

I, myself, have never been wild about memorizing parts of the human anatomy. But when clients learn how their brain actually pulls off this amazing feat, they pay more attention to mind mapping, so a brief summary of this model should be helpful. But rather than giving you an anatomy lesson, I'll give you a functional picture of how your brain works. (For those who are interested, the anatomy appears in Appendix A.)

To begin with, your brain's mind-mapping system consists of three physically distinct but interacting parts, which aligns nicely with the triune brain model I mentioned earlier.

First, there's a "representational" component located in the reptilian (rear) portion of your brain. This is where your mind maps initially arise. Your brain constructs two maps for you and two for the other person. One is a "cognitive" map of thoughts, beliefs and knowledge. The other is an "emotional" map of feelings and hard-wired as well as socially-based emotional reactions.[5]

Secondly, there's an intermediate "attribution and information processing" component in the mammalian portion (middle) of your brain. This is where the process of attributing mental states to yourself and others occurs. For instance, this is where your brain decides you're a good person; but the guy walking down the street is a creep, and his hot girlfriend would be better off with you.[6]

Finally, there's an "application/execution" component in your prefrontal neocortex, the uniquely human region of your brain located in your forehead. This is where your brain makes decisions about implementing or acting upon your maps of yourself and others.[7] In other words, here's where you decide that even though you're probably smarter than this dude, you'd rather not have him pound you into the cement for coming on to his girlfriend.

These three areas are interconnected by neural pathways and brain chemicals that relay signals between neurons ("neurotransmitters"). Recent discoveries reveal a complex system that staggers my imagination. It turns out dopamine and serotonin work together to form an integrated regulatory system that innervates all three mind-mapping regions.[8] These two neurotransmitters work in concert to develop your mental maps and assist in updating and maintaining them. In other words, this dopamine-serotonin neurochemical system turns on and off to form your initial map of another person's mind and then modifies it based on actual errors in predicting their actions and behaviors.

This is pretty amazing information. But you can't really appreciate the neurochemistry of mind-mapping without a more sophisticated description of its neurophysiology. So bear with me for a moment while I regale you with a more complex picture of what's taking place.

1. There are actually two separate mind-mapping networks. One is the cognitive network which maps your and other people's thoughts, knowledge, and

beliefs. The other is the affective mind-mapping network, which maps your and other people's feelings and emotions. Together they form the larger mind-mapping system by your brain's integrating these cognitive and emotional maps into an integrated whole.

2. These cognitive and emotional mind-mapping networks share connections in the reptilian (rear) region of your brain. This is where your brain actually creates representational cognitive and emotional mental maps, and your brain distinguishes maps of your own mind from those of other people.[9]

3. Previously I said your ventral and dorsal attention networks track different streams of information regarding your own versus other people's mental states. Your brain alternately turns these two systems on and off to direct information to the correct areas of the self-other mind-mapping system in your reptilian brain. It also uses this mechanism to shift your attention from focusing on your own mind, to focusing on other people's minds, and then back and forth endlessly. This is how your attention "decouples" from your current focus and allows you to refocus on something else.

4. Norepinephrine, another neurotransmitter, plays a critical role in this ventral-dorsal switching process which distinguishes maps of your own mind from maps of other people's minds. Norepinephrine is a third part of the neurochemistry of mind-mapping, and how it interacts with the serotonin-dopamine system is still being discovered. Norepinephrine also promotes vigilance, increases your arousal and alertness, enhances memory formation and retrieval, and mobilizes your brain and body for fight-or-flight reactions. It's pretty cool that this neurochemical not only helps you track other people's minds, it also prepares you to act on what you're mapping.

If you want the whole incredible story of how the brain produces mind mapping, all the neurobiology and neurochemistry, and what interacts with what, it's succinctly laid out in Appendix A. Researchers continue to discover new neural connections between brain regions in the mind-mapping network, and incredible secrets of its neurochemistry are still being revealed. At this point, for example, scientists can only speculate about research findings which indicate that oxytocin (a neurohormone) greatly improves healthy subjects' mind-mapping abilities when sniffed![10] But some things aren't likely to change. For instance, growing up blind doesn't change the neural basis of mind mapping.[11]

MIND MAPPING SHAPES YOUR LIFE STORY

I've just given you a rough idea of how your brain maps your and other people's thoughts and emotions, and maintains or changes these maps over time. Here's how it stores and retrieves these maps: Researchers have long known the human brain is functionally subdivided into left and right hemispheres, with each side contributing specialized functions. What researchers haven't known until recently is the extent to which the left and right hemispheres of the brain are differentially involved in mind mapping.

Although your left and right brain hemispheres are integrated and work together, their respective abilities differ. Left-brain thinking is verbal, analytical, and deductive. It processes information sequentially, first looking at details, then, after putting them all together, coming to conclusions. Left-brain abilities include logic, linear and verbal thinking, mathematics, and an astute memory for facts and song lyrics. Your left brain uses deductive logic and reasoning built on facts, ideas, and recalled memories. Your ability to use language comes from this part of your brain. Your left brain talks to itself in words. When you read this book as text, you're using your left brain.

Right-brain thinking, in contrast, is nonverbal, intuitive, and holistic. It relies on pictures rather than words, and processes information in an instinctual, instantaneous way. It first looks at the whole picture, then the details. It focuses on feelings rather than facts. Right-brain abilities are holistic thinking, nonverbal communication, imagination, intuition, daydreaming, art skills, rhythm, and remembering the tunes of songs. Visualize your first kiss. Remember the way you felt the moment your lips touched. In doing this, you just engaged your right brain.

Mind mapping involves both brain hemispheres, working back and forth in powerful combination. In addition to mapping out your and other people's thoughts, beliefs, and knowledge, your left brain stores your "autobiographical memory"–the story of your life based on remembered past events including your mental maps of your and other people's mental states at the time. This information is actually distributed in various parts of the brain rather than stored in a single location. In contrast your right brain, which maps out your and other people's feelings and emotions, is also responsible for retrieving and re-assembling your autobiographical memory.

Here's why this is important: Many of my clients are stunned when a completely different picture of their lives emerges, having spent hundreds of hours (and thousands of dollars!) on prior psychotherapy. The flaws in their previous understandings become obvious. They realize insight-oriented talk therapy can only go so far, leaving them with the mistaken impression that they comprehend their past. Because their understanding is based on deductive logic, words, and remembered details, they often fail to see the bigger picture, which has much greater power and meaning. This awareness typically arises when I help them recognize gaps in their autobiographical memory of pivotal events in their lives. From years of experience, I know exactly where to look to expose this gap.

Without fail, what's missing in clients' autobiographical memories of prior troubling experiences is their mind maps of the people involved. As I'll explain in Part Two, a right-brain retrieval error often occurs when you try to remember past traumatic events. Common stereotypes suggest people don't remember prior bad experiences at all–that they are repressed in the unconscious. But my experience indicates complete amnesia is unnecessary because the absence of mind-mapping information about the people involved sufficiently changes the meaning of things to make remembering the event itself tolerable.

A variant of this involves instances where you can remember an event and you can retrieve your maps of the self-perceptions, knowledge, ideas and thoughts of the people involved. However, you still misinterpret their mental states because maps of their emotions and feelings (a right brain process) are absent. When clients put the mind-mapping information back into their autobiographical memories, the meanings of their life stories change significantly. Digesting these new pictures can be upsetting.

Part of the shock comes from realizing your mind-mapping information has been absent from your memories. Most clients had no idea anything was missing! This realization encourages re-examining other memories through the lens of mind mapping, trying to retrieve what their nemesis (the person in their memory engaging in cruel, destructive, or traumatic behavior) was thinking and feeling towards them at the time.[12] After the initial shock of recognition wears off, clients feel more relaxed and grounded.

Getting to this point can be challenging. When clients try to fill in the gaps in their autobiographical memories, without fail, they attempt this using their left brains. For instance, they try to work off historical facts like "My father used to take me fishing as a child." Using deductive logic gives them a (false) sense of moving forward, but unfortunately, they're using a method that's guaranteed to fail. Their left brains are simply incapable of retrieving the missing mind-mapping data.

Clients usually have no difficulty retrieving detailed memories using right-brain methods I'll describe in Part Three, including those who initially reported remembering virtually nothing of their childhoods. Their story often changes to something like "My father took me fishing so he could get away from my mother. There was little conversation between us. I got to tag along while Dad did what he liked to do."

You only get part of the picture using your left brain to retrieve missing mind maps. But I've come to the conclusion this mistake is purposeful. Left-brain attempts to access missing mind-mapping information are actually a defense mechanism that prevents your brain from becoming dysregulated by seeing upsetting things. This way is like looking at something using only one eye. Sure, you can see *something*, but it's only two-dimensional. Using both eyes produces stereoscopic (three-dimensional) vision, which contains much greater details.

Whenever clients try to grasp the meaning of their past intellectually—for example, by offering theories they've polished in years of prior therapy—I interfere and push them to describe what they actually *see* when they look at their memories. People who are avoiding the emotional impact of their left-brain, intellectual theories always fail. They say, "I understand it intellectually, but I don't get it emotionally."

I reply, "Well, if you don't get it emotionally, then you don't get it! And if you want to get it, stop telling me your theories and tell me what you *see!*"

To produce a more accurate autobiographical memory by retrieving mind-mapping information, you have to visualize the picture and watch what's happening rather than intellectualizing, rationalizing, or speculating about what things means. This, as I've explained, is a right-brain process.

Be aware of this as you read this book and sift through your autobiographical memories. Mind maps of your antagonists are probably missing in your memories, whether they involve your spouse, children, parents or siblings, employer or co-workers. To retrieve these maps, use your right brain to visualize what's happening in a particular remembered event. Keep watching your memory as if you were studying a videotape, watching the action develop. You'll see things that aren't evident in your verbal descriptions and summaries. If you keep at it, eventually you'll see the meaning of the entire picture suddenly come together all at once. You'll get an entirely different picture of the event and perhaps your whole life. I'll show you exactly how to do this in subsequent chapters.

MIND MAPPING IMPACTS YOUR BRAIN

Mind mapping lies at the heart of your connections with other people. Mind mapping constructs your understanding of people and events as they occur in real time. Likewise, your ability to accurately retrieve mental maps of other people and yourself greatly shapes your autobiographical memory and the story of your life. How you think and feel, what you desire, what you refuse to see in yourself, all this and more dramatically impacts those around you for better and for worse. The same holds true for other people's thoughts, feelings, and behaviors profoundly affecting how you feel about yourself, how you see the world, and what you say and do.

It's less obvious that mind mapping also changes your brain as well. But mind mapping profoundly interconnects us all, down to the level of interpersonal neurobiology. Traumatic interpersonal experiences can have far-reaching negative neurobiological impacts on your brain. I'll explain this in Part Two where we'll dive into the murky world of *traumatic* mind mapping.

These negative impacts, however, are not set in stone. In Part Three, I'll show you how to reverse damage done by traumatic mind mapping and heal yourself in the process. Moreover, you don't have to wait until then to use mind mapping to create *positive* neuroplasticity for yourself and those around you. It's actually quite simple. Do something positive that's completely out of character for you. When you violate other people's mental maps of you, it grabs their attention and they'll start re-mapping you to make sense of what's going on. If you keep this up, their maps of you and how they interact with you will change.

Do you want your aloof adolescent daughter to be less adversarial? Want your husband to stop glancing at the football game and have a real conversation with you? Wish your employees were more motivated? Just do something positive that challenges their map of you. You'll not only have their undivided attention, you'll be changing your brain and their brains for the better. Most people are amazed to discover the breadth and depth of mind mapping's impact on their lives. This particularly happens when they learn about children's mind mapping abilities. Once you know how early this starts, and the full extent to which they can map you, you'll never look at your kids the same way again. This is where we're headed in our next chapter.

3

Mind Mapping in Children

Mind mapping is such an intrinsic part of being human that it shouldn't surprise us how quickly it emerges in children. But once you realize the full extent of this, you'll never look at kids the same way again. Especially if you're a parent. Knowing your kids are tracking you all the time makes you realize what most parents want to avoid: You're living under the spotlight of constant scrutiny. How you respond determines the sort of relationships you'll have with your children throughout your lifetime.

Implicit Mind Mapping

Signs of mind mapping emerge soon after birth. Babies begin sharing eye contact ("mutual gaze") with caretakers when they're three months old. This interpersonal communication is an important precursor of actual mind-mapping ability. Another shared experience is directing their parents' attention toward objects that interest them ("declarative pointing"). Using gestures and mutual eye gaze to deliberately establish "joint attention" with parents are milestones in a baby's social development and important precursors of mind mapping. By their first birthdays, children understand pointing is an intentional act and readily share joint attention with others while gazing at objects within their visual field. (In other words, a one-year-old child has better mind-mapping ability than a four-year-old chimpanzee.) This is the basis for creating powerful "moments of meeting" (mutual mind mapping) in childhood and later in life.

At 18 months old, babies' rudimentary abilities take a phenomenal leap. They can follow someone's attention to objects *outside* their own visual field. Babies can figure out that there's an object they can't see, hidden behind a screen, by monitoring the eyes of someone who can see the object. In other words, by this age a baby can track something in another person's *mind!* Like I said, mind mapping ability starts incredibly young.

From this point until they're about four years old, children demonstrate "implicit" mind-mapping ability. This means they clearly have mind-mapping ability but lack the verbal skills to reliably demonstrate it. Researchers study implicit mind mapping by showing children a cartoon of a wall with two mouse holes. One hole has a piece of cheese in front of it. The researcher asks the child, "Which hole is the mouse most likely to come out of?" and watches the child's eyes.

Children who have implicit mind-mapping ability refer to what they've "mapped" out about mice in cartoons and stories. What do mice want? They want cheese! By making language unnecessary, scientists can tell if children have mind-mapping ability by tracking where they look.

Explicit Mind Mapping

As children's verbal skills improve, they demonstrate "explicit" mind-mapping ability. They can now report exactly what's going through their minds. This makes it possible to accurately assess the extent of their abilities.

For the last 30 years, scientists have been conducting ingenious experiments on children of widely ranging ages and cognitive aptitudes to measure their ability to discern the intentions of others—so much so, that in 2001 a meta-analysis of 178 studies involving over 4,000 children was conducted.[1] (For those wanting more in-depth details of children's mind-mapping abilities than I can mention here, I highly recommend Martin Doherty's *Theory of Mind: How Children Understand Other's Thoughts and Feelings*.[2]) Rest assured that everything I tell you here has lots of research to back it up.

Researchers were particularly eager to figure out exactly when children's mind-mapping ability emerges. So in 1983 Heinz Wimmer and Josef Perner, professors of cognitive neuroscience at the University of Salzburg, devised an experiment involving what's known as a false-belief test and gave it to children ranging from three to nine years old. They told each child a story about a boy named Maxi who sees his mother put chocolate in a blue cupboard before he goes outside to play.[3] (Take a moment to visualize what I just described. Do you see the blue cupboard where Maxi's mother put the chocolate?)

While Maxi's gone, his mother takes the chocolate out of the blue cupboard to bake a cake. When she's finished, however, she doesn't put the remaining chocolate back in the blue cupboard where it came from. Instead, she puts it in a green cupboard on the other side of the room. (Can you see the green cupboard where Maxi's mother put the chocolate?) When Maxi returns, he's hungry and starts to look for the chocolate. Where does Maxi look first?

Most three-year-olds flunked the test, saying Maxi would look in the green cupboard. They don't really understand that other people can have perspectives and knowledge that differ from their own. Since they knew the chocolate was in the green cupboard, they assumed Maxi did too. They were incapable of understanding Maxi's perspective differed from their own or that Maxi could be mistaken.

Four-year-olds passed the test more than half the time, demonstrating they understood other people's minds can hold false beliefs about the world. Older children demonstrated even better comprehension of false beliefs and greater ability to understand how someone else's perspective could differ from their own. These are essential components of accurately mapping other people's minds.

Probably the best known false-belief test is the "Sally-Anne test" developed by Simon Baron-Cohen, Alan Leslie, and Uta Frith in 1985.[4] Rather than telling children a story, the experiment is depicted in comic-strip fashion: Frame 1: Sally and Anne are playing with a ball. Frame 2: Sally puts the ball away in a box and walks out of the room. Frame 3: While Sally is gone Anne moves the ball to a closet and also leaves the room. Children are then asked, "Where will Sally look for the ball when she comes back into the room?"

No matter what form these experiments take, the results are amazingly consistent, as explained below.

Starting around age four, most children display mind mapping ability. They are able to create mental maps of other people's minds. They understand other people's minds may be inaccurate pictures of reality, differing from how they themselves see things. They recognize other people's behavior can best be predicted by knowing how these people see things and what they believe–even if their perceptions and beliefs are wrong.

This isn't an isolated phenomenon reserved for well nourished, well educated children in wealthy countries. Nearly all children acquire this ability around age four, regardless of gender, race, religion, geography, or education. And as parents around the world soon find out, it isn't long before your kids start using your existing false beliefs, and trying to implant new ones, to manipulate you.

Enjoy Your Brief Moment of Omniscience

Very young children think their parents are omniscient, meaning they know everything. Before the age of four, children can't imagine their parents' minds aren't perfect pictures of reality. I learned this from personal experience one day as I walked through a mountain meadow with my daughter, Sarah, when she was three and a half. We stopped to eat lunch. I'd made two sandwiches, one for me and one for her. I wasn't very hungry and decided to only eat half of my sandwich. I offered the other half to Sarah, but she refused to eat it.

"Why don't you want to eat the sandwich?" I asked her. "I thought you liked bologna. Look, I'm eating it too."

Sarah began to get angry. "You know why!"

"No, I honestly don't," I said, mystified.
"Yes, you do!" she insisted and started to cry.
"No, I don't" I said desperately. I ran through a mental checklist of what sin I might have committed and came up blank. I felt stupid. I started begging. "Sarah, please tell me. I don't understand. Why won't you eat this?"

"You *know* why I don't want to! You can read my mind!" She was emphatic, no doubt in her mind.

"No, I can't, honey." I almost started laughing but managed to keep myself under control.

"Yes, you can! You can read my mind!"

Like many other proud parents, I thought Sarah's assertion was incredibly cute. She was absolutely certain that my mind was perfect and I saw *everything*–including what she was thinking! Sarah was clear I should know exactly what she wanted.

As it turned out, Sarah didn't want half of *my* sandwich, she wanted *her own* sandwich. (I should have known!) Once she got what she wanted, she settled down. We enjoyed eating our *own* sandwiches together, and I remember thinking, *I don't want this moment to end. You think I know everything, sweet Sarah. But ten years from now you're going to think I'm a complete idiot! And if you're a typical teenager, you're not going to want me knowing anything that goes through your mind!*

FOUR-YEAR-OLDS: BUDDING LIE DETECTORS

A child's cognitive development explodes around the age of four. They understand what people do is directly connected to what's in their minds. They know that, if they can figure out what's going on in your head, they can predict what you're going to do.

As we saw in the Sally-Anne test, four-year-olds can also detect false beliefs. They can pick up on big lies as well as little "white" ones. For instance, let's say your four-year-old daughter accompanies you to the shopping mall; and while you're there you walk past a pet store with rabbits in the window for sale. When you get home, your daughter asks if she can have a bunny rabbit. You already have three dogs, two cats, and a hamster.

The last thing you want is another pet, so you tell her, "No. I'm sorry, honey, we can't. I have no idea where to get one."

Lots of adults think they can get away with telling lies to small children, but they can't. Four-year-olds understand that if someone sees something, he or she has knowledge of it, so they know what you know if they know what you've seen. In other words, your four-year-old daughter knows you're lying. She knows what you know because she saw you looking at the rabbits in the pet store.

By the time children are six years old, they can detect the difference between their parents' lying and playing make-believe. Take a moment to contemplate just how sophisticated a discernment this is! Most parents don't want to believe their six-year-old knows when they're lying, because it means they have to face issues they'd rather not address if they want to have a healthy relationship with their child.

When you realize how little you can get away with once your kids are four and older, it challenges how you approach parenting. You can't get away with as much as you thought you could. Once you grasp young kids can map your mind and understand what you're doing, it impacts the way you act around them.

For instance, children as young as four understand that someone can want something mean, hurtful, and socially wrong to happen to someone else. More specifically, kids this age can recognize when their parents want something bad to happen to *them*. Kids can see when their parents enjoy hitting or yelling at them and wanting them to feel bad.

My clients are typically shocked, sobered, and oftentimes ashamed when they realize this. And the ones who want to be good parents start cleaning up their acts.

LITTLE LIE, BIG MOMENT

Even if we don't realize what we're seeing at the time, every parent witnesses scientific proof that their child now has mind-mapping ability. It's that moment around the age of four when your kid tells a fib for the very first time. Researchers consider the onset of telling lies to be positive proof children's mind-mapping ability has arrived. Cute and adorable as it might seem at the time, this also means your child realizes you aren't perfect.

For example, while your four-year-old is in another room, he drops a glass of water, which breaks. You subsequently give him ice cream you promised him earlier because you don't know about the broken glass. This is a red-letter day in your child's mind-mapping ability because he now knows your mind is capable of a false belief. Before long he arrives at the stunning realization he can use this to manipulate you. He'll spend the next decade or two implanting false beliefs in your head to acquire bigger and more expensive things than ice cream.

Once kids know your mind is capable of a false belief, their mission seems to become deliberately implanting them. However, that doesn't mean they're good at it (yet). In fact, they're terrible liars at first. They giggle self-consciously and draw attention to themselves, advertising the fact they're up to no good. Or they can't stop laughing because they're having a grand old time trying to fool you.

Kids lie in a variety of ways, including stating untruths, omitting important facts, hiding evidence, and creating misleading appearances. For instance, the boy who dropped the glass of water could lie several different ways, depending on how he sizes up his parents. He could do this:

1. By commission. ("I didn't break the glass. Sally did.")
2. By omission. ("Oops, I forgot to tell you I broke the glass.")
3. By misdirection. ("Mom, have you seen my glass of water?")

END OF YOUR PRIVACY

You are probably just as stunned as I was to learn that four-year-old children (and older) can map your mind. Even if your kids are now fully grown, you start reviewing memories from their childhoods and reconsidering your interactions with them now. Once you realize you're under constant surveillance, any privacy you thought you had evaporates. Like it or not, your kids are mapping you all the time. If they look like they don't see you at all, that often indicates they see you all too well.

If you're a good parent, this realization forever shapes how you act around your kids. You start looking at past and present events differently. You realize startling things you never recognized before because you didn't understand the full extent of your children's mind-mapping abilities. In particular, this makes you less complacent about problems at home.

For example, four-year-olds understand interpersonal conflict in forced-choice situations. They recognize two people can have discrepant desires, and only one of them is going to be happy about the outcome. In one experiment, for example, researchers tell children a story about two sailors on an island who want to row their boat to the mainland; but each sailor wants to go to a different place. Four-year-olds understand one sailor is going to be unhappy about the outcome, whereas three-year-olds do not.

In other words, you're kidding yourself if you think your children don't understand that you and your mate are fighting, even if you don't argue directly in front of them. If you're constantly bickering, your kids are monitoring your arguments. Children often have greater understanding of what these fights are all about, and how you and your spouse feel about each other, than most parents realize. This increases as they get older.

Around age 11 a child's mind-mapping ability reorganizes into adult form, allowing the child to understand sexual motivations and complex social interactions. The child closely monitors a parent's physical displays of affection and understands things like extramarital affairs. Little, if anything, goes over a child's head. And you thought your sex life was private!

One client told me how he learned how observant kids are and how ignorant of their vigilance we can be. Gregory was getting his five-year-old ready for school one day when his son asked, "Daddy, what's moly?"

Gregory said, "I don't know. What are you talking about?"

"*Moly!*" his son said. "What is *moly*!?"

"I have no idea," Gregory replied.

Gregory's son started to get upset. "Yes you do, Daddy! You

know what moly is! Why won't you tell me?!"

"I'm sorry son, but I don't know what moly is. Where did you hear this? On TV?"

"No! You say it! You say moly all the time"

"What? I don't say moly. I don't even know what moly is. You must have gotten this from a cartoon or something."

Gregory's son got increasingly agitated and started to cry. It took a while to calm him down enough to send him to school. The experience was upsetting, but Gregory told himself his son was getting upset about nothing important.

A few days later, Gregory offered to make breakfast for his wife and son. He made pancakes because he knew his kid loved them, and together they both usually made a big deal about it. This day was no different. His son was almost jumping out of his chair as Gregory delivered a freshly cooked stack to the table. The boy's enthusiasm was infectious, prompting Gregory to exclaim, "Holy Moly! Who wants pancakes?!"

His son pointed at him and said, "See, Daddy, you *do know* what moly is! What is *holy* moly?" Gregory's eyes lit up with recognition. *That's* where the word came from! It had come from him!

From mind mapping, Gregory's son knew him better than Gregory knew himself. Gregory thought *Wow! This kid is soaking up **everything** I'm doing, so I'd better watch what I say and do from here on out!*

SEE THE WORLD THROUGH YOUR CHILDREN'S EYES

Once children are 11 or 12, their ever-expanding mind-mapping ability reorganizes into adult sophistication. If you're not completely honest with them, they're going to pick up on it. (Just don't expect them to be honest in return.) Teenagers' disrespect toward their parents often stems from disappointment seeing how their parents lie and act blind to themselves.

It's hard to beat kids' mind-mapping radar. Realizing how early children's mind mapping starts, how fast it develops, and how much they understand lights a fire underneath many parents. This motivates them to deal with long-festering personal, marital, or family problems. Many clients can't face their children knowing that their kids see them for who they really were and recognize their shortcomings. Unfortunately, this isn't true of all parents, whether they know about mind mapping or not.

A client named Josh told me about the many ways he tried to prevent his alcoholic mother from drinking during his childhood and adolescence. She'd go on drinking binges for days at a time. He'd beg her not to drink, and she would agree. But then he'd find her empty liquor bottles hidden under the couch.

When Josh was 12 years old, he tried to enroll his mother in Alcoholics Anonymous. His father refused to support his efforts to keep his mother sober. One day Josh found his mother passed out on the living room floor. She was unresponsive to his attempts to rouse her, and broken furniture suggested she'd fallen. Panicked and afraid she was dead, Josh ran to tell his father to call an ambulance and then raced back to his mother's side. Josh couldn't believe how long it took his dad to casually amble into the living room.

Josh's face was wet with tears and flushed with fear. He looked up at his father and asked, "Is Mommy dead?!"

Josh's father prodded his wife's motionless body with his foot and smiled at Josh. He said, "Nah, she's just sleeping. She's tired. Let her be. She needs her beauty rest," before ambling back out of the room.

In that moment, Josh realized his father was lying. His father knew his mother wasn't sleeping because she was tired. She was drunk. Josh's father refused to help his wife or help Josh deal with her drinking problem. Josh carried this memory for 30 years without fully understanding it. The part that stood out in his mind was his father's smile. Why was he smiling about his wife's passing out drunk?

Josh didn't think this terrible memory could get worse–until he finally made himself visualize the living room scene and map his father's mind. That's when he realized his father wasn't smiling about his wife; it was about Josh. Josh's father didn't respect Josh's efforts to help his mother. Instead, he was enjoying Josh's distress. He enjoyed toying with Josh and lying to him about what was happening. This brought up more anger about his past, but it allowed Josh to finally see his childhood for what it was and move on.

So to summarize, there are three important reasons for understanding children's mind-mapping abilities. One involves your children. Another relates to your own childhood. And the third involves your present life. First, understanding children's mind-mapping abilities lets you do more than simply mitigate negative impacts on them. You'll gain a whole new appreciation and enjoyment of your kids. Parenting becomes a richer experience for you and a much better experience for them. You'll recognize things you've overlooked and savor precious moments of your children's childhoods before they're gone. It doesn't matter if your kids are now fully grown and have families of their own. They're still mapping you. Once you view their childhoods through the lens of mind mapping, you'll understand how they came to be who they are.

Second, looking at your own childhood through the lens of children's mind-mapping abilities helps you understand how you became who you are too. This forces your mind to grapple with difficult things you previously told yourself you were too young to comprehend.

Third, and finally, learning about children's mind-mapping abilities helps you deal more effectively with difficult people and situations in your current life. It lays the groundwork for understanding mind mapping in *adults*, which we're going to focus on in the next chapter.

4

DOES EVERYONE HAVE MIND-MAPPING ABILITY?

Discovering the extent of children's mind-mapping abilities should inspire you to open your eyes and take stock of your life. It can feel like the walls are closing in on you because there's no room to lie, have affairs, or take out your frustrations on your kids without negatively affecting them.

This shocking realization usually triggers others: If your four-year-old has mind-mapping ability, how far does this go? Does everyone have mind-mapping ability? Who doesn't? Or more to the point, does your seemingly oblivious spouse, parent, boss or coworker have mind-mapping ability too? This is not an innocuous question because the ramifications of the answers are powerful and far-reaching. Because of this, we'll approach this in a series of steps that build towards a final answer.

DOGS HAVE MIND-MAPPING ABILITY

First off, just how widely distributed is mind-mapping ability—not just among people, but among other animals? In Chapter Two, I said the scientific evidence for chimpanzees' mind-mapping ability is equivocal. The same is true for elephants, dolphins, and whales. (Notice these are all mammals.) But one animal has proven mind-mapping ability akin to humans, and you won't be surprised which one it is: dogs. Let me also be more precise about what I'm saying. Dogs not only have mind-mapping ability with other dogs, they have some ability to map humans.

Dogs have a very particular place in our hearts because they have mind-mapping ability. If you've had a dog, this comes as no surprise to you. But you may not have been aware this is the source of your connection with them. And did you know advertisers deliberately exploit your dog-connection to manipulate you into buying their products? Watch how many TV commercials have a dog thrown in, although Fido has nothing to do with the product being pitched. Once you understand how mind mapping hugely endears dogs to us, you appreciate your connection with them all the more. It makes you feel all warm and fuzzy inside to realize that your four-legged pal has you mapped. It may also make you glad your dog can't talk and tattle on you.

If you were impressed by children's mind-mapping abilities, get this: Recent research reveals dogs have enough mind-mapping ability to take their owners' sides when their owners have *minor* social conflicts with other people. At Kyoto University's Companion Animal Mind Project, researchers divided 154 dogs and their owners into 3 groups. In every group, each dog observed an interaction between its owner and two strangers, during which the owner struggles to open a box. In the first group, the owner asks one of the strangers for help. This stranger snubs the owner ("bad stranger") while the other stranger looks on ("neutral stranger"). In the second group, the strangers help the owner open the box. In the third group, the owner doesn't actually interact with the strangers at all.

After each interaction, the strangers offered the dog a treat to eat. Dogs in the first group that witnessed their owners getting snubbed, avoided the "bad" stranger in favor of the "neutral" stranger. They wouldn't accept something to eat from someone who refused to help their master. Dogs in the other two groups accepted treats from neutral strangers or those who assisted their owner. Researchers concluded dogs' mind-mapping abilities prompted them to side with their owners during negative social interactions with other people.[1] Pretty good for a non-human, isn't it?

Studies at the Canine Cognition Center at Yale University further investigated a dog's ability to pick up on human social cues. It's long been known dogs respond to pointing. If you get a dog's attention and point to a spot, the dog will go there (whereas a chimpanzee will not). But it turns out you don't even have to point. In one study, dogs who observed researchers glancing at a spot where a treat was hidden were typically able to find it. Other research, similar to the Kyoto University study, used human puppets to show that dogs can tell the difference between puppets that are "mean" and those that are "nice"—and when given the choice, dogs prefer to interact with nice puppets.[2]

These and other studies demonstrate that dogs are constantly observing humans, watching our subtle behaviors, and paying attention to what interests us. This is one of the main reasons we feel so connected to dogs. They're mapping us every step of the way!

When humans look at each other, our eyes automatically shift slightly to the left. This is called left gaze bias. You probably haven't noticed this, but it happens without fail. We look at the right side of each other's faces first (supposedly because the right side gives a more accurate representation of what's going on inside of our minds than the left side).

Well, dogs' eyes display left gaze bias when they look at humans faces! Dogs are the only animals that can see and understand our emotions by looking at our faces.[3] Dogs and their owners both experience an oxytocin surge when they gaze into each other's eyes, which humans normally experience in romantic encounters or after childbirth.[4] Dogs process meaningful words in the left hemisphere of their brain, and analyze tone and intonation in their right hemisphere, just like humans do.[5]

Just as you now look at four-year-olds differently, knowing dogs have mind-mapping ability gives you an entirely new appreciation for humankind's best friend. You can better appreciate what's going on behind those soulful eyes whenever a dog looks at you. I love dogs, and knowing what they're capable of sharing with me increases my appreciation of what's going on in their minds.

IF CHILDREN AND DOGS HAVE MIND-MAPPING ABILITY, DOES EVERYONE?

So now that we've canvassed mind-mapping ability across the animal world, let's focus in on things directly relevant to *your* life: If children and dogs have mind-mapping ability, what about the difficult people in your life?

For instance, what about a partner who seems oblivious to your needs and who repeatedly does inconsiderate or hurtful things? What about a teenage daughter who embarrasses you in front of your friends? Or a mother who often tells you, "You're an attractive girl, but you'd be a real knockout if you'd lose another ten pounds. I have some pants that would fit you because they're too big on me." Or, what about a co-worker who is always making cruel, sarcastic remarks? Do these people have mind-mapping ability? Because, if they have mind-mapping ability, then what they're doing become entirely greater offenses.

This is why I said earlier that the question of who has mind-mapping ability is not an academic matter. It's a highly emotionally-charged issue: Like most people, you probably console yourself by thinking these people must not understand the impact they're having when they say or do hurtful things. It's much easier to tolerate poor treatment by telling yourself your antagonists don't know what they're doing. The excuses we make for these people are often quite creative, because we're really making excuses for *ourselves*. Telling ourselves other people don't know what they're doing excuses *us* from having to really deal with them.

The three all-time most popular ways of accomplish-ing this are: *They don't understand the impact they're having.* Or, *They're in so much pain or so self-absorbed they're not paying attention to anyone else.* Or, *They're just trying to protect themselves by using an outmoded psychological defense mechanism from their past.*

But currently, the most fashionable excuse is *They're somewhere on the autism spectrum. They probably have Asperger's Syndrome.* The exploding number of autism and Asperger's diagnoses is fueled, in part, by attempts to avoid the difficult realization that people do what they do because they have mind-mapping ability, not because they lack it.

I told you about children's and dogs' mind-mapping abilities to prepare you for realizing just how pervasive mind mapping is. Your mind will probably want to escape the difficult realization that most people have pretty good mind-mapping ability–and there are lots of mean or cruel people walking around. I refer to meanness and cruelty because they can see their impact in other people's minds.

Mind-mapping ability can be impaired to varying degrees. But, by and large, most people have enough to engage in complex social interactions that require recognizing and responding to subtle expectations and social cues. Mind mapping isn't something you can turn on and off like a light switch. You can't be a titan of industry or a shrewd negotiator at work and be clueless about interpersonal relationships at home. You can't repeatedly build up a person's expectations and then let that person down unless you have mind-mapping ability. The same holds true for parents who pressure kids to maintain appearances in public to control what other people think of the family. Their concern for what's in other people's minds demonstrates mind-mapping ability.

So the next time your bitchy sister-in-law gives you yet another gift you really can't stand, perhaps you should stop criticizing her for not putting much thought into her gift. She may have spent a lot of time envisioning your face when you open her gift. From what you now know about mind mapping, she may know exactly what she's doing.

Parents' Impact on Children's Mind-Mapping Abilities

Deciding who has mind-mapping ability and who doesn't is a highly volatile issue because the answer determines how you view people's actions and their accountability. And as I said, your mind wants to err in the direction of believing people don't have mind-mapping ability when they actually do. One way this surfaces is in the common belief that people who come from troubled homes don't develop mind-mapping ability. So let's consider how a person's childhood family impacts his or her mind mapping. The answers may surprise you.

Lots of research has explored how parenting influences children's mind-mapping development. Many of these studies look at precocious development, meaning what makes kids have rudimentary mind mapping sooner than their same age peers. This started in the 1990s with a landmark study by Judy Dunn and associates on how family environment influences children's mind-mapping abilities. By actually observing and recording families at home, they discovered children were more likely to succeed on mind-mapping tasks during their third and fourth years of life if their families discussed feelings, if they used mental state language, and if mothers frequently attempted to control older siblings' behavior. [6]

Another leading researcher, Elizabeth Meins has also spent years teasing out how parenting affects mind-mapping development. She initially reported that children who had been securely attached to their mothers during infancy performed better at age four on the "Maxi moves the chocolate" task (described last chapter) than their insecure counterparts.[7] However, Meins subsequently explained her findings differently, proposing that the real link was mothers' mind-mindedness–their proclivity to treat their infant as an individual with a mind, capable of intentional behavior.[8] This arose from discovering that mothers of securely-attached babies were more likely to focus on their child's mental attributes (rather than physical appearance or behavioral tendencies) when given an open invitation to describe their children.[9]

Meins suggested that children who are securely attached in infancy outperform their insecurely attached peers on mind-mapping tasks because secure-group mothers treat their kids as individuals with minds.[10] She subsequently demonstrated that maternal mind-mindedness could be observed in infant-mother interactions during the first year of life, and individual differences between mothers were identifiable.[11] Mothers demonstrated mind-mindedness by appropriately interpreting their infant's behavior and saying things about their baby's thoughts, desires, intentions, and memories. Mothers' mind-minded comments turned out to predict attachment security better than maternal sensitivity, which has traditionally been regarded as the best predictor. [12]

Mothers' mind-minded comments when babies are six months old are the earliest facilitators of precocious mind mapping in children. This takes effect when babies are just learning to control their bodies, before they develop language or form attachment relationships. For instance, Dunn observed six-month-old infants and mothers engaging in free-play, and rated each mother's mind-mindedness. More than three years later, this was the best predictor of children's mind-mapping ability when Dunn tested them at age four.

However, Dunn found that only a mother's mental state language, which appropriately reflected their infant's state of mind, facilitated mind-mapping ability later on. Comments mismatched to an infant's mind state (inaccurate or inappropriate mind-related comments) had no effect.[13] Mothers' accurate mental state talk during infants' earliest months has a "scaffolding effect," allowing kids to understand their own behavior in terms of their underlying mental states.[14]

I find this information pretty amazing. But please remember: These studies DON'T prove that children who lack mind-minded mothers never develop mind-mapping ability or that their abilities are necessarily impaired in adulthood. The importance of this clarification will become clear momentarily.

What If Your Mother Wasn't Mind-Minded?

So what if you didn't have a mind-minded mother? Or parents who read books to you that talked about feelings? What if you came out of a troubled home full of conflict? To answer this, we have to consider Judy Dunn's subsequent ground-breaking research. Elizabeth Meins ultimately concluded that her work agreed with Dunn's studies linking general exposure to mental state language and mind-mapping ability.[15]

Dunn proposed certain family interactions promote mind-mapping ability because they confront children with conflicting views of the world. She specifically focused on discussions involving mind-related topics, sibling conflict management, joint play, shared jokes, and moral reasoning. She discovered conflict facilitates a child's understanding that reality is represented in a person's mind, and it can be (or often is) misrepresented.[16] There's also an "apprenticeship" aspect where children's interactions with older family members provide informal tutoring in mind mapping.[17] Just having an older sibling facilitates your mind-mapping ability.[18] You may also get mind-minded comments from people other than your mother (e.g., grandparents, aunts, family friends), or overhear comments made by you mother to other people, particularly your siblings.

You also get exposed to mind-minded comments outside the house. Contacts with older children and adults beyond your nuclear family promote your mind-mapping abilities.[19] You get a lot more experience with conflict when you start going to school with other kids. This is why Dunn's studies included entire families and children's interactions with peers, and she analyzed conversations recorded at home during unstructured observations.[20] So, even if your mother wasn't particularly mind-minded, most people go through plenty of experiences at home and in other relationships sufficient to develop their mind-mapping abilities.

For instance, Dunn and colleagues studied the participation in family conflicts by two-year-olds. They focused on how children come to understand other family members' feelings and intentions, and the family rules and punishments. They also looked at how this knowledge influenced kids' emotional functioning. As children understood other family members and the family rules, this increasingly shaped their teasing, supportive and prohibiting actions, and comments about violated rules. Children reacted differently to emotions displayed by various other family members and the topics of family disputes.[21]

Becoming a family member involves developing relationships with other people and understanding the relationships among them. You have to learn the family's routines, rules, expectations, jokes, games, prohibitions, and sanctions. Participating in family interactions has huge emotional significance to children, especially during family disputes. Intense encounters with siblings and parents over conflicts of interest and rule transgressions expand your understanding of other people's intentions and feelings, as well as conventional rules and morality.

Dunn found that, by two years of age, most children refer to feeling states in themselves and others, and discuss the causes of their feelings in a variety of contexts (including pretend games). Moreover, if mothers and older siblings made feeling-state comments when children are 18 months old, kids were more likely to be discussing feeling states when they were 24 months.[22] Other research confirms that most two-year-olds respond when other family members become distressed or angry.[23] Children at 33 months of age who participate in family discussions about feelings and arguments have better social understanding, emotion-labeling, perspective-taking, and false-belief identification at 40 months.

Family conversations about people, interactions with siblings, and relationships with and between other family members develop your mind-mapping ability and social understanding.[24] Talking with your mother and siblings about mind-related things when you're 36 months old makes you better able to understand other people's perspectives on emotionally-laden issues when you are age 6. If your family frequently talks about feelings in arguments and rule violations and punishments, it helps your subsequent ability to recognize emotions, but sheer frequency of family conversations doesn't make a difference.[25]

At 40 months, a child's ability to explain someone's behavior in terms of false beliefs isn't connected to his or her rudimentary understanding of emotions. But by the time the child reaches kindergarten, that child's emotional understanding encompasses mixed emotions and moral sensibility. You probably wouldn't think of this, but precocious understanding of other people's minds often leads kids to have negative initial perceptions of school and sensitivity to teachers' criticisms.[26]

Four-year-old kids increasingly talk about mental states in conversations with siblings and friends, and more so than with mothers. They can recognize a variety of emotional expressions (e.g., happy, sad, angry, neutral) and they understand external events cause emotions.[27] Those who engage in more mind-minded talk are better able to understand false beliefs. And if both your parents frequently mention mental states, you are more likely to have cooperative interactions with siblings and friends.[28]

Preschool-age children face daily moral issues and transgressions like acts of unkindness, failure to share, exclusion from play, teasing, and rule-breaking. Children as young as four can distinguish between different kinds of cultural breaches.[29] Dunn studied children's views on permissible transgressions against friends and their justification. She found children's justifications involving interpersonal reasons (e.g., friendship quality) utilized their understanding of mental states and emotions.[30]

Your first year of full-day schooling at age five is one of the most significant developmental milestones of early childhood, and mind mapping is hugely important.[31] For example, your mind-mapping ability at this age predicts your academic achievement two years later at age seven.[32] Once you start school (or preschool), your opportunities to develop your mind-mapping abilities expand exponentially.

As this occurs, kids increasingly use mental state terms in informal conversation with playmates.[33] But, it turns out there's a specific link between mutually reciprocated friendships and mind-mapping ability. First evidence of this surfaced when investigators discovered that four-year-old preschoolers who had at least one mutual friend were better mind mappers than peers who had none.[34]

Reciprocated friendships hone your mind-mapping abilities. Close relationships require becoming aware of how someone else's perspectives differ from your own. You get to explore other kids' thoughts, feelings, and reactions through the intimacy of reciprocated friendships. Conversely, mind-mapping ability facilitates mutual friendships when you're starting full-day peer contact in school. This allows you to attract friends and understand their thoughts and feelings well enough to keep relationships going.

But what if you weren't the most popular kid at school? What if you didn't have a lot of friends growing up? Children lucky enough to have at least one mutual friend are somewhat buffered against social isolation and peer rejection.[35] This is true throughout childhood and adolescence.[36] Many kids who are disliked or ignored by their peer group have at least one best friend, and they are satisfied with this situation. Conversely, if you're one of the "popular" kids at school but you have no mutual friends, you're more likely to feel lonely and dissatisfied.[37]

Other research explored how mind-mapping ability predicts mutual friendships from age five to seven years. Many group-rejected children had a mutual friend (53%), whereas three percent of those highest in group status did not. Five-year-olds with a mutual friend had better mind-mapping ability than their friendless peers. In contrast, chronically friendless seven-year-olds displayed lower mind-mapping ability. The better your mind-mapping ability when you start school, the more you're able to develop friendships and maintain them over time.[38]

But what if you're not the first kid on your block to understand false beliefs? What if you don't develop mind mapping sooner than most kids? What if you're just average? Or maybe even a little slow? Are you screwed for life? Not by a long shot.

Remember, mind mapping, first and foremost, is a survival skill. Humans probably evolved mind mapping in competitive situations in which resources were limited, and understanding your competitors provided a selective advantage.[39] Even other primates exhibit some surprisingly sophisticated mind-mapping abilities in competitive situations.[40]

So, does your mind mapping stop developing if you're a loner? Nope. You still have plenty of contact with other kids at school. Even if you're not usually picked when your peers choose up teams at recess, you watch which kids get selected and figure out why. You monitor the playground bully for signs he's coming after you. If you're engaging in small acts of vandalism, you're looking over your shoulder to see who's watching you. If you're playing up to your teacher for attention, you're using mind mapping to figure out how to do it. If you're depressed and moping about your house, you're mapping out whether or not your parents are paying attention and whether they care or not.

When you become a teenager, you start dating. You "solve your loneliness problem" by getting into relationships with people of your preferred sexual orientation. First you have a "crush" on someone. Then you graduate to "boyfriend and girlfriend" (or girlfriend and girlfriend, etc.). Eventually you start having sexual relationships. Maybe you're having sex with someone older. Maybe you're dating someone your own age (or younger). Even "robbing the cradle" involves huge amounts of mind mapping and implanting false beliefs. This probably isn't what researchers envision when they imagine studies of mutually reciprocated relationships.

Adolescence is the epitome of mind-blindness coinciding with constant mind mapping. Over the course of several partners–and some painful experiences–your mind mapping improves. If you're insecure and anticipating rejection and you desperately need approval–like most teenagers–you're scanning your girlfriend or boyfriend and your peer group all the time.

Then you get a job or go to school. (Remember your first day at work or college?) You're mapping everyone full time, trying to figure out where you fit in. *Who are the alpha males and females? Who are their surrogates and supports? Who is smarter or prettier than you? Who can you trust? Who should you watch out for?* How do you get a promotion or bonus, or get your professor to raise your grade? *Is that pretty girl smiling at you because she wants you to ask her out or because she thinks you're pathetic?* By reading her eyes and mouth—and blundering through lots of awkward and embarrassing moments—your mind mapping continues to sharpen.

Eventually you get into an ongoing committed (or uncommitted) relationship, or settle into single life and the dating scene. Do you think mind mapping stops when you get married?

The whole next chapter explores mind mapping in adult love relationships. How about meeting prospective partners from a dating website? Are you kidding? Is this person a serial axe murderer? Whether you're scrutinizing candidates for "partner for life," or trying to figure out the one you've got, loads of things are going to develop your mind-mapping radar.

WHAT IF YOU COME FROM A TROUBLED HOME?

But let's get down to the really tough question: What if you came from a troubled family? Comments reflecting other people's mental processes nurture your mind-mapping ability whether you're a child, a teenager, or an adult. When you're young, caregivers' and other adults' tendencies to explain someone's behavior by referencing their mental states develops your understanding of how beliefs and desires determine behavior. When you're older, if your backbiting parents and siblings are always bad-mouthing other people and attributing nefarious motives and agendas, you develop good mind-mapping ability at the same time you learn not to trust them.

What if your parents use mind-minded language inappropriately or inaccurately—like suggesting you're thinking or feeling things that you're not, or telling you that you are stupid? Remember back to Meins' study that found exposure to mental state language wrongly (incorrectly) reflecting an infant's state of mind had no impact on their subsequent mind-mapping ability at age four. I'm not saying parents who mess with your mind have no impact. But if your parents repeatedly use mind-minded language to screw with your head, there's a good chance you'll be a great mind mapper (and mind masker) with low self-esteem. I've seen this pattern in my clients a lot.

Many people—including therapists—believe that children don't develop mind-mapping ability if they grow up in really bad homes, places where parents constantly distort reality, family members manipulate each other, and lying is routine. Their faulty reasoning proposes that children don't develop mind-mapping ability if they're not mapping a coherent mind, or they don't receive straight answers. This doesn't take into account children's abilities to detect false beliefs. Children from troubled homes learn to take their parents' distorted thinking into account as their mind mapping develops. *Not a single study supports the widely held supposition that bad childhood parenting precludes mind-mapping ability in adults.*

Judy Dunn's observational studies reveal when family members most often talk about their own or other people's thoughts and feelings. The stereotype is this happens when parents discuss their kids' feelings or read books to them. But Dunn discovered this happens most often when family members are arguing and fighting! In tension-filled moments, people say things like *The way you stared at that younger woman made me really angry!* Or *You didn't congratulate me on my job promotion, which makes me feel like you really don't care!* Or *How dare you spend so much money without checking with me first! You think you can do whatever you want!* Witnessing arguments is unpleasant for kids, but this sharpens their mind-mapping abilities rather than hampering them.

This shouldn't surprise you. Once again, the primary purpose of mind mapping isn't to facilitate bonding, it is survival. Growing up in an unstable environment with little or bad parenting often produces excellent mind mappers. If your father is an explosive man who goes from calm to violently angry—bam!—just like that, you'll hone your mind-mapping abilities to improve your chances of predicting when he might blow up. Likewise, if your mother is a stumbling-down alcoholic, you're going to map her all the time to identify triggers that push her to drink, so you can steer her away from them.

As adults, people from troubled homes often tell themselves *I really didn't know how bad it was, because I was immersed in it. It was all I knew!* As convincing as this may sound, they're lying to themselves. Kids know whether what's happening at home is good or bad. They go to school, watch TV, talk to friends, and play at other kids' homes. They have many ways to figure out if what they're experiencing at home is normal or not. This shows up in staying away from home as much as possible or being embarrassed to have friends over.

In Germany where I often give lectures and workshops, there's a phrase that describes how kids figure out what's happening at home is wrong: *Strassen Engel, Haus Teufel,* which means "Street Angel, House Devil." This refers to parents who act much better outside the home than in it. When children see their parents deliberately present themselves differently in public, they know their parents know that what the parents are doing at home is wrong. This is how children determine their parents aren't doing the best they can.

There's probably no better example of this than the family court judge whose learning-disabled teenage daughter filmed him violently beating her at home.[41] The fact that the girl was prepared to film her father with her smartphone meant she mapped his mind. She knew he was going to beat her again, and he wouldn't want other people to know this.

Next, consider a mother who is a complete disaster inside her house. She's always screaming, drinking too much, and making vile, offensive comments. But the second she walks out her front door, she acts like Mother Teresa or Miss Manners. Her two-faced self-presentation doesn't confuse her kids' mind-mapping abilities. To the contrary, her children unfortunately conclude at a very young age that their mother is a fraud. Watching her get away with this leads them to conclude most (if not all) adults are frauds too and not to be trusted. They know they better look out for themselves because their mother can't be counted on. Her kids become shrewd mind mappers, who mask that they're mapping other people all the time.

People from dysfunctional families often fly under other people's radar by portraying themselves as emotional Neanderthals. This is particularly true of men. When they get married, they set their spouse up as the *de facto* relationship expert, while they pretend to be knuckle-dragging brutes. It's easy for them to get away with this with women who like to believe they're better at relationships than men. (Ladies, you know guys are emotionally stunted, insensitive creatures to begin with, don't you?!) This subterfuge bolsters popular myths that people from troubled homes don't have mind-mapping abilities. But folks from problem families are often more strategic in handling relationships and more manipulative than their partners ever imagine.

Women are often shocked to realize their seemingly dense hubbies not only have excellent mind-mapping ability, they also have mind-masking ability. (We'll discuss mind masking in depth next chapter.) Men's capacity to conceal their thoughts, feelings, and agendas often beats their wife's radar. You can avoid this pitfall if you remember the evolutionary origins of mind mapping. It doesn't stem from our mammalian urge to bond. It arises from our reptilian urge to survive. And from what I've seen, the more you are a human reptile, the more cunningly you employ mind mapping.

WHERE DO SCHIZOPHRENIA, AUTISM, AND ASPERGER'S SYNDROME FIT IN?

As mentioned earlier, mind-mapping ability isn't an all-or-nothing affair. Lots of brain studies have tested people suffering from brain damage and various disorders.[42] Three problems have received particular attention: schizophrenia, autism, and Asperger's Syndrome. Let's look at each one to help you gain perspective.

Some types of schizophrenia interfere with mind-mapping ability, but not all. If you suffer from catatonic schizophrenia, spending your day in a fetal position in the closet with a blanket pulled over your head, you're not going to be mapping other people. On the other hand, if you're paranoid schizophrenic, you're mapping other people *all the time* and thinking they're all mapping you. Why do some paranoid schizophrenics wear tinfoil on their heads? It's to prevent Martians or government agents from mapping their thoughts and stealing them! Mind mapping is a full-time preoccupation if you're paranoid schizophrenic. In fact, some researchers conceptualize this disorder as a problem of overactive unmodulated, mind-mapping ability.

Similarly, some forms of autism and Asperger's Syndrome interfere with mind mapping, but it's not true that autism and Asperger's *de facto* inhibit this ability. This is an important point because of the unfortunate explosion of autism spectrum diagnoses in recent years. I've seen a corresponding increase in the number of clients who've told me their prior therapist diagnosed them as having mild autism or Asperger's Syndrome, or they've diagnosed themselves this way. According to a useful article in *New York Magazine* entitled "Autism Spectrum: Are You On It?" it is fashionable to see yourself as having some level of autism.[43] The diagnosis is everywhere.

Psychologist Myrna Siegel, director of the autism clinic at the University of California, San Francisco since the 1980s, says she sees more false-positive assessments by clinicians than ever before. Therapists often misattribute someone's callous or cruel behavior to a supposed inability to anticipate the impact of his actions, and misdiagnose sadism masquerading as disability.

Likewise, people often refer to themselves as having "a touch of Asperger's." I'm quoted in this article as saying having Asperger's-like symptoms does not mean you have Asperger's any more than having a big belly necessarily means you are pregnant. Autism is the new ADHD. Lots of people say they have a little of it. But having difficulty learning doesn't necessarily mean you have attention-deficit hyperactivity disorder.

But rather than getting caught up in who has autism or Asperger's Syndrome and who doesn't, here's what you need to know: Even if you have autism, it doesn't automatically mean you lack mind-mapping ability. Many high-functioning autistic people have good mind-mapping abilities. And for those who don't, there are well-developed programs that can teach them how to develop it. People who actually have Asperger's Syndrome often have mind-mapping ability too. I've become particularly wary of folks who self-diagnose themselves to have these ailments. *If you make (real or imagined) autism or Asperger's Syndrome an excuse for your bad behavior, you're actually demonstrating the mind-mapping ability you claim not to have.*

Research indicates high-functioning autistic children can be taught mind mapping well enough to understand false beliefs, distinguish fantasy and reality, and comprehend the relationships between beliefs, desires, and reality. They can also recognize five levels of mental states involving pretense and emotions, discriminate between four basic emotions (happy, sad, angry, and afraid), and understand the principles underlying these fundamental mental states. They actually accomplish mind mapping differently than you and I do, but they can still do it.[44] High-functioning autistic people often develop sufficient mind-mapping abilities to function productively in society.

So what explains the rapid rise in autism diagnoses? Better reporting and more inclusive diagnostic standards may account for some of this.[45] But some of this surge comes from psychotherapists who emphasize "safety and security," core components of the most popular "attachment-based" therapies today. These therapists avoid identifying cruelty in clients—particularly when it's directed at a spouse or a child—because it interferes with the picture they're trying to create and might potentially destabilize a marriage or family.

By assigning autism spectrum diagnoses and believing people with autism or Asperger's Syndrome lack mind-mapping ability, therapists are able to construct an alternate reality for clients where people who repeatedly do inconsiderate or downright hurtful things supposedly don't know what they're doing. An autism spectrum or Asperger's diagnosis unfortunately becomes a "Get-Out-of-Jail-Free card" that superficially stabilizes troubled relationships but keeps real problems from being addressed.

Certainly some children *do* have autism or Asperger's Syndrome, and early diagnosis and intervention are terribly important. I've worked with families with autistic children, and I know how badly social services need to be expanded to help them as their kids become adults. Moreover, some grown adults today should have been diagnosed "on the autism spectrum" when they were younger. But misusing and abusing autism and Asperger's diagnoses trivializes and disrespects people who actually suffer from these disorders.

The difference between being cruel and making a mistake is whether or not you anticipate or recognize the impact you're having. This is why the question of who has mind-mapping ability and who doesn't is such a big deal. If your partner actually has the mind-mapping ability of a fish, then your tendency is to accept his behavior and forgive repeated hurtful actions. But if he's tracking you and continually pushing the boundaries of what he can get away with... well, does the name Lorena Bobbitt ring a bell? [46]

MIND MAPPING IN CHILDREN WITH AUTISM

Many therapists' textbooks contain flat statements that people with autism spectrum disorders don't have mind-mapping ability.[47] I remember a mother getting angry at me for saying autistic children don't have mind-mapping ability. She was quite proud of the social skills her son worked hard to develop. It turns out that the answer to the question "Do autistic people have mind-mapping ability?" boils down to how you look at mind mapping.

According to a scientific review of the literature, the false belief tests I described earlier became the litmus test in research and biased scientists' answers to this question.[48] Other socially relevant aspects of false-belief comprehension are rarely considered, such as sharing pretense,[49] communication,[50] and sensitivity to criticism.[51]

The "mind-mapping deficit" hypothesis of autism began in the mid-1980s with the discovery that children with autism fail to infer another person's false belief.[52] This led to proposals that autism's cardinal social-communication impairments could be explained by a specific deficit in cognitive meta-representation.[53] This changed the future of autism research for the next two decades. Subsequent studies demonstrated that autistic children also had difficulty understanding the knowledge,[54] intention,[55] and complex emotions of others[56] as well as deception,[57] jokes and lies,[58] irony, white lies, and double bluff.[59]

However, further research on pretend play in autism indicated that lack of spontaneous imaginative activity could not be attributed to a failure in cognitive decoupling and meta-representations. Children with autism could engage in pretense when prompted.[60] Children with higher verbal mental age showed spontaneous pretend activity of a stereotyped nature.[61]

So it turns out this mother was correct. Her son did indeed have mind-mapping abilities, although he couldn't pass the false-belief test. Distinct components of children's mind-mapping abilities emerge at different stages of development, and show up in their real-life social experiences and competencies.[62] Sweeping over-generalizations about people with autism or Asperger's Syndrome lacking mind-mapping ability are unwarranted.

If you want a list of all the diseases and disorders that can impact mind-mapping ability, you'll find this in Appendix A. Just keep in mind that the studies that make this compilation may be subject to this problem of how they operationalize mind-mapping ability.

WAIT! MY HUSBAND HAS MIND-MAPPING ABILITY?!

Earlier I challenged the misguided notion that women are much better than men at intuiting other people's thoughts and feelings. According to brain science, there's no gender difference in mind-mapping ability. But both men and women promote this erroneous idea for different reasons. Women do it to support the picture that they're more sensitive and more into intimacy and bonding. Men do it because it lets them get away with a lot more things.

I'm particularly fond of a cartoon that illustrates what I'm talking about. A woman walks into her kitchen to find her husband applying nonstick spray to the *outside bottom* of a cookie sheet, instead of the upside where the cookie dough actually goes.

"What are you doing?" she asks him.

"Your dad and I wanted some chocolate chip cookies, so I'm making a batch," the husband explains. "The recipe says to spray the bottom of the cookie sheet with nonstick spray, so that's what I'm doing."

The wife's bulging eyes say it all: *When it comes to anything kitchen-related, men are such idiots!*

"Let me do it," she says. "I'll bring the cookies to you and Dad when they're done."

The man walks out of the kitchen and returns to the den where his father-in-law is watching a football game on the television. The husband exclaims, "It worked! She's making us a batch of cookies!"

His father-in-law grins. "I told you it would!"

In other words, the father had his daughter mapped. He knew she carried a mental picture of men being completely helpless in the kitchen, and he utilized it to manipulate her. People often use mind mapping to convince others to do what they want them to do—especially by implanting false beliefs of being incompetent or unable to do things. This cartoon offers a great example of men having mind-mapping ability but trying to appear as if they don't. When women realize this, their pictures of their relationships with men instantly change—but rarely in the direction of more tolerance.

When you think your partner is a stereotypic, emotionally ignorant guy, it's easier to put up with his behavior you don't like. But when you realize he can map you just as well as you can map him, and he's using this ability to control you, this makes you much less accepting of the status quo. Men no longer get away with manipulating women who are willing to give them the benefit of the doubt when they act oblivious to what they're doing.

This isn't limited to heterosexual couples. If you're a gay

man whose partner likes to act clueless, you might handle things differently the next time he says, "I had no idea you'd be upset if I had sex with somebody 15 years younger than you!" Just because you're a guy doesn't mean you're immune from buying into this stereotype about men being obtuse.

So in summary, here's the answer to the question posed in this chapter's title: The vast majority of people have mind-mapping ability. These abilities are far more sophisticated and work much better than you might infer from the way people treat each other. As you'll see in the next chapter and throughout the rest of this book, this has enormous ramifications.

5

MIND MAPPING IN ADULT LIFE

Thus far we've reviewed the scientific research about mind mapping. We've seen how children's mind-mapping abilities are amazingly robust, emerging early in childhood and transitioning into adult sophistication in early adolescence. Even kids growing up in dysfunctional families develop mind-mapping and mind-masking capabilities, and men are as good in this regard as women. This is why most adults are good mind mappers.

However, we've only scratched the surface of how mind mapping shows up in daily life. In this chapter we'll look at it from the practical perspective of adult love relationships. Appreciating its complex role in common relationship difficulties can help you better understand what's going on in your own life and turn such problems around.

MIND MAPPING IN MARRIAGE

Mind mapping plays a central role in healthy satisfying relationships as well as in troubled, destructive ones. The more you value good relationships in general, the more important it is to understand how mind mapping plays out for better or worse. This particularly holds true for marriage. It's hard to find better examples of how mind mapping underlies the best and worst of times.

When you map strangers on the street, your assumptions about their agendas will likely have very little impact on your life. When you map your partner, however, the impact can be life-altering. Let's say, for example, your wife prepares a meal you love; but she herself doesn't like. If you're in a good, long-term marriage, you might attribute her kind and thoughtful gesture to her continuing generosity and consideration. You could thank her for going out of her way to do this nice thing for you, or you might not give it a second thought because you take your wife for granted. Either way, your wife's behavior fits your map of her that you've developed over time.

If you're newly married, however, you might not be sure what to make of this, because you're still mapping out your wife. You'll probably err on the side of optimism and attribute her making this meal to her thoughtfulness, how nice she is, and how lucky you are to have married her. When you picture her cooking the meal, you envision her thinking *I don't like this food, but I know he does. I know this will make him happy, and it makes me happy to do this for him.*

As much as you like eating the food–it's your favorite meal!–this accounts for only a small part of your pleasure. You derive particular satisfaction from the meaning encoded in the mental picture you've made of your wife, happily humming as she prepares your food. Even though she won't be eating any of it, she's delighted to please you. If the picture you've created of your wife's mind turns out to be accurate, you're a lucky man. This is a good example of how mind mapping can play a positive role in your romantic life.

But what if your mental picture of your wife involves her angrily shoving pots and pans around? What if she cooked this meal, which she hates but you love, begrudgingly? Are there strings attached? Does she think you owe her something? You'll use your mind-mapping ability to figure this out. If you decide the image of your wife throwing pots is correct and she now feels you are indebted to her, your favorite meal won't be nearly as satisfying.

Alternatively, let's say you've been married for ten years, and you and your wife have grown indifferent—if not hostile—to each other. Your initial reaction won't be to attribute this to her kindness. You'll probably think *Did she do this because she wants that new car she's been talking about? Or is she feeling guilty about something? Did she spend too much money at the store again?!*

If you and your wife haven't been getting along, you won't let her unusual and unexpected behavior pass without mapping her out. She's violated your map of her, and you're not going to relax until you figure out what's going on. You'll ask yourself *Why did she do this? What's her motivation? Is she making a peace offering? Has she finally seen the error of her ways and realized what a great guy I am? Or is she preparing to spring something new on me that I'm not going to like? Is she trying to disarm me and butter me up? Or has she poisoned the food so she can collect on my life insurance?*

With the stakes potentially so high (hopefully, your wife isn't trying to kill you!) you are going to map her very carefully. You may not eat a single bite of your favorite meal; and if you do, you probably won't enjoy it very much.

The good news is you won't be starting from scratch. You'll come to a decision fairly quickly. Your brain recognizes familiar things much faster than it does novel ones. By working off your time-tested map of your wife, you can quickly narrow down her intentions in cooking this meal. The bad news is your existing map predisposes you to interpret novel situations in line with your expectations and overlook the possibility that your partner is doing something new to make things better (or worse).

Maybe you should give her the benefit of the doubt. Maybe she's intentionally violating your map of her in order to grab your attention. Perhaps she didn't cook your favorite meal just to please you. Maybe she did something nice to set the stage for a long overdue conversation about how you ignore her and never talk to her and how unhappy this makes her. Does her map of you say you're going to take her words as a rebuke and you're going to get huffy and defensive? Was cooking your favorite food her attempt to soften your reaction, by planting the idea in your mind that she still cares about you?

Something motivated your wife to cook your favorite meal, but whether it's good or bad remains to be seen. Deciding whether she's being underhanded and trying to manipulate and control you or she's making a good-faith effort to save your marriage could very well shape the rest of your life.

In every scenario you can imagine, from making you feel loved to trying to kill you, your wife is mapping you and using her map to deal with you. Asking her why she cooked your favorite meal will only get you so far–especially if she's trying to kill you for the insurance money. You'll still have to interpret her response, and this will depend on the accuracy of your pre-existing map of her, your openness to updating this map, and your ability to map her mind in the moment. Once you understand how mind mapping constantly occurs in adult love relationships, seemingly straightforward interactions such as cooking a meal become intense dramas and mind-mapping storms.

There's another important lesson here: Intentionally violating your partner's map of you can be especially helpful when you're experiencing serious difficulties in your relationship and trying to turn things around. In a hostile environment, subtle positive gestures often go unnoticed. Partners tightly grasp the negative maps they've made of each other. Sometimes doing something constructive that's out of character and a bit flashy can be extremely beneficial. The same holds true when you're trying to remedy boredom in a long-term relationship. As long as you act in ways consistent with your partner's map of you, he or she won't pay close attention to you. He might nod his head and say, "Yeah, sure, alright," but he won't *really* be listening to you. I'm not just referring to mealtime conversations. This often applies to sex too.

Mind Mapping and Sexual Desire

Most couples experience sexual desire problems at some point in their relationship. Here's why: Even if partners start off with the same level of sexual desire, eventually one partner wants sex more often than the other. There's always a "higher desire partner" (HDP) and a "lower desire partner" (LDP).[1] HDPs are neither "better" nor necessarily healthier than LDPs.[2] Some LDPs know more about sex or like sex more than HDPs. HDPs are simply the ones who more frequently initiate sex.

However, LDPs always control sex, whether they know it or like it or not—but not because they want to be withholding. This happens because they decide which of the HDPs' initiations they're going to accept, and this determines when sex happens.[3] This result of evolutionary biology creates ongoing tension and conflict in relationships, much of which occurs through mind mapping. Every time they go to bed, partners map each other trying to gauge the other's sexual interest. Even if neither partner says a word, there's still an explosion of mind mapping.

HDPs constantly study LDPs, looking for signs of receptivity to a sexual advance. *Does she seem like she's in the mood? Is she relaxed or tense? Is she wearing sexy lingerie (which means I'm going to get lucky), or her flannel pajamas (which means no nookie tonight)? Did she light a candle, or is her nose buried in a book?* Mind mapping provides answers to these all-important questions.

LDPs similarly map their HDPs for cues of sexual interest. Scanning and warding off sexual advances often start long before bedtime. The longer couples go without having sex, the more LDPs monitor their partners. LDPs become adept at masking their mind mapping, so they may look like they're deeply engrossed in a book when they get into bed, but that's a facade. They're mapping their HDPs from behind those pages, looking for signs of sexual initiation they'd rather not see.

Simply by touching their partners' arms and glancing at them inquisitively, HDPs implant their message: *You know I want to have sex. Well? What about it?* In some relationships, this triggers a withering defense from LDPs. They'll shoot their HDPs a scathing look that says *I really don't feel like it!* like they've been offered an opportunity to kill baby seals. In other relationships, however, LDPs refuse to acknowledge the HDPs' overture. Their nonverbal response—*I have* zero *interest. Don't even think about it!*–is no less contemptuous. Some HDPs turn away, acting like they never made a sexual approach. Others pout or mope in ways that say *You've hurt my feelings. You're a selfish person!* Regardless of the specifics, these HDPs and LDPs initiate a tsunami of mind mapping, and none of them will be having sex anytime soon.

As the sexual drought increases, so does couples' mind mapping–as long as LDPs think their HDPs still want sex. When enough time elapses without HDPs threatening divorce, LDPs take this to mean their partners have given up and reluctantly accepted staying together without sex. At this point LDPs' mind mapping may decrease because they think there's no reason to remain on high alert. However, sometimes HDPs have implanted a false belief that they are still monogamous, when in fact they're not.

This plays out differently in less polarized couples. When LDPs map out that their HDPs are losing interest, they know it's the time to initiate sex. The LDPs aren't suddenly overwhelmed with passion. They realize it's no longer safe to continue refusing sex because their HDPs might satisfy their sexual needs elsewhere. By having sex, LDPs implant the false belief in their HDPs' minds of more sex to come. In reality, this gives LDPs another six months before they have to initiate sex again. This pattern revolves around mind mapping, which some couples continue for years.

When some HDPs recognize what's going on, they realize their partners' sexual disinterest isn't going to change and they ask for a divorce. This prompts LDPs to furiously mind map their HDPs. *Are they bluffing? Is this just a threat? Are they really willing to leave?* When LDPs see that HDPs are serious and ready to go, this can trigger a crisis that improves relationships because this is when many couples seek professional help. However, if LDPs propose treatment to buy time and implant the false belief they're open to changing, this eventually surfaces in therapy.

In Part Three, I'll show you how to use mind mapping to resolve sexual boredom and enhance sexual desire, but here's a quick tip for now. There's method to the madness of getting buck naked, putting on a tie, draping a towel over your arm, and greeting your partner in bed with the announcement, "I'm Carson, the butler (or Anna, the Lady's maid), and I'm here to serve you!" You can figure out your own variations, but the goal is the same—to step outside your partner's map of your eroticism.

Don't be put off by your partner's laughter or a negative reaction that says *You have to be kidding!* Your partner will probably be awkward and anxious too, not knowing how to respond. And that's exactly what you want. Using mind mapping, you've suddenly created the same circumstances that makes sex interesting at the outset of a relationship: You don't know what's going to happen next. You have to pay attention.

As further proof, visualize the best sexual encounter(s) you've ever had. When you visualize your partner, you're probably seeing more than his (or her) body. You're seeing him (or her) looking at you. You're actually seeing a lot more than the two of you having sex. You are mapping his (or her) mind.

MIND MAPPING AND SEXUAL DYSFUNCTIONS

Virtually everyone experiences sexual-performance difficulties at one time or another. Some women have never had an orgasm in their lives. Others have no difficulty by themselves, but they can't reach orgasm with a partner. Some can have an orgasm receiving oral sex but not during intercourse, and for others it's the other way around.

Now imagine you are one of these women—take your pick—and visualize what's going through your mind while receiving oral sex from your partner. You better believe you'd be mapping him (or her)! You want to know *Is he getting tired? Is he bored? Does he think I'm taking too long? Is he starting to get lockjaw? Is he enjoying himself, or does he want this over with already? Will he be angry or think less of me if I don't have an orgasm?*

On the other hand, one out of three men struggle with erectile difficulties, and another third have premature ejaculation. One out of twelve men has difficulty having an orgasm (just like one out of three women). When these difficulties arise, men react the same way women do.

Imagine you are one of these guys, and your dysfunction shows up. Whether you can't get it up or keep it up, or you ejaculate too quickly, or you never orgasm at all, you start furiously mapping your partner's mind rather than relaxing and having a good time. You want to know *How disappointed is she? Is she angry at me? Is she taking this personally? Should we just stop, or does she want me to stimulate her another way? Does she think I'm less of a man because of this?*

To show you how pervasive mind mapping is during sex, let's say you're *the partner* of a man who has erection or orgasm problems. You develop elaborate mind mapping and mind masking to accommodate his difficulty. You have to look like you're interested, even enthusiastic because you don't want to damage your partner's self-esteem. You moan and groan at all the appropriate times to implant the idea that you're having a good time.

But at the same time, you don't let yourself get turned on because that only leads to more frustration in the end. You have to mask your mind so your partner doesn't see your disappointment because that makes him feel even worse. In the name of love and intimacy, you shut your mind to your partner. You don't want him reading what you're doing, even if you're trying to be kind. Unfortunately, hyper-vigilant guys who just "failed in bed" will likely interpret your mind masking as emotional withdrawal.

Couples struggling with sexual dysfunctions often turn off the lights when they attempt to have sex. Visualize yourself trying to mask your disappointment in the dark when things don't go well. Imagine how lonely and painful it is to hide from the person you love while you're doing what's supposedly the most intimate thing partners can share. You end up having sex by yourself in the company of your partner. It's terribly sad when sex highlights how far apart you and your partner really are, but darkness offers no refuge. If you and your partner know each other, there's no place to hide.

If you're having sexual problems, the first step in turning things around involves allowing your partner to map your mind. It's easier to get your genitals working when you're not putting on an Oscar-worthy performance to keep your partner from seeing you're a nervous wreck. Stop masking your mind from your partner. This makes your partner assume the worst. Instead, deliberately open your mind, so your partner can read you. Speak openly about what you're doing and what's happening between you, and invite your partner to do the same. This will bolster your connection and allow the two of you to relax in each other's presence.

These simple steps are often enough to resolve some sexual dysfunctions. But even when they don't, you will have solved a large part of your problem. You won't feel as lonely, and you'll have a collaborative alliance with your partner. My prior book, *Resurrecting Sex*, offers specific techniques to resolve most common sexual problems, but none of them work well when partners mask their minds from each other.[4]

Mind mapping also plays a crucial role in the best sex you'll ever have. I'll show you how to put this into action in Part Three of this book. However, good genital functioning doesn't guarantee good sex. Some people look like sexual athletes who can go all night having multiple orgasms, but they never let their partner into their mind. If you're interested in sexual intimacy, regardless of what your body is doing, there's no better way to create this than through mind mapping.

Mind Mapping in Affairs

As mentioned previously, mind-mapping allows people to deceive and manipulate each other. Many evolutionary anthropologists believe we developed our ability to detect lying, deception, misdirection, and manipulation around the same time we learned how to pull off these shenanigans. Does this facilitate stable, long-term relationships? The complex answer is yes and no. No one wants a relationship with someone who lies and deceives. However, deceiving and detecting deception have allowed people to keep families together long enough to build complex civilizations.

Most people acknowledge the importance of "emergency lies,"[5] even if they've never heard the term. These are "little white lies" designed to spare another person's feelings. For example, what do you say when your partner asks what you're thinking about while you're having sex? The "correct" answer is, "I'm thinking about you, dear!" even though your mind has drifted to work issues or the pretty young barista at your favorite coffee shop. You are masking your own mind and trying to implant a false belief in your partner's head.

People differ on what constitutes an "emergency" requiring an emergency lie. As originally intended, the emergency is an urgent need to avoid hurting someone else's feelings, and the lie is relatively minor. But for people in committed relationships who also have affairs, the emergency is usually you just got caught and you want to save your own ass. In such cases, your lie is a whopper like *I swear I was working late in the office by myself. My secretary went home hours earlier!*

The sad truth is many of us can only keep our relationships going by lying, deceiving, and manipulating. Our transgressions don't have to involve affairs. It could be how we handle money, share information, or make decisions. To get away with manipulation, you have to be a good mind mapper and mind masker. For example, you have to figure out what your partner knows and suspects. You have to implant false beliefs in your partner's head to cover your tracks. You have to know if your partner buys your lies or not. If you succeeded, you need to shut up because you're more likely to get caught if you keep talking. If you haven't succeeded, you need to keep going. If you partner discovers holes in your story, you'll need to implant more false beliefs to plug them.

Your career as an adulterer hinges on your ability to conjure up excuses your partner will buy. "I was late because my car ran out of gas" works better than "I forgot to call because the battery in my phone died" because your partner will pick up on the bad logic of your statement. If you say, "My cell phone doesn't work at night," you should rethink your Casanova-like ambitions because you're a terrible liar. When you get caught, saying "I didn't mean to hurt you" doesn't get you very far. Although it's counterintuitive, you're often better off saying "I figured you'd be hurt, and this was okay with me because I was angry with you." Your partner is more likely to believe you.

Deception precludes having a true partnership, but it sure keeps lots of relationships going. This doesn't bother many of us who don't really form collaborative alliances. We take prisoners instead. We keep our partners perpetually in the dark or confused because we know they'll be out the door if they ever get an accurate picture of us. Wrapping your mind around this changes your picture of what drives many marriages. Mind mapping lies at the core of all human relationships; but unfortunately, many marriages can't survive without mind masking.

I'm certainly not advocating affairs or lying to your spouse. I'm saying you can be certain your partner has mind-mapping ability if he (or she) has managed to have an affair for some time. This is important if you really want to understand what has happened and with whom you're dealing. Having worked with numerous couples, I'm acutely aware how affairs erode relationships and frequently lead to divorce. However, this doesn't necessarily doom a marriage because that same mind-mapping ability used for deception can also help heal the rift.

If you and your partner force yourselves to be open with each other and allow your minds to be thoroughly mapped, this provides a solid foundation on which to rebuild your relationship. If, however, you continue masking your minds, your relationship is more likely to end in divorce. (The more mind masking that occurs, the nastier the divorce tends to be.) To have any chance of finding peace in your relationship, remember this: Your partner's mind won't rest until everything makes sense and all the details fit together. Continuing to lie is self-defeating, unless you think staying together is more important than being happy together.

SINGLES AND MIND MAPPING

What about single adults? If you're looking for a mate, mind mapping is an indispensable part of the process. You need to penetrate the deceptive self-descriptions people post on dating websites. If you take the next step and decide to meet face to face, you need to use your limited time together to best advantage. After ruling out that he or she is a deranged serial killer, you have to track whether this person is interested in what you say or in having sex with you, or both, or neither.

But there's more to finding a partner than selecting the right person. You have to manage your interactions to lead to a second date. You have to gauge how much personal information to disclose, what to reveal, when to disclose it, how to present it, and what to hold back. *Are you talking too much? Does your date find you fascinating? Are you laughing too much at his jokes? Should you claim a sudden headache and run away as fast as you can?* You'll be mapping potential partners in search of answers while appearing to be innocently chatting away.

THERE'S A WHOLE LOT MORE GOING ON THAN YOU THOUGHT

Here's a riddle for you: How can you tell the married couples eating in a restaurant? They're the ones not talking to each other! If you've ever seen discomfited couples, who look like they have nothing left to say to each other, then you know what I'm talking about. Both partners stare at their food amidst a stony silence lingering between them. If you're single, these couples make you glad you're dining alone.

In contrast, partners on their initial dates look into each other's eyes and chatter away, forming maps of each other's minds. The issues discussed are relatively easy: *Where did you grow up? What's your occupation? What's your favorite color? What movies or sports do you like? What political party do you vote for? Are you liberal or conservative? Where do you go on vacations? Are you vegetarian?* Everyone is on their best behavior, laughing at each other's jokes. Neither partner rolls his or her eyes like the joke's been heard a million times.

Partners who've been together for years have mapped each other thoroughly across a variety of situations. They know each other backwards and forwards. This is why married couples in the restaurant aren't talking to each other. It's not like they have nothing left to say. They've discussed all the easy subjects new couples talk about. All that remains are difficult issues neither partner wants to discuss, like *Why don't you stick up for me with your mother?* Or *Why do you expect me to take out the garbage all the time?* Or *Are you having an affair with your yoga teacher?* And how do married couples know their partner doesn't want to hear what they have to say? Even if you and your partner are at odds and "not speaking to each other," you're still communicating via mind mapping all the time.

No matter where you look, from the best to the worst things emotionally committed relationships have to offer, mind mapping will be at center stage. In Part Two, we'll take a closer look at mind mapping in adult love relationships through in-depth case studies. But for you to really appreciate them, you have to understand more about mind masking. That's the focus of our next chapter.

6

Mind Masking: Defeating Mind Mapping

Now let's plunge deeper into the depths of mind mapping. To help you organize everything you're learning, I've developed a framework outlining four levels of mind-mapping ability, wherein each of the levels includes those preceding it. Thus far we've been discussing people at the first and most basic level: They have the ability to map other people's minds. I refer to these folks as "mind mappers" or "trackers."

Second Level of Mind-Mapping Ability: Mind Masking

The second level of mind-mapping ability involves mind masking, which heretofore I've only touched on briefly. *Mind masking is the ability to screen your mind from other people,* making it more difficult for them to accurately and rapidly detect your inner mental states. This could involve shielding what you actually want, know, think, feel or believe. People at this level–known as "maskers"–can map your mind while simultaneously masking their own. They might look like they're hardly paying attention to you while they're figuring you out. Or they might look like they're being open and honest with you while they're lying through their teeth.

Mind masking of some sort is an integral part of all human interactions. But good maskers excel at keeping you from mapping them. As noted in Chapter Three, children start engaging in deliberate deception around the age of four. So it shouldn't surprise you that most people are not only proficient mind mappers, they're skilled mind maskers too.

WHY DO PEOPLE MASK THEIR MINDS?

Just as your brain has the ability to make a mental map of another person's mind, it can attempt to keep other people from mapping what's going in your own mind. Mind maskers are perpetually vigilant for signs that other people are trying to map them. They scan every interaction for indications someone is trying to get into their heads.

For some people, mind masking is a way of life, a way of being in the world. They can seem affable and charming—as long as you keep your psychological distance and don't complain, question, or challenge them. But if you trigger their mind-mapping radar and make them feel like they're being probed, they quickly become hard-edged and send you a not-so-subtle signal to back off. For instance, you better not complain that it's easier to decipher ancient Sanskrit texts than it is to map your partner. Henceforth you'll probably find it easier to get into a government top-secret facility than to get into your partner's head.

Mind masking always entails some degree of deception, but that doesn't mean it's always antisocial. *Sometimes* it is, but not always. Here are some important prosocial examples of mind masking:

1. *Mind masking creates tolerable interpersonal space.* There's a saying from several decades ago that might sound out of place in today's "tell-all" society but which I believe is still applicable: "Marriage is improved by the two or three things not said each day." What if you couldn't get away with "emergency lies?"

2. *Mind masking allows you to hide what is important to you or what makes you happy.* For instance, if

your parents take pleasure in your unhappiness, letting them know you're eagerly anticipating something guarantees that's the first thing they'll take away as punishment. Similarly, how could you negotiate good contract terms if you couldn't mask your mind?
3. *Mind masking helps you avoid being a target of emotional or physical abuse.* Oftentimes it's important to hide what you feel about someone who is annoying, frightening, or abusing you. If you weren't able to do this, your own thoughts and feelings would invite more abuse.
4. *Mind masking provides necessary camouflage in certain situations.* What if you couldn't mask your sexual thoughts and feelings at parties, enabling everyone to read you? Without mind-masking ability, you'd be forced to endure a lot of awkward silences (or slaps in the face) every time you rode in an elevator with someone you found attractive!

There are times where mind masking facilitates positive intent, like when you've received bad news and want to wait until you and your partner have a private moment to share this. Or you've given someone a gift wrapped in a box, and you want her to guess what it is before she opens it. Or your father-in-law repeatedly tries to goad you into a political debate at family events. Or a dangerous-looking stranger tries to get your attention as you walk by.

If people could read everything going on in everyone else's heads, we'd all be in a world of trouble. If you couldn't mask your mind, it would be virtually impossible to maintain a serious relationship, hold a job, or engage in common daily interactions. Simply put, people who can't mask their minds can't function in society.

In adult love relationships, mind masking is a fact of life. Most people depend on getting a positive, reflected sense of self from others. We present ourselves to others as we would like to be seen, and this usually differs from who we really are. Maintaining this facade requires constantly masking your mind. You may think of this as preserving your privacy, but you're intentionally hiding your true self. In actuality you're using mind masking to deal with your fear of being rejected if you were ever really known. And without mind mapping, how else would you convince other people that you don't really see them?

- How could you persuade your girlfriend that you love her for who she is, including her infantile obsession with all things Kardashian, and not just for her smoking-hot body?
- Conversely, how could you convince your money-hungry, morally bankrupt boyfriend that his Wall Street salary actually means nothing to you?
- What if these two people got married? How could they stay together if they couldn't mask their minds?

Mind masking flourishes in families with adolescent children. Parents don't want to be mapped—remember they're not omniscient to their kids anymore. Teenagers don't want their sex lives, drinking, and drugging to be mapped either. Mind masking allows them to avoid parental scrutiny and control. It's how they defy their parents' authority and prove to themselves that parents don't know everything. How else could you convince your mother that you're spending the night at your friend's house while you're out partying with your girlfriend or boyfriend?

People sometimes believe their thoughts, feelings, and emotions are more apparent to others than is actually the case.[1] The "illusion of transparency" is a real or perceived inability to mask your mind to an observer.[2] There are loads of times when people want to mask their minds—from anxiety over public speaking to approaching a potential romantic partner, to feelings of disgust over food at a dinner party or over someone's bad behavior.

Research suggests you're probably better at masking your mind than you believe. People overestimate the extent to which their internal states "leak out" in negotiations[3] and potential emergencies.[4] Research subjects who were asked to tell lies believed they revealed their deception more than they had and overestimated how much other people could detect their falsehoods.[5] Other people who sampled foul-tasting drinks in view of an observer thought their disgust was more apparent than it was.[6]

PEOPLE FROM BAD HOMES DEVELOP TERRIFIC MIND-MASKING ABILITY

As noted earlier, people from troubled homes tend to have the same *mind-mapping* ability as those who come out of warm, stable environments. Where they differ is their proficiency at *mind masking*. If you're growing up in an unstable household with irresponsible, untrustworthy parents, you get good at mind masking out of necessity. You learn to map other people while looking like you're not paying attention to them. When you grow up in a household with level-headed, loving parents, there's less incentive to develop mind-masking ability because you don't feel compelled to guard your back at all times.

Let's say your father is a violent man who lashes out at the slightest provocation. Not only are you going to map him all the time so you can anticipate his next explosion, you're also going to hide this because it's likely to set him off if he recognizes this. Let's say you're mapping your father and thinking *You're such a jerk! You're a loser!* If you don't mask your mind, the chances of your next beating will have skyrocketed.

Mind masking, like mind mapping, is a survival skill. For example, let's say you're a young child and your family demands that you maintain a positive facade in public about the mayhem happening at home. Masking your mind becomes second nature. Everyone in your family masks their minds, so doing it makes you part of the group. It's how people in your family interact with each other. This doesn't make you feel secure with your untrustworthy kin, so mind masking serves multiple purposes.

Before you know it, mind masking becomes a standard way of relating to everyone. It shapes your personality and controls your relationships. You learn how to present yourself as an open, honest person while you're flying under everyone's radar. You act like you're more naive or socially inept than you really are, so most people don't realize you're masking. Unfortunately this contributes to your not trusting anyone, because you can't trust the people who don't see you, and you can't trust those who do.

What Does It Take To Mask Your Mind?

If you grew up in an unstable, dysfunctional family, you don't need lessons on how to mask your mind. But let's say you grew up in Disneyland, you had the world's most perfect family, and you never masked your mind in your entire life. What if you didn't know how to hide what you were thinking and feeling and needed lessons? I know you know how to mask your mind, but it's instructive to see that it's teachable. Moreover, knowing how it's done helps you detect mind-masking in other people.

If you're good at masking your mind, even accomplished mind mappers may not figure out exactly what you're hiding. However, quite often they'll be able to detect you're masking *something*. To thwart this, really good mind maskers hide the fact they're masking in a variety of ways. Here's how they do it:

1. *Substitute false content.* For example, if you stayed out late and your wife is wondering where you were, you have to do more to mask your mind than simply refuse to talk about it. You have to "fill the holes" in your story. You might say something like "I understand you're wondering where I was last night. I drank too much, but I did the right thing under the circumstances, which I knew you'd want me to do. I got some coffee and waited until I was sober enough to drive."
2. *Don't tamper with more of the truth than you need to.* "I know I didn't come home until 4 a.m."
3. *Appear to be forthcoming, even eager to comply.* "You deserve a truthful answer. I know I didn't use good judgment and should have called a cab."
4. *Show a picture of your mind you know people want or expect to see.* When your wife doesn't simply accept

your story, you can respond "I'm getting defensive, like I always do whenever you question my answers. You know I don't like to be cross-examined."
5. *Create plausible deniability.* "I had to wait a long time at the diner where I stopped because there was only one waitress."
6. *Take umbrage when you're caught withholding important information.* "I told you about the waitress at the diner. It was late and the diner was closing, so I offered her a ride home. I wasn't withholding anything. Now you're accusing me of lying!"
7. *Talk obliquely or abstractly and gloss over details.* When the full picture emerges, which is usually quite different, you can claim, "That was what I was talking about!"
8. *Be convincing or confusing. You don't need to be accurate.* "How would you like it if every time you come home an hour late, I cross-examined you like this?!"

All of these methods contribute to one overall goal: breaking the link every four-year-old understands—what people do is a reflection of what's in their minds. Mind maskers attempt to defeat this by claiming they've misspoke...or they didn't mean what they said or did...or they didn't get a chance to say what they really meant because you interrupted them. In other words, they're saying you can't read their mind through their behavior. Once you understand this strategy, it's easier to identify.

Why am I teaching you how someone masks his mind? I'm not trying to help you be a better mind masker. I'm more concerned with helping you identify another person's mind-masking ability. Knowing these common techniques will make it easier for you to recognize when this occurs in your presence.

MIND MASKING IN ADULT LOVE RELATIONSHIPS

Detecting when your partner masks his mind is particularly important if you're in a committed relationship. Mind masking frequently occurs when couples experience serious difficulties that could destabilize their relationship like medical or financial problems. Partners put on "smiley faces," hoping to hide their anxieties and keep from further burdening their mate. Couples locked in repetitive arguments also mask their minds to avoid further conflicts.

Unfortunately, these well-intended tactics often backfire. When you realize your partner is masking, it makes you even more insecure. Naturally you wonder what he or she is hiding. Couples often become so enmeshed in mind masking, it creates an emotional gridlock in which neither partner reveals his or her mind, and both partners are suspicious.

The darker sides of marriage involve mind masking too. Extramarital affairs quickly come to mind. But I've encountered couples where one partner never seems to learn or remember how their mate likes to be touched or kissed. One husband tortured his wife for 25 years while giving her oral sex, by shifting his technique when she was about to reach orgasm. This expert mind-masker covered his tracks by insisting on weekly "relationship check-in" meetings to track his wife's frustration level.

Many couples go from unhappy marriages to unhappy divorces, continuing to shield their minds and arguing over their mutual refusals to be seen accurately. Beyond signing legal documents and changing where they live, their divorce doesn't change anything.

Third Level of Mind-Mapping Ability: Implanting False Beliefs

We've covered two levels of mind-mapping sophistication: mind mapping and mind masking. The third level is the ability to implant false beliefs in other people's minds, i.e., lying. All too many people lie constantly throughout their lives just for the sheer pleasure of it. Mind masking is like a shiny new toy for people with a strong antisocial streak. They just can't wait to play with it. It shouldn't surprise you that sociopaths have excellent mind-masking abilities. They'll implant a false belief in your mind—like they're your friend and they really care about you—and mask their minds as they do it. If you could look inside their minds, their cover would be blown and you'd run in the other direction.

Successful lying and manipulation requires mind-mapping and mind-masking abilities. This means once you've determined someone's an accomplished liar, you can be confident you are dealing with a mind mapper and masker. A common example is your partner having an affair you didn't know about. Your partner had to mask his or her mind to hide the affair and map your mind to successfully implant a false picture in your head.

Why care about establishing whether or not your partner is capable of implanting false beliefs when, according to brain science, you yourself have been doing this since you were four years old? It's because people tend to give their partners the benefit of the doubt and soft-pedal the fact that he or she is willing and able to lie right to their faces. Whether it's your partner's willingness to lie or it's her sheer brazenness, realizing your mate stuck a false picture in your head usually blows your map of your partner to pieces.

Besides removing pertinent information such as their true motives, good liars introduce misinformation to lead you to erroneous conclusions. These dual abilities allow liars to embed highly nuanced pictures in your mind. As complex as this sounds, most of us do it automatically. We don't even have to think about it. Lying, it turns out, is easy. Detecting lies is hard. (I'll help you get better at this in Part Three.)

GOOD THERAPISTS NEED TO BE GOOD LIARS

When I'm conducting training workshops for therapists, I ask them if they've ever lied to their own therapists. Typically, nearly everyone raises their hand. Then I ask, "How many of your therapists detected your lies?" Far fewer hands go up this time. I follow up by asking, "What happened when you beat your therapist's radar and he or she doesn't know you're lying?" The answers are always the same: Even if you continue seeing your therapist, you lose respect and trust for him or her. Your treatment basically ends right there.

Then I'll ask, "How many of you are good liars?" Only one or two brave souls will raise their hands. The rest of the audience smiles innocently like a bunch of goody two-shoe characters. Then I deliver the bombshell: "Well, if you're not a good liar, then you're probably a lousy therapist because you're not going to detect when clients are lying to you!" When I ask the question again—"How many of you are good liars?"—hands shoot up all around the room.

If you're a mental health professional, it's so important that you be a good liar. This doesn't mean you should lie. Being a good liar makes you adept at recognizing when clients are lying to you. Don't delude yourself that your clients are going to speak honestly with you when they first meet you. It's far more likely they'll mask their minds and offer you a distorted picture. When you build treatment on a false picture of your clients and their situation, they will map out where you're blind to them. And their treatment will be over before it really gets started.

Fourth Level of Mind-Mapping Ability: Mind Twisting

People at the peak of mind-mapping abilities are the mind twisters. These people map, mask, and lie with such finesse it puts them in their very own category. Mind twisters maneuver people, not by *concealing* their minds but by *revealing* them. Unlike mind maskers, mind twisters want you to map their minds because that's how they manipulate you.

Another reason mind twisters deserve a class by themselves is that, unlike mind mapping, mind masking, and lying, mind twisting has no prosocial uses. For example, mind twisters can tie you up in knots by telling you something they know contradicts what they previously told you, while acting like they don't see the contradiction. The outcome satisfies their dark urges and motivations at your expense.

Here's a minor example of how a mind twister operates. Let's say you catch your partner in a lie. "I feel terrible about this," your partner says tearfully. "I can't believe I lied to you. I don't know why I did it. I was having a bad day. I'm so sorry. I love you so much!"

Your partner looks so devastated that, instead of being angry, you start offering consolation. "Don't beat yourself up about it," you say. "It's not that big of a deal." In this case, your partner manipulated you by showing you a false picture of her mind being full of remorse. She wanted you to map her, knowing you would feel sorry for her.

Another mind-twisting method involves revealing what you want without directly asking for it. For instance, mind twisters will offer you a map of their minds that shows they want you to loan them money, give them a gift, or let them borrow your car. This ruse lets them act like grateful recipients of your unexpected generosity who owe you nothing in return because they (technically) never asked for it. If this doesn't work because you refuse, they deny asking and act like they don't know what you're talking about.

I've worked with clients whose ne'er-do-well parents sponge off them for years using this method. I remember one client, Bradford, who bought his parents a comfortable home and provided them with a monthly stipend. Bradford's parents never saved a nickel, spent his money freely, and expected more whenever they ran out, which happened frequently.

One day Bradford's mother approached him with what appeared, at first glance, to be good news. She said, "Your dad's thinking about going out and looking for a job! He's drinking less and trying to get back on his feet. I think he'd have a better chance if he had some new clothes and a few bucks in his pocket. At the very least it would boost his self-confidence, and he hasn't had much of that lately. You know I wouldn't ask for anything for myself, but this is for your father."

Bradford immediately recognized what was happening from what he'd learned in therapy. His mother was implanting the idea *You need to give us more money!* She wanted him to map her mind, so he could see what she wanted. She was speaking to him in a code she knew he understood.

Bradford responded using similar code. "I'm really glad to hear Dad's doing better," he said. "It's been a long time coming, and I know it would mean a lot to both of you if he succeeded. But I think giving him more money is more likely to undercut him rather than help him. If he's going to make it, he's got to make it on his own. I'm interested in seeing him accomplish that." When translated this meant *No, I'm not giving you more money!*

"Oh, well, I just thought you might like to know," Bradford's mind-twisting mother replied. This meant *Forget I asked! This never happened. I didn't want your money anyway, you ungrateful little shit!*

"I appreciate that," Bradford said. Translation: *I see what you're trying to do, Mom.*

Bradford and his mother were communicating on two levels simultaneously. On the surface, they were discussing a new start for his father. Below the surface, they were having an argument. Bradford's mother was shaking him down for more money, and Bradford was telling her *That's it! No more!*

Liars and mind twisters have much in common, wherein both try to manipulate you to do what they want. One tries to accomplish this by masking his or her mind while the other wants to be mapped out. Mind twisters rely on more contrived tactics than good liars employ. But the biggest difference is the amount of squeeze in their manipulations.

We've now looked at mapping other people's minds from a wide range of perspectives. We've covered four tiers of mind-mapping ability as well as some mind-masking scenarios. As you can see, there are many ways people prevent others from mapping their minds.

But another aspect of mind mapping involves knowing what's going on in your own mind. Often these abilities are unrelated. You can be masterful at mapping out other people and shielding your mind from them while simultaneously having no accurate knowledge of who you really are. We'll delve into mapping out your own mind in the next chapter.

7

Do You Know Your Own Mind?

"**Who** am I?" Humans are singularly preoccupied with this question.

The answer to this inquiry is your "self," your essential being, which distinguishes you from everyone else. The object of this introspection is more than just your personality, your core identity, your response tendencies, or your thoughts and feelings. The human self is an amazing organism.

For one thing, the human self has both elasticity and continuity. You have a sense of yourself existing throughout your life as a single entity, and yet you have been constantly changing, growing, and adapting. You're not the same person you were when you were ten years old, but you connect that "self" with who you are now. You have many sides to yourself at any given point, which makes it difficult to pinpoint exactly who *you* are. This elasticity and continuity makes discovering "who you are" a difficult proposition. But if you're really interested in answering "Who am I?" it helps to understand how your brain and mind operate. It's more complicated than you'd think.

Our Selves Are Connected

Most animals have just a simple self, which is centered in their body. If animals with simple selves can be said to think, one principal thought would be: *Everything inside me, including my skin, is me; everything outside my body is not me.* In this way animals with simple selves are capable of some measure of self-consciousness. They are sentient beings capable of experiencing pleasure and pain. Monkeys, rats, birds, and snakes all have simple selves.

Humans, however, have complex selves because we are more than just our bodies. Our brains allow us to have a remembered past. *When I was young, I played sports all the time. That was a long time ago, but I'm still that same person, even though I don't play sports anymore.*

We also have anticipated futures. *I'm going to learn how to sail. Then, when I can afford it, I'm going to buy myself a boat.* This amazing flexibility *and* continuity of self is rooted in words and language. In other words our "self," which we experience existing within the confines of our skulls, actually exists because we share a language and interact with other people.

Your "self" exists, not just within you, but in the interpersonal space we share with others. You wouldn't have a remembered past or an anticipated future if you lived in a world devoid of other people. The implications to this amazing realization often go unrecognized. This means our minds are inherently interconnected, not just through mind mapping but through our complex self as well. Thus, it shouldn't surprise us that our mind affects the brains of the people around us, and vice versa. Our sense of self and our brain's ability to function effectively often hinge on what other people think and feel about us.

Mind-mapping is an essential element of your complex self. There's more to this than the fact that other people are a large part of your remembered past and projected future. It turns out that the same parts of your brain that picture other people's minds also allow you to map your own mind. Mind-mapping not only allows you to predict other people's behavior, it helps you figure out "who you are." It's how you come to know yourself. As you'll see, this isn't as simple as asking yourself *What am I thinking? What am I feeling? What do I want?!*

SOURCES OF SELF-KNOWLEDGE

Self-knowledge is the answer to the question *Who am I?* This requires cultivating self-awareness and developing a "self-concept" built on attributes you believe yourself to have. Self-knowledge informs your mental pictures of yourself, including your map of your own mind.

You actually view yourself through an ever-changing lens over the course of your life, but the way you see yourself at any given moment is your "current self-representation." *(I'm a father who has a young daughter.)* Your current self-representation shifts depending on the role you wish to play in a particular situation. *(I own a small company, and I'm about to fire my bookkeeper.)* Research demonstrates that the way you see yourself at any given moment greatly influences your thoughts, emotions, behavior, and the way you process information.[1]

There are several ways you can gain knowledge about yourself. One is through messages you receive about yourself from other people. Another popular way is through introspection. A third method involves watching what you do and how you act (like you would watch someone else), and inferring your mental states from your behavior. As insightful as these common strategies might seem to be, they're not as reliable as you may have thought.

REFLECTED SENSE OF SELF

From the time in infancy when you first became self-aware, you've looked to other people to figure out who you are. All of us start off dependent on other people's perceptions of us. That's because human consciousness emerges much earlier in life than our ability to support our own "self." Unfortunately, many grown adults continually pander to get positive feedback from others. Some of us struggle with this our entire lives.

If this is you, your reflected sense of self comes from how other people respond to you. You imagine how you appear in their eyes, how they judge and evaluate you, to garner some idea of how they feel about you. This allows you to see yourself from outside yourself and leaves you feeling good or bad depending on what you map out. Your self-concept hinges on your perceptions of their appraisals, which may differ from how they actually see you.

Now, your self-concept is not a passive thing. Once it takes shape, your self-concept influences what you attend to and how you interpret new information. Unfortunately, this creates a self-perpetuating cycle that promotes unwarranted and shortsighted pictures of yourself and interferes with your knowing who you really are. This is how you can grow up thinking you're God's gift to the world, or you're worthless and don't deserve to take up space on the planet.

Mind mapping is pivotal in getting a positive reflected sense of self from others because this tells you how to present yourself to get their approval. Even when you get this, it's a lost cause as far as developing an accurate self-concept is concerned. Other people aren't put on earth to be accurate "mirrors" for you. More often than not, they will send you a picture of yourself that suits their own purposes. If you are desperate for them to like you, many people will map this out and manipulate your picture of yourself to their own ends.

Throughout my career I've focused on the many problems that arise when you depend on getting a positive reflected sense of self. When people start new relationships, they look at themselves through their partner's eyes and soak up as much validation and acceptance as they can get. This approach inevitably fails–and gets you into relationships you soon regret–because it's happening during the "can-do-no-wrong" stage, which never lasts. They expect their partner's validation to last nonetheless.

Unfortunately, that's not how adult relationships work, and that's not how you develop a solid sense of self. The validation new partners feed each other inevitably dries up. When your relationship is on the rocks and your partner wouldn't validate you if you were the last person on earth, *that's* when you find out who you really are. When faced with other people's disapproval--when they *refuse* to accept and validate you–that's when you're forced to relinquish their picture of you and hold on to your perception of yourself. *That's* how you develop a solid sense of self.

Mapping out how the important people in your life see you at such moments is often shocking and upsetting, particularly when you realize it says more about who *they* are than who *you* are. We'll explore this in greater depth in Part Two. But if you're familiar with my prior books, you'll recognize I'm describing how the natural process of differentiation occurs (becoming a solid person and developing a unique "self" by being in relationships). Starting in the next section I'll show you how differentiation and interpersonal neurobiology are inextricably interlinked.

Perceived Specialness of Introspection

Besides depending on a positive reflected sense of self, the most common way we attempt to know ourselves is through introspection. Introspection is the process of looking inwards and reflecting on your feelings, thoughts, attitudes, and behaviors. Sometimes this results in meaningful self-knowledge, but often it does not.

If, like most people, you think your introspection is infallible, blame the French philosopher Rene Descartes. In 1637 Descartes presented his famous line of reasoning, "I think, therefore I am," as verification of his own existence. He championed the idea that our minds are like theaters, present-ing ongoing shows that can only be viewed by us.[2] Descartes' view coincided with another seventeenth-century philosopher, John Locke, who proposed introspection is something like an inner sense.[3]

Introspection seems to place us in the unique position of knowing exactly what we're thinking and feeling at any given moment. Your view of your inner world seems more reliable than your perceptions about the external world and other people's mental states. More importantly, most of us believe our pictures of ourselves are more accurate than other people's pictures of us. Researchers call this notion "privileged access." It's the erroneous idea that you're positioned better than anyone else to know what's going on inside yourself.

Does my challenging the idea that you are the ultimate authority on yourself surprise you? Most of my clients buy this notion of privileged access without question. Their lack of skepticism is hardly surprising. It seems so intuitively true, many psychotherapists embrace this assumption without reservation. In fact, over the last 30 years it's been fashionable for therapists to demonstrate their clinical acumen by claiming that clients are the ultimate authorities on themselves. (I understand this makes no sense when you stop and think about it. If clients understand themselves so well, what do they need a therapist for? What would a therapist have to offer?)

This actually harkens back to Descartes' next book, his celebrated 1641 *Meditations on First Philosophy*, in which he claimed introspection is infallible. Moreover, he said our minds are "transparent" and "omniscient;" and whatever happens within them is always available to us through introspection.[4] In other words, Descartes proposed that you can't be wrong about what you think you think and feel.[5]

Here's the trickle-down effect of all this: You very likely believe other people can be wrong about your mental states, but you cannot be wrong about yourself because you're "closest to the data." Most of us operate under the delusion we have privileged access to our own minds. We think introspection is immune from error because it (supposedly) doesn't involve inference or deductive thinking, the way this comes into play when we're mapping someone else. This give us the (unwarranted) confidence to shout at our loved ones (and therapist), "How dare you tell me what I'm thinking or feeling!" Like a lot of intuitive truths, however, this one fails under scrutiny.

Fallibility of Introspection[6]

You're probably not aware of this, but scientists have questioned the accuracy of introspection since the 1970's. Critics point out introspection relies too heavily on previously established pictures of yourself, which are often inaccurate. Introspection rarely involves self-confrontation, and because of this it readily perpetuates inaccurate self-perceptions that lead to overly positive or negative self-evaluations.

Another problem is oftentimes we're not really aware of what we're feeling or why we feel the way we do. When you introspect, you tend to "fill in the blanks" with what you presume you must have felt, based on your preestablished (and often inaccurate) self-image. In many instances where you think you're introspecting, you're actually theorizing.[7] I've found the more that people are seriously out of touch with themselves, the more certain they are that they know themselves better than anyone else, the more they confabulate what they must have felt, and the more their confabulations are self-aggrandizing.

Loads of studies document the fallibility of introspection.[8] The notion of privileged access is uniformly rejected by researchers and philosophers alike. It turns out people are notoriously bad judges of what they're feeling. Your introspective judgments are often mistaken due to three well-known phenomena in perception research: expectation, context, and memory errors. Your expectation of what you expect to see and feel strongly influences your self-report. Context (like comparisons, for example) hugely influences how you perceive things.[9] Introspective memory errors also lead to erroneous pictures of your experiences.[10]

The sad truth is your access to your own mind may not be as good as you thought.[11] You probably often inaccurately perceive and describe your own introspective processes.[12] Research indicates children can't remember their own recently held false beliefs (mistakes), even within a few minutes. According to kids' recollection, they always had it right. You've probably seen adults have this same difficulty.

Your mind isn't transparent or omniscient either. Just as you can mistakenly postulate what's *in* your mind, i.e., what you must have thought and felt about something, you can also be wrong about what's *not* there, i.e., what you're sure you *didn't* feel. The great nineteenth-century American psychologist William James argued that, even if your immediate awareness of your mind is perfect, you're faced with the fallibilities of observation and memory a moment later.[13] Many scientists have concluded the evidence is "consistent with the most pessimistic view concerning people's ability to report accurately about their cognitive processes."[14] (I understand your temptation to hide this fact from your partner.)

DISTORTIONS IN MAPPING YOUR OWN MIND

A line in the *Talmud* refutes Descartes' belief that introspection is the key to self-knowledge. It says, "We don't see the world as it is; we see the world as we are." Meaning: How we see things says more about *us* than external reality. Unfortunately, this plays into our need to feel good about ourselves. We think of ourselves in favorable terms to maximize our feelings of self-worth and enhance our self-esteem. This leads us to seek information that confirms our positive qualities and negates our shortcomings.

Counterbalancing this is our need to know ourselves *accurately*. This makes us willing (to varying degrees) to know the truth about ourselves, regardless of whether what we learn is positive or negative. Some of us want the benefits of making better choices through accurate self-knowledge. Others feel a moral obligation or existential drive to truly know ourselves.

Unfortunately, these motivations are often far outweighed by our need to see ourselves positively and our tendency to protect our self-concept from change. Once you've developed a picture of yourself, you welcome information consistent with what you believe to be true about yourself, and reject and avoid things that reveal inconsistencies or inaccuracies. Like most people, you probably feel better when you believe other people see you the same way you see yourself. This means you seek out people who give you self-verifying feedback and avoid those who show you that your self-image is wrong. This "confirmation bias" increases your self-esteem, reduces your anxiety, and justifies whatever you are (or aren't) doing. Your need for accurate self-knowledge often gets lost in the shuffle.

WAYS OF FOOLING YOURSELF

As a therapist, I'm frequently confronted with clients' unflagging belief in the infallibility of their introspection. I've put together a dozen ways you can fool yourself about who you are. Here are the four biggest pitfalls:

1. *Emotional reasoning:* You believe that what (you think) you feel must automatically be the royal road to the truth. For instance, if you feel hurt by someone, that person must have done something hurtful to you, otherwise why would you feel hurt? Coming up with plausible motives for being hurt, you convince yourself that you've been wronged. (An alternative possibility is you're overly sensitive or predisposed to feel victimized. This is the flaw in couple therapy focused on partners' emotions.)

2. *Cognitive dissonance:* To relieve the uncomfortable tension you feel when you recognize inconsistencies between your thoughts and beliefs and your actions, you revise your picture of your thoughts and feelings to support your actions. This leads you to believe your own self-justifying stories. For example, when parents punish children excessively for minor transgressions, they magnify the misbehavior in their minds or develop a "higher motive" for doing what they did.

3. *Confabulation:* You manufacture a plausible story to account for your behavior or gaps in your explanations. People unknowingly or deliberately fill in holes in their autobiographical memory with fabrications they believe or prefer to be true. Who hasn't confabulated *Oh, I meant to tell you this, but you didn't give me a chance....* when we had no such intention at all? It's more like we think *A good person would have told you in advance. I'm a good person. Therefore, I must have meant to tell you.*

4. *Filtering:* Selective attention obviates a balanced objective assessment of all the evidence. This can involve dwelling on a single detail and making it a bigger deal than it really is—or ignoring, dismissing, or denying alternative viewpoints.

You'll find eight additional well-researched introspective errors in this endnote.[15] But perhaps what's most telling is the need to prove to you that introspection about what you're thinking and feeling is not perfect. This masks another reality I've discovered from working with my clients: Repeated bad mind-mapping experiences with other people can leave you seriously unaware of what you're thinking, feeling, and doing. Introspection can lead you to be seriously out of touch with yourself, and you won't be aware of it.

OTHER PEOPLE KNOW YOU BETTER THAN YOU KNOW YOURSELF

Although scientists reject introspection as being privileged self-knowledge, they don't claim that attaining self-knowledge is impossible. You have to start by accepting there's nothing special about your own judgments concerning "what makes you tick." You're *not* in a better position to know your own mind. Given humans' well-documented difficulties looking at themselves objectively, sometimes other people know you better than you know yourself.[16]

Realizing your self-perception isn't accurate can be quite upsetting. It's hard to give up the notions that introspection is infallible and you know yourself best. Some clients experience intense anxiety and embarrassment. This is a fairly natural response when your mind begins to see its own errors. *Wait a second,* your mind says, *There's a hole in my radar? I can't accept this! If I can't trust my own perceptions, then what can I trust?!*

Regardless of whether or not you trust other people, brain science says they can often see you better than you can see yourself. That's because it's easier for the human brain to make an accurate mental map of another person's mind, than it is to map one's own mind. This is how your self-image and the way others see you can be as different as night and day. Growth comes when you stop believing everything in your mind. The next step involves accepting you may not know your own mind as well as other people know you.

YOU CAN BE RIGHT ABOUT YOUR PARTNER BUT WRONG ABOUT YOURSELF

Like I said, it's easier for your brain to map another person's mind than it is to map your own. This fact of life has far-reaching real-world consequences. Like most people, you probably think you see your partner more accurately than your partner sees himself, and in many cases you'll be correct. Unfortunately, this often leads to an incorrect presumption: *If you see your partner accurately and his view of himself is wrong, then when he tells you something about yourself that you don't agree with, you erroneously presume he's wrong about you too.*

Here's the bad news: *You can be correct about your partner but wrong about yourself; and moreover, this is often the case.* The false confidence that comes from accurately mapping your partner's mind can blind you to your own mind. This creates loads of problems in adult love relationships. For example, you can be right about your husband's underhanded motives in an argument while you're wrong about your own. You can see he's arguing just to prove he's right, but as far as you're concerned you're not arguing at all. You think you're simply trying to have a conversation and keeping the record straight.

According to brain science, it's far more likely that you and your partner are right about each other and wrong about yourselves. You can avoid lots of heartache and solve many arguments if you accept that your partner probably knows you better than you know yourself. My friend, renowned Swiss psychiatrist Jürg Willi, said your partner's criticisms are usually accurate and should be taken as contributions to your growth. However, few of us ever say "Thanks for sharing!"

More often we refuse to abandon the flattering pictures we've created of ourselves. We continue to insist no one else could possibly know our feelings and motivations better than we do. To defend our intransigence, we devise theories as to why our partners enjoy telling us terrible untruths about ourselves. We then use this to justify discounting their observations and lashing out at them. You might say, "You don't understand me at all! Don't presume to tell me why I did what I did! How could you know what my motivations were?! Don't dump your projections on me!"

While your mind has no problem seeing your partner's limitations and contradictions, it doesn't like seeing them in you. When you finally accept this, your relationships will improve. You'll make better use of feedback from others. You'll be better able to learn from your mistakes and avoid misunderstandings. You'll be better able to compensate for your shortcomings and blind spots. In the long run, other people will find you much easier to love.

DID YOU MARRY A COMPLETE LUNATIC?

You might think that, if you depend on getting a positive reflected sense of self from others, you'd take their feedback seriously when it challenges your self-perceptions. Unfortunately, human nature doesn't work this way. In fact, it's usually the exact opposite, which causes loads of relationship problems: You're more likely to cling to idealized self-images, and the more tenacious your grip will be. People who lack a solid sense of self forged through difficult self-confrontation tend to say things like *I know exactly who I am. I'm the type of person who always does _____ and would never do _____.*

Earlier I noted that one of the incredible things about the human self is how it can change over time while maintaining its continuity. A solid sense of self is actually flexible and permeable rather than rigid and brittle. You can accept unflattering feedback and give up glorified images of yourself. You can adopt new ways of living and being in the world without losing your identity or sense of direction. You can change roles and ways of interacting that no longer fit you.

Unfortunately, these aren't the people I see as clients. When I'm listening to spouses bicker, one often says to the other, "Who the hell are you to tell me what I'm thinking and feeling?!" This usually succeeds in backing the partner off.

As the sparring continues, I'll often intervene by asking, "Did you marry a complete lunatic?"

The answer is usually the same, although the delay in

responding varies. "No," they'll say.

I'll continue, "Well, if you haven't married a complete lunatic, can your spouse really be wrong about *everything* about you, like you keep insisting?"

Typically, there's another long pause and perhaps a shrug. "Probably not," they'll begrudgingly respond.

"Well then," I'll reply, "instead of trying to prove your partner is wrong, why don't you figure out how your partner is right. I know this isn't as immediately gratifying as proving your partner doesn't know what she's talking about. As you mature, it is easier to give up idealized self-images. However, the process of letting go is like a lobster shedding its shell: It allows the lobster to grow, but it's not a very secure feeling."

Figuring out how your partner is right, rather than how you are right, uses a different part of your brain. Your thinking shifts, and when your partner sees this, it has a powerful impact on him or her. Don't reject feedback because you don't think they know who you are. They don't have to be experts on themselves to be an authority on you.[17]

We'll conclude this section by clearing up what might appear to be a contradiction: For 30 years I've underscored how the common quest for a positive reflected sense of self creates problems in relationships. Here I'm saying your partner may have a more accurate picture of you than you do. How do these two ideas fit together?

There's a difference between accepting that your partner probably sees you more accurately than you see yourself and depending on him or her to make you feel good about yourself. I'm also not suggesting you take everything your partner says about you as gospel truth. Being a slave to your partner's feedback is no better than reflexively rejecting it. There's no substitute for self-confrontation because a solid sense of self develops from inner conflict rather than getting another person's approval. Interacting with other people and challenging ourselves to examine our true outcomes allow us to relinquish false pictures we've created of ourselves. This is how we create a healthier world for ourselves and for everyone around us.

Part Two

Problems of Mind Mapping

8

TRAUMATIC MIND MAPPING

When people initially hear about mind mapping, they tend to think about prosocial forms which facilitate bonding, like comforting, care-taking, and teaching. But antisocial forms like teasing, mocking, withholding, lying, and cheating are as much a part of human nature as being kind, empathic, and supportive. For every prosocial application of mind mapping, there's an antisocial one as well. This section of *Brain Talk* focuses on the latter because ignorance puts you in peril. And if you come from a highly troubled family, and you want to improve your life, pay good attention to what you're about to read. It can change your world.

If you were surprised by what you learned in Part One, I promise Part Two will be even more of an eye-opening experience. Part One laid the groundwork for understanding how your mind and brain are interconnected with the minds and brains of other people. Parts Two and Three go further into the underlying interpersonal neurobiology of how we affect each other's brains for better or worse. Here we'll look at the negative brain impacts of mind mapping in everyday life. In Part Three, I'll show you how to use mind mapping to help your own brain and other people's brains for the better.

Antisocial Applications of Mind Mapping

Whenever someone commits an antisocial act—something that shows no regard for the rights or feelings of others—the first question that enters our minds is *Does this person understand what he's doing and the impact he's having?* Whether we realize it or not, our mind is asking *Does this person have mind-mapping ability?* The answer strongly influences how we understand what he (or she) is doing, how we feel about it, and how we respond to it. *Determining whether or not someone has mind-mapping ability is one of the most important things to know in dealing with the other person.*

For instance, if someone is cruelly teasing you, then you can be certain she has mind-mapping ability. Psychotherapists have known for years that, if you tell yourself your antagonist has no idea what she's doing or she's just joking, you'll respond more cordially. This is why some therapists constantly "reframe" troubling vignettes to encourage clients to presume another person's good intent. But if you want to deal effectively with people who are tormenting you, realize that for someone to tease you effectively, he or she has to know where you're vulnerable and see that you're cringing. There's no fun in teasing you if she can't tell she's struck a nerve, and she can't know this for sure without mind-mapping ability.

Biting sarcasm—using words that mean the opposite of what you really want to say in order to insult someone or show irritation—is another good example of antisocial mind mapping ability. How your brain actually maps out sarcasm is well documented, and it's pretty amazing. First, your left brain uses deductive logic to realize that the speaker's statements are not to be taken literally. Then your right brain decouples from reality (the spoken words) to grasp the speaker's intentions and to mentally represent the speaker's perspective.

In other words, detecting sarcasm requires sophisticated social thinking involving mind-mapping ability, including the ability to identify emotions. Biting sarcasm requires these exact same abilities. You have to recognize your recipient's beliefs and knowledge of the current situation to successfully transmit a double-message. Someone who's being sarcastic actually uses your ability to detect sarcasm against you.[1]

Likewise, if you want to frustrate or thwart other people, one of the best ways involves anticipating where they're heading in a conversation, and cutting them off by abruptly changing the topic. You can't pull this off unless you have mind-mapping ability. The same goes for successfully mocking, stealing, lying, having affairs, messing with somebody's reality, or emotionally abusing people. If someone does any or all of these things (the list goes on), your antagonist has mind-mapping ability.

There's a common misperception that bullies act the way they do because they lack social skills. Supposedly they use their hands because they can't rely on their brains or their mouths. This is incorrect. Research indicates many bullies possess above-average mind-mapping abilities–this is what allows them to get away with bullying for extended periods.[2] This pushes you to consider less forgiving reasons why bullies abuse other people. The possibility we're least open to considering is that they *like* doing it. They take pleasure in being cruel.

Con men are great mind mappers. Sociopaths positively excel at it. People commonly think these folks are indifferent to other people's needs, feelings, and best interests, which allows them to act solely in their own self-interest. It's an unfortunate truism among therapists that sociopaths lack empathy (the ability to intuit other people's feelings). But the truth is sociopaths and con men are extremely sensitive to other people's feelings and agendas. They're adept at using other people's emotions to their own advantage, which makes their deliberate exploitations more cruel and reprehensible.

We've already explained how you can't be a good liar if you don't have mind-mapping ability. Liars must be able to withhold true facts (mind masking) while offering false information, and these two abilities allow them to implant the exact picture they want in other people's minds. But as harmful as lying is, it's not the lies themselves that often cause the most damage. It's the liar's refusal to admit what he or she has done.

For example, if you ask your wife if she's having an affair

and she says yes, that's going to be extremely painful. But there's exponentially greater impact if your wife denies this and you subsequently see her leaving a hotel room arm in arm with someone else. It happens again when you confront her with the evidence and she still refuses to confess. "You're wrong," she responds. "I'm telling you the truth. If you insist, I'll lie to you and say I'm having an affair. Will that make you happy?" Her attitude is *I don't care what proof you have. If I say it's not true, I expect you to approach this as if it's not true.* In this case, your wife's willingness to deliberately lie to your face and distort reality causes far more damage than the actual affair, although both produce traumatic mind mapping in you.

I often hear people say, "People who do cruel things don't understand their impact." This is a typical response to cruelty. We assume people who do cruel things don't have mind-mapping ability because it's easier on our own brain. It boggles our mind that people deliberately act this way when they're aware of their negative impact. But such people *do indeed* have mind-mapping ability. Some don't give a damn about their effect on others, while some particularly enjoy treating other people badly.

When we think someone is deliberately trying to hurt us, it blows our mind and our mind-mapping ability shuts down. We don't want to look into a mind that rationalizes doing harmful things *to us*–even if we're fascinated by mass killers and criminals. Rather than deal with it directly, we prefer to keep ourselves in a gray area where we remain unsure of that person's true motivations. But this isn't as simple as our reluctance to see what we don't want to acknowledge. Your brain aids and abets this by ceasing to operate effectively. You get "spaghetti brain."

Traumatic Mind Mapping

Everything we've discussed thus far about mind mapping is based on voluminous, well documented research. What we're talking about now–traumatic mind mapping–comes from working with my clients.[3] In a nutshell, traumatic mind mapping is a mental and physical impairment that occurs when you're mapping someone else's mind; and what you see is so terrible and upsetting that it makes your own mind and brain fall apart.

All therapists experience traumatic mind mapping at some point during therapy. They get "spaghetti brain" from mapping out their clients' minds or listening to terrible stories from their clients' past or present. It gets hard to think straight. They don't pick up obvious clues or contradictions. They don't ask relevant questions, especially ones that would increase their own or their clients' anxieties. Like everyone else, therapists prefer to believe their clients (or their clients' adversaries) did whatever bad things they did because they were emotionally clueless or suffering from autism or Asperger's Syndrome.

Let me give you three examples of what traumatic mind-mapping looks like in action. You can decide for yourself whether the people in these vignettes know what they're doing or not.

In our first example, you are four years old and sick with a fever. Looking for comfort and reassurance, you go to your mother, who's sitting on the couch reading a book, and you cling to her leg. However, your mother shakes her leg to wiggle out of your embrace, kicks you out of the way, and yells at you to leave her alone. In disbelief you run back to your mother, and this pattern repeats several times. You dissolve into tears, crying out for her. In response, your mother walks out of the room while telling you she's sorry she ever had you.

In our second example, your father is a clergyman who is discovered having an affair with a woman in his congregation. This creates a huge scandal in town, and he gets thrown out of his church. Your family relocates to a new town where your father starts up with another congregation. Your father's girlfriend from the prior town moves there too, and he openly continues his affair with her. The first scandal was bad enough, but it blows your mind that your father–supposedly a man of God–hasn't learned his lesson. When you ask your mother about this, she acts like she doesn't know what you're talking about.

For our third example, you've had a difficult relationship with your self-centered overly, controlling mother for most of your life. It's difficult for you to confront her about her manipulations, but you finally get up enough nerve to do it. You expect her to get upset and either deflect the topic or start talking over you. However, she actually listens in silence for ten minutes. Occasionally she nods her head to show she understands what you're saying. At several points she smiles, seeming to agree with you. You're shocked and delighted by your mother's response and begin to doubt yourself. Maybe you were wrong about her. You're so overwhelmed with such a great sense of relief, you thank your mother several times for being open minded, which she graciously accepts. But, as she's rising from her chair at the end of the conversation, your mother looks at you sadly, shaking her head from side to side, and says, "I don't know what you're talking about."

As varied as these examples are, they have something in common. The generic traumatic mind-mapping experience involves discovering that your antagonist has mind-mapping ability and knows what he or she is doing. In these particular examples, the antagonist was a parent, but it could be your spouse, child, boss, co-worker, or a stranger on the street. When you realize this person is tracking you, and he knows what he's doing while he's doing terrible things, it gives you traumatic mind mapping–particularly if you see he's enjoying your discomfort and disappointment.

As you learned in Chapter Three, two-year-olds can intuit when caregivers are negatively disposed towards them. By the age of four, children can understand that parents can want them to suffer, and kids can recognize when their parents enjoy hitting or emotionally torturing them. However, many astute, experienced therapists suddenly lack the mind mapping ability of a four-year old *because* they've intuitively understood what's going on in their clients' stories. This gives therapists traumatic mind mapping long before clients wake up to what they're describing–and clients who appear to be "walking and talking in their sleep" increase their therapist's spaghetti brain.

I'm describing how seasoned professionals who are accustomed to hearing lots of bad things still develop traumatic mind mapping from dealing with *other people's* lives. But when it comes to their own personal lives, therapists are no different than you. When you realize your spouse has mapped your mind and is tracking you and using this information to manipulate or torture you, it blows your mind and gives you "spaghetti brain." Next thing you know, you're playing amateur neuropsychologist and diagnosing your loved ones with autism-spectrum disorders.

It changes your picture of your parents (and subsequently your entire life) when you realize they have mind-mapping ability, and they're using it while behaving badly. You suddenly grasp that your parents aren't emotionally blind or indifferent like you wanted to believe. I can't tell you how many people have told me, "My parents didn't care about my feelings!" like this was the worst thing in the world. In the worst cases, they're tracking everything you d;, they have you mapped; and you are *very much* on their minds. Rather than ignoring your feelings, they are *very* invested in making you feel (or do) things against your wishes.

Traumatic Mind Mapping Impairs Your Response to Stress

Once I recognized traumatic mind mapping in my clients, I wanted to know what kind of physical impact this had on them, particularly their brains. To accomplish this, I started studying clients who reported extreme traumatic mind-mapping events. I wanted scientific proof that their brains had been impacted, and I wanted to know what that impact looked like.

With their permission I began monitoring my clients' physiological responses in the midst of psychotherapy using small, wireless biometric recording devices that didn't interfere with therapy. Being naturally inquisitive, I measured my own physiological responses during these sessions as well. In addition to recording our responses, my equipment also provided real-time physiological readout of the three of us, which I monitored on a small laptop computer by my side. I also videotaped these sessions and subsequently overlaid the recorded physiological measurements. This allowed me to review what people were saying and doing—and what was happening inside them—at any point in our sessions. Here is a picture of what couples therapy looks like when it is all put together (Figure 1).

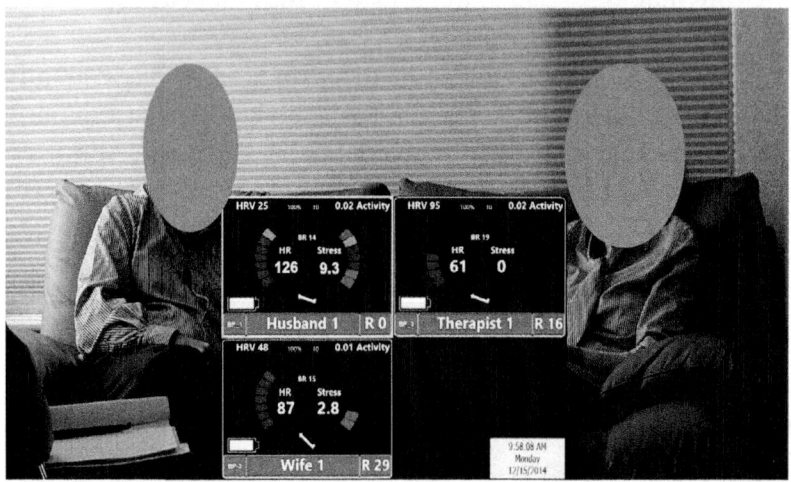

Figure 1. Display of in-session psychophysiology readings of clients and therapist.

Existing research gave me some place to start. I focused on clients' heart rates and heart-rate variability during intense psychotherapy sessions discussing anxiety-provoking issues. Prior studies document something called the hypothalamic-pituitary-adrenal (HPA) axis, which is a brain-based system that regulates one's response to stress. If you are healthy, your brain increases your heart rate in high anxiety situations, especially when you enter "fight or flight" mode. However, if you've been over-exposed to extreme stress, your stress-response wears out. For instance, first responders, emergency room doctors, and veterans of prolonged combat show impaired HPA responses.[4]

I also studied clients' heart-rate variability because the time between heart beats of a healthy person is highly variable (as much as 50 to 100 milliseconds), whereas people who are unhealthy show less heart-rate variability.[5] Sadly, neonatal intensive care units know which babies won't survive because their heart-rate variability flattens out to single digits.[6]

Figure 1 shows a couple therapy session with the husband, Michael, sitting on the left and his wife, Irene, on the right. The physiological measurements for the three of us are superimposed. Michael's readings are in the upper left box, Irene's measurements are in the lower left box, and mine are in the upper right. Now look at any given box. For each person, heart rate is on the large gauge to the left, stress level is on the large gauge to the right, and heart-rate variability is displayed numerically in the upper left corner. Stress level is a useful, calculated index (on a scale from zero to ten) derived from heart-rate variability. Low heart-rate variability reflects greater physiological loading.

Notice Michael's extreme heart rate and stress level (upper left), while Irene's levels (lower left) and my levels (upper right) show far more moderate responses. This pattern was typical for each of us throughout most sessions. Irene's household was no picnic when she was growing up, but Michael experienced extreme emotional and sexual abuse by his mother and father throughout his childhood and adolescence.

I also conducted several sessions with Michael and Irene and their two teenagers. Figure 2 contains a picture of Irene (left side) and Michael (left side) in family therapy, facing their children. In this recording, heart-rate variability is centered between gauges for heart rate and stress.

As you can see, even their kids were monitored during the session. Teenagers are notoriously resistant to family therapy. But if you pop a set of gauges on them, they want to watch their own reactions—as well as those of their parents—as the session progresses. This completely changes the dynamics of therapy. Figure 3 illustrates the self-contained wireless device I use to take the measurements displayed here. It attaches to the chest with two disposable electrodes.

Here's what I discovered from what I've observed and recorded: People who report repeated, extreme, traumatic mind-mapping events exhibit two distinctive patterns of physiological response. One group, like Michael, has difficulty

Figure 2. In-session psychophysiological monitoring during family therapy.

Figure 3. Zephyr™ BioModule™ system used by the author.

keeping their heart rates under control. The normal resting heart rate for a healthy adult is usually between 60 and 80 beats a minute. During stressful sessions, the heart rates of many clients can spike to 100 or 120 beats per minute. But clients who report repeated, extreme, traumatic mind mapping can have heart rates as high as 145, together with *continual* heart rates as high as 135! Average heart rates this high are not conducive to living to a ripe, old age.

Similarly, heart-rate variability (the intervals between heart beats) of 22 milliseconds or higher is considered normal and healthy, and it can go as high as 100 milliseconds or more. However, in Figure 1, Michael has a heart-rate variability of nine milliseconds, and I've seen it get as low as four or five milliseconds. As I said, heart rate variability is a good index of health: If a baby in a neonatal intensive care unit showed such low heart-rate variability, its chances of survival would not be good. Twenty months later during a highly stressful family session, Michael and Irene show much better heart-rate variability, lower heart rate, and less stress (Figure 2).

Now, earlier I mentioned I've observed a second physiological pattern. This other group of clients has a completely different reaction to anxiety and stress. They can tell me the most horrific stories about traumas they've experienced, and their heart rates won't change a bit. They'll stay level the entire time, while my heart rate will be going up and down like a rollercoaster. This is the "flat line" response I described earlier that is commonly found in people who work in extremely stressful

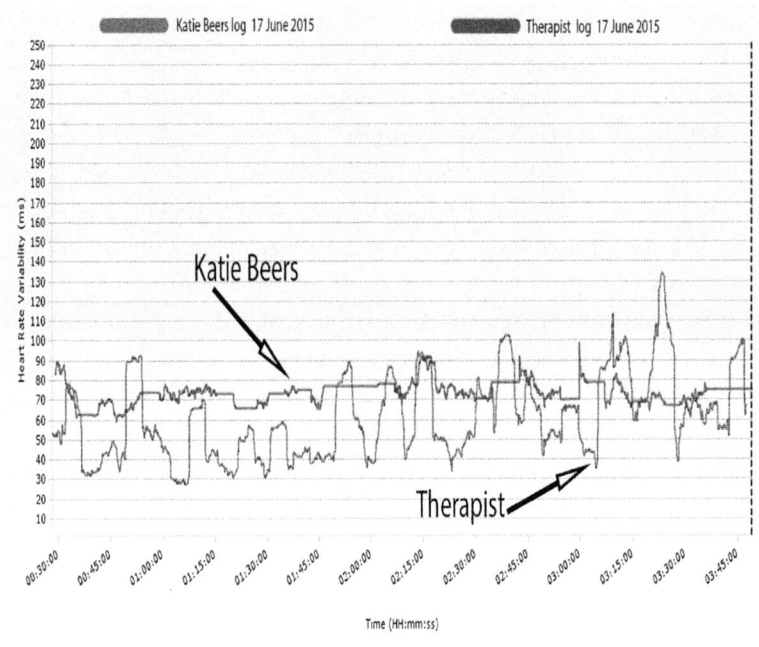

Figure 4. Heart rate variability readings during psychotherapy session.

environments such as first responders, emergency room doctors, and soldiers in combat zones. Their hypothalamic-pituitary-adrenal (HPA) axis, which regulates their response to stress, gets overwhelmed and stops responding as it should.

Figure 4 shows the heart-rate variability of a very special woman as well as mine during a psychotherapy session. Ten-year-old Katie Beers received national attention in 1992, when she was discovered being held underground for two weeks by a neighbor in a six-foot-by-seven-foot bunker concealed underneath a two hundred-pound concrete trap door. Upon her release, Katie said the daily sexual abuse she endured underground wasn't as bad as the sexual abuse she received from her godparents, with whom she lived across the street.

Katie now goes by a different name in her current life. She lives with her husband and children. She's very productively employed, and she tirelessly advocates for childhood victims of sexual abuse. I'm telling you about Katie's session with her permission. We both hope what you read here makes you more aware of the scope of this problem and encourages you to seek treatment if this has happened to you.

In the session depicted in Figure 4, we were discussing the details of Katie's experiences. As you can see from the many tall peaks and sunken valleys in my heart rate, I was having massive physical responses as I listened to her. A few times I asked Katie to give me a moment to get my heart rate down and to digest what she was telling me. Katie, however, showed minimal heart rate changes as she described the most disturbing experiences you can imagine!

In-session physiological monitoring is still cutting-edge practice, and a small number of therapists do this. Until recently it simply wasn't possible unless you were part of a university-based research study using millions of dollars of equipment. However, technological advances have put this within the grasp of private practitioners, and this will become standard procedure as brain-based therapies develop.

Studying heart rates and heart-rate variabilities of clients reporting repeated, severe, traumatic mind mapping has convinced me of several things:

1. Traumatic mind mapping is very much a real phenomenon and not something I imagined.
2. Traumatic mind mapping negatively impacts your brain in ways that affect your body and general health. It exhausts you physically and mentally and impairs your cognitive and emotional functioning. Judging from research on the impact of stress on your body's immune system, it's very likely that traumatic mind mapping suppresses your resistance to disease and maybe even shortens your life. [7]
3. It appears possible to improve your heart rate variability once it has been compromised. You'll be glad to know Michael's heart-rate variability and his ability to control his heart rate are improving. In fact, I first noticed improvement in his physiological metrics before I recognized his improved ability to keep his thoughts focused and organized, and his emotional reactivity under control in the midst of important confrontations from Irene.

THE CLOSER THE RELATIONSHIP, THE BIGGER THE IMPACT

Here's an example of how you can go from enjoying a nice, sunny day and feeling perfectly fine to witnessing something so horrific your mind collapses. Visualize yourself walking down the street on your way to meet a friend for coffee. Suddenly, a man bolts out of the coffee shop, nearly crashing into you, and runs down the street.

Another man emerges from the store with a gun; and standing right in front of you, he takes his time and calmly shoots the running man in the back. Then he walks over to the wounded man lying on the ground and bends over to say something. Then the shooter smiles as he finishes off the helpless man and strolls away, stopping to tell a woman with a baby carriage to be careful crossing the street. As horrific as the actual violence is, mapping the shooter's mind makes it much more traumatic.

You think *How could he do such a thing and be so calm?!*

He actually talked to his victim before finishing him off! And then he stops to tell a mother to be careful with her baby? What kind of person can do that! You instantly imagine how the shooter's mind works. How does he think? Where are his normal emotions? What are his beliefs? The traumatic mind mapping this produces makes your mind fall apart, and your body starts shaking. You realize you easily could have been killed had you gotten in the way.

This scenario involves a couple of strangers. Now imagine the impact if you actually knew the man who got shot or the shooter. What if this turned out to be the friend you were meeting? The more important the people involved are to you, the bigger its impact on you. This is why traumatic mind mapping so commonly occurs with extreme marital discord, domestic violence, and sexual abuse from someone you know. Things difficult to deal with under any circumstances become *highly* traumatic events when they involve people who are supposed to take care of you.

Traumatic Mind Mapping in Troubled Homes

During my 20 years of teaching at Louisiana State University Medical Center in New Orleans, I witnessed the effects of Louisiana's extreme poverty. One client, Billy, told me that his childhood ambition was to own a bicycle. Unfortunately, his parents were extremely poor and couldn't afford to buy him one, so Billy got a newspaper route delivering *The Times-Picayune.*

Billy saved his earnings in a jar hidden under clothing in his dresser drawer, and every week he added a little money to his stash. He nearly had enough to buy that bike when he came home one day to find the jar with his money was gone! Confused and upset, Billy asked his parents if they knew what had happened to his money.

Both parents said the same thing: Each one was certain the other stole it, but neither did anything about it. By the next day, they acted like nothing had happened and told Billy to stop moping about. Shortly thereafter, Billy's parents replaced the air conditioning window unit in their bedroom. Given the family's impoverished income, it wasn't hard for Billy to guess where the money came from.

Billy's traumatic mind mapping was deep and lasting. The people he trusted most in the world had not only stolen from him, they had lied to him too. They knew how important getting this bike was to him and how hard he worked for it. They saw his mounting excitement and happiness about almost having enough money. They were positioned to know how disappointed he was going to be by their actions. Accepting that the important people in your life have mind-mapping ability can be as unsettling as realizing the Earth is round when you've believed it is flat. But Billy never understood or got over this event until he let himself realize this was the case.

Of course, traumatic mind mapping isn't limited to poverty-stricken homes. Another client, Conrad, told me about the day he watched his father, a business executive, being dragged by police across their well-manicured lawn. He was arrested for embezzling money from the company he worked for. It turned out Conrad's family's lavish life-style was more than his father's pay grade afforded.

Conrad's first thought was *How could my father keep telling me with a straight face, 'If you work long and hard, your efforts will be rewarded,' when he was stealing money the entire time? I thought he was an upstanding guy, but he turns out to be a crook! Was he secretly laughing at me when I believed him? He's embarrassed us in front of the entire community!*

Images of his father's being dragged away in handcuffs stuck in Conrad's mind until he came for treatment. As I listened to Conrad, I spotted another familiar pattern: He told his entire story as if he only had one parent. All his attention focused on his father. I wasn't surprised because I've seen this pattern repeatedly in people who've experienced severe traumatic mind mapping. It's like they have one "bad" parent and one "good" one, and they just focus on the "bad" one. Having an identified "good" parent seems to pacify their mind. Unfortunately, the "good" one often turns out to be no better than the other.

To help Conrad, I asked him about his mother's reaction when the police dragged his father away. Conrad told me he never thought about it. I pointed out that a child would look at his mother at such times to make sense out of what was happening. He said his mother was standing right next to him, but he couldn't remember the look on her face.

Conrad's next thought was *Did my mother know what my father was doing? She used to be a bookkeeper, and she paid the household bills, balanced their checkbook, and did their taxes. Did she ever ask him where all our money came from? Did she care?! She was always saying we needed to keep up with our neighbors to maintain appearances.*

There were two possible answers and neither was good. The first was Conrad's mother didn't know, and she was negligent in not asking. The second was she knew Conrad's father was bringing home way more money than he should. This was the more plausible picture. But either way Conrad was having traumatic pictures of *both* his parents' minds. Repairing this hole in his autobiographical memory was a necessary step in leaving this all behind. In Part Three, I'll show you how I helped Conrad figure out which picture of his mother was most accurate.

SEXUAL ABUSE MOST OFTEN OCCURS THROUGH TRAUMATIC MIND MAPPING

According to the National Sexual Violence Resource Center, one out of four girls and one out of six boys will be sexually abused before they're eighteen years old.[8] This alarming statistic becomes even moreso when you realize this only accounts for reported offenses. The vast majority of child sexual abuse goes unreported.

When people think about sexual abuse, they focus on improper physical contact. They're looking for "reportable offenses," evidence that would stand up in court and lead to a conviction. Victims know that's how the police and prosecutors look at things. In the absence of incontrovertible proof, most victims don't self-report, whether it's to the police, teachers, or their own parents. This leads to drastic under-reporting of sexual abuse in homes and workplaces and creates misleading appearances that disguise the pervasiveness of this problem.

Unfortunately, a lot of sexual abuse doesn't occur via touch. After treating victims for over 30 years, I've concluded that the majority of sexual abuse occurs through traumatic mind mapping. For instance, imagine you're six years old and you innocently walk into the living room where you discover your father having sex with the family maid. Your father has her bent over the back of the couch, and he's having sex with her from behind, doggy style. When your father sees you, he turns towards you; and without ceasing to thrust, he looks at you and smiles.

Beyond the damage caused by watching your dad have sex—and he's doing it with someone other than your mother—this pales in comparison to his smiling at you as he's doing it. By doing this, your father has included you in the sex act and put you in a terrible position with your mother. He's also turned this into an interpersonal event between the two of you: He wants you to map his mind and see that he's having a good time.

Another client, Karen, told me about being a dutiful daughter and bringing her first boyfriend home to get her father's approval. Not long into the visit, Karen's father asked the boy, "So, is my daughter any good in bed?" These eight words were more than enough to create sexual abuse and cause brain damage through traumatic mind mapping. Karen was confronted with her father's sexual interest in her, and she never got over it. The picture of this encounter was emblazoned in her mind. Her relationships with men were a disaster, especially around sex.

I helped Karen glean things from her picture that her mind refused to see: One was that her father didn't bother to mask his sexual interest in her; he wanted her to know it. With this in place, it doesn't take much for abusers to retraumatize their victims. All it takes is a secretive smile in the midst of a crowd of people. This helped Karen finally understand why her father's smiling at her at the dinner table, in front of her mother, could leave her feeling so violated. On the surface there was nothing to report. But Karen's father sexually abused her repeatedly through her adolescence by the use of suggestive acts that instilled traumatic mind mapping.

This disturbing revelation was closely followed by another

when I asked Karen about another disturbing aspect of her story: Either Karen's father wasn't worried about her mother's response to his inappropriate question to the boyfriend, or he was sure Karen wouldn't tell her mother (which she didn't). Either way this suggested Karen's father thought he had his wife and Karen mapped, and his assessment was that he had nothing to fear. Apparently he felt he was in control of the situation and thought he could get away with sexualizing his relationship with Karen.

Let me give you one final example, so you can see how far this goes: Another client, Penny, had slept in the same bedroom with her sister ever since they were little girls. When she was 16 and her sister was 14, their older brother began coming to their room in the middle of the night and having sex with both of them. Being repeatedly raped at night, and receiving knowing looks from her brother during the day, created constant traumatic mind mapping for Penny. But years of prior sexual abuse treatment focusing on the rapes failed to help her resolve this. I've found this often occurs when people have a myopic view of traumatic experiences, which is more the norm than the exception. The full picture of Penny's circumstances–the family dynamics that promoted these experiences–was missing.

To broaden out Penny's understanding of what had really happened to her, I asked her why her brother wasn't worried that their parents would give him the beating of his life. This might be the case in homes where there was only one daughter and it was her word against her brother's. But given there were two daughters who could corroborate each other's accusations, most boys would be pretty sure they wouldn't get away with this. However, Penny's brother obviously thought he could handle the situation if both sisters told their parents. How could he be so sure?

My questions prompted Penny to map her brother's mind. More specifically, she started looking at this through her brother's mind-mapping ability. That's when Penny realized her brother's complacency stemmed from his map of their parents. He knew they wouldn't do anything even if the girls said something. Penny's realization triggered a torrent of memories she hadn't previously mentioned—like how her father was a chronic philanderer who was frequently unemployed because of drinking on the job. Her mother, who had no female friends, never taught Penny or her sister a thing about sex or menstrual hygiene, except that sex was dirty. Her parents argued all the time and never intervened when the three kids fought. Penny and her siblings were expected to settle disputes amongst themselves.

With these new pieces, can you visualize the world Penny grew up in? Without minimizing the impact of the sexual encounters, do you now see they were part of something far worse? Penny grew up with a lot more traumatic mind mapping than her description of her problem initially suggested. But once the whole picture came out, Penny was able to digest her childhood and get on with her life. If there's a "silver lining" to such stories, that's it: As damaging as sexual abuse and traumatic mind mapping can be, they're actually highly treatable using this method.

PTSD FROM TRAUMATIC MIND MAPPING

If you aren't familiar with Post-traumatic Stress Disorder (PTSD), you've probably at least heard about it. It's an anxiety disorder caused by very frightening, stressful events. More than three million new cases are diagnosed in the United States each year.[9] Like many therapists, I've treated lots of clients who have it. But what particularly interests me about PTSD is the role traumatic mind mapping plays in who actually develops PTSD and who doesn't.

For instance, in a recent study, 61 percent of more than 6,000 teenagers (ages 13 to 17) reported experiencing traumatic events.[10] Some lived through natural disasters like a tsunami, a hurricane, or a tornado. Others survived life-threatening illnesses or injuries. Some experienced the death of a loved one like a parent or sibling, while others endured physical or sexual abuse.

However, not everyone who experiences trauma develops PTSD, so researchers set out to explore this. It turned out that, overall, only four percent of the trauma-exposed teens actually developed PTSD, which is a much lower percentage than you might have guessed. But, as it turned out, the kind of trauma greatly determined who got PTSD and who didn't. PTSD occurred in less than ten percent of those who'd survived a natural disaster, serious illness, or death of a parent or sibling. In a marked contrast, 39 percent of teens who were raped, and 25 percent of those who were physically abused by caregivers developed PTSD.

What could explain such large differences? My own interpretation is physical and sexual abuse cause PTSD more frequently because they involve traumatic mind mapping, whereas the other traumas do not. When your father is beating you, part of the time you may watch the belt in his hand, but you'll monitor his face too. You'll also study his voice. You're going to map his mind, trying to decipher his intentions. You'll be thinking *How far is he going to go this time? Is this one of his usual disciplinary beatings, or is he completely out of control? Why does he think I deserve this? Or is he just taking out his frustrations on me? Is he going to seriously hurt me, and if so, how badly? Can I submit to this, or do I have to get the hell out of the house?*

If someone breaks into your house in the middle of the night, climbs into your bed, and starts raping you, I guarantee you won't focus all your attention on his penis. You'll focus on his mind. *Why is he doing this to me? Did he pick me at random or does he know me from some place? Is he enjoying this? Is he doing this because he's angry at me? Or is he a sadistic rapist who gets turned on by the sight of blood or seeing terror on his victim's face. Is this all he wants, or is he going to kill me?* Even if you "disassociate" (your mind "goes somewhere else") this happens after you have traumatic mind mapping, not before.

A Princeton University study of low-income mothers in New Orleans during Hurricane Katrina sheds more light on the role of traumatic mind mapping in PTSD. Nearly 33 percent of these mothers had PTSD. Some of them were forced by rising floodwaters to climb into the attics of their houses and chop holes in the roof to escape. Some lived for days without food or water. But this didn't produce PTSD as frequently as watching vigilantes beat a homeless man to death or seeing people get shot over a loaf of bread. I think it's because these incidents involved traumatic mind mapping, whereas dealing with flooding itself didn't necessarily involve another person's mind.[11]

MIND-MAPPING DISTORTIONS FROM TRAUMATIC MIND MAPPING

Traumatic mind mapping negatively impacts your mind-mapping ability. Mind mapping itself is by no means perfect, even when it doesn't involve traumatic mind mapping. Distortions and inaccuracies can occur. For example, the well known defense mechanism of "projection" involves attributing your own (often unacceptable) motivations, thoughts, and feelings to someone else. In other words, your map of someone else's mind is actually determined by what you refuse to see in your own mind.

Then there's "Stockholm syndrome," wherein hostages express irrational empathy, sympathy and positive feelings toward their captors, sometimes to the point of defending them.[12] The Federal Bureau of Investigation's Hostage Barricade Database System indicates roughly eight percent of victims demonstrate this pattern.[13] Victims seem completely mind-blind to their captors. However, this doesn't require a typical Hollywood hostage scenario. All it takes is "strong emotional ties that develop between two persons where one person intermittently harasses, beats, threatens, abuses, or intimidates the other."[14] If you stop envisioning bank robberies and international terrorism, you'll recognize this goes on in families and marriages all the time.

I've documented five main types of mind-mapping distortions created by traumatic mind mapping:

1. *Gaps in your mind-mapping radar.* This mind mapping distortion has a unique pattern: You are hyper-vigilant *and* blind. This means you have "narrow-gap blindness" when mapping specific people who are very important to you, and it's vital to your peace of mind to see them in favorable ways. This occurs in the context of your being hyper-vigilant everywhere else in your life. You

look mind-blind to select people at the same time that you are overly suspicious of everyone else. You are also blind to your blindness. You're not aware you're missing anything. You have no doubt you're seeing everything to be seen.

2. *Holes in your autobiographical memory.* You remember very little of your childhood, and the mind-mapping data for events you do remember is missing. When you recall events, your mental maps of other people's agendas and motivations (at that time) is absent. You don't retrieve what you thought they were really doing, and without this your recollection of events tends towards overly positive interpretations of what happened. When you finally recall your mind-mapping information, the meanings of your memories can change drastically. This is another way you can look mind-blind about your antagonists.

3. *You attribute positive motivations to your antagonist.* One client told me about how, in the middle of playing a football game in high school, his father ran down to the sidelines after he dropped a pass and screamed, "You stink! You ought to quit! You're an embarrassment to your team!" Initially this man told me his father was actually trying to encourage him, demonstrating an unwarranted attribution of goodwill.

4. *You minimize your antagonist's negative motivations.* I often hear battered women say, "He didn't mean it. He didn't want to hit me. He just lost his temper. It's my fault. He wouldn't have done this unprovoked." Even four-year-olds can figure out that repeated negative behaviors are reliable indications of people's intent. But these women are so traumatized they deny it, just like lots of kids who, when beaten by their parents, tell themselves this exact same litany.

5. *You dismiss the negative impacts of your antagonist's bad behaviors.* People often tell themselves *It's*

not so important. Just let it go. Or they try to tell themselves things turned out worse than their antagonist wanted. Unfortunately, it often takes dire and unavoidable circumstances to force us to wake up and deal with interpersonal situations we'd prefer to avoid.

These common distortions explain how seemingly intelligent people end up staying in dysfunctional relationships long after they should have left. Remember my example of catching your wife leaving a hotel with her lover, which she still vehemently denies? I've had clients who caught their mates having affairs *three or more times.* You'd think they would have said *Wait a second, we've been through this before* and gotten divorced, but they didn't. Instead, they doubted the accuracy of their overwhelming evidence.

When you encounter folks like these it makes you wonder *What's the matter with this guy? He's smart; he's an intelligent professional; but, when it comes to dealing with his wife, he can't think straight. Can't he put two and two together and see she's having multiple affairs?!* The answer is: No, he can't. And even if he could, he doesn't want to. When you have traumatic mind mapping, you often don't put two and two together—even if you're a therapist.

EXAMPLE OF HOLES IN MIND-MAPPING RADAR

A woman, Phyllis, once told me she moved 15 times and went to 15 different schools while she was growing up. She herself was a therapist. When I asked her why her family moved so many times, I fully expected her to say one of her parents was in the military. However, Phyllis said she didn't know. Our conversation went like this:

"What do you mean you don't know?" I said.

"They never told me," Phyllis replied.

"You moved 15 times and neither of your parents ever told you why?"

"That's right. My father would just come home one day and tell us we were leaving. He'd say he bought a new house in a different city, and we'd start packing."

I was stunned for a moment. Then I asked Phyllis about her mother's education and employment. Phyllis told me her mother was a college-educated high school teacher, who also received training as a therapist. Hearing this stunned me even more. I asked, "Why do you think your mother never explained why your family moved so often when she clearly had the ability to understand and anticipate your feelings and reactions?"

Phyllis replied, "Our relationship has gotten better in the last few years, so recently I felt comfortable enough to ask her why she never told me. But I still had to work up to it. It wasn't an easy conversation. My mother said she never had time to explain why we moved because she was too busy packing or unpacking or trying to get us kids registered in a new school."

Phyllis' apparent blindness to this illogical and implausible explanation left me speechless, and I wasn't the only one who saw it. Phyllis' husband's mouth dropped open and his eyes were bulging as he witnessed her cognitive functioning seemingly evaporate into thin air.

ANTICIPATORY TRAUMATIC MIND MAPPING

Although mind mapping isn't distortion-free, it's amazingly accurate when it's working. It's good enough for humans to carry out complex social interactions. Our brains' ability to detect deception and mind masking is good enough to force philandering spouses to work overtime to get away with it. You might not always know when a stranger is trying to con you or when a friend is pulling the wool over your eyes; but the better you know someone, the better you'll be at spotting deception instantly—*unless you have traumatic mind mapping!*

Remember, mind mapping is a survival skill. It's more like an army tank than a delicate flower. It's designed to operate in stressful situations and keep you alive. Over the course of our species' evolution, it's worked well enough to make us the most adaptive animal on the planet. We couldn't do this if life experiences constantly knocked out our mind-mapping ability.

However, repeated traumatic mind mapping creates another problem, which we briefly broached when we touched on sexual abuse. Once you've mapped your abuser's mind, you map him even when he's not doing something bad because you want to see the abuse coming. This creates what I call *anticipatory* traumatic mind mapping.

When you suffer from anticipatory traumatic mind mapping, you envision being traumatized by something or someone before it actually occurs. You stay in a perpetual state of high alert, constantly looking for signs of things you don't want to face. You live in a world of constant danger and expected disappointment. Living like this negatively impacts your mind, brain, and body.

For instance, another of my clients, Amy, grew up with emotionally abusive and physically violent parents. Living with them was an ongoing series of traumatic mind-mapping experiences that left Amy perpetually on the lookout for signs of sudden danger. Amy's father liked to spank her bare buttocks. Amy's mother enjoyed slapping her face, and breaking her things. Amy could never relax in their presence, even when they sat together in the living room calmly watching television. Amy always kept one eye on them and one eye on the exit, prepared to run on a moment's notice.

Family sexual abuse delivers a similar double whammy. Not only do you get traumatic mind mapping during the actual experience, you also develop anticipatory traumatic mind mapping. You imagine future encounters with your abuser in hopes of avoiding them. When you're with your abuser, you watch for traumatic moves even when they're not occurring. So even if physical sexual abuse occurs infrequently, anticipatory traumatic mind mapping puts you in that state constantly.

Lots of people grow up like Amy and like Karen (with her father) and Penny (with her brother), all of whom I've mentioned earlier. You may have too. If so, you probably experience frequent difficulties in love relationships, particularly allowing your partner to hold you. You might be married, but it still can be hard to relax around your spouse. You're always anticipating the worst and keeping an eye out for trouble.

Anticipatory traumatic mind mapping can be debilitating, but it's fixable. In Part Three, I'll show you how to take care of this problem. The methods I'm sharing with you have helped countless troubled couples and singles relax and find peace. But before we get there, we need to explore other dark things that create traumatic mind mapping. I'm referring to things like deliberate cruelty and enjoying the suffering of other people. That's where we're headed next.

9

ANTISOCIAL EMPATHY

We're about to take a brief tour of even darker sides of humanity. I don't do this lightly or without reason. I hope to focus attention on ailments afflicting millions of people caused by severe traumatic mind mapping, ailments that are nonexistent as far as most people are concerned. Everything I'll describe here is treatable and reversible, which I'll explain in Part Three. Hang on to this fact while you read about things no one wants to believe actually exist, like parents who are so bad they trigger involuntary reactions in their children's brains.

I'll provide case examples to illustrate these problems. You'll probably be inclined to approach them intellectually, as if you're just reading text. However, I recommend you stop and visualize each vignette to get a visceral understanding of what's happening. Some vignettes may make you shake your head in bewilderment and ask yourself *Where does empathy fit in? Don't the people in these examples have any empathy?* I'll tell you in advance that some folks I'll describe are *incredibly* empathic. This may not make sense when you first envision them. But by the end of this chapter, I guarantee it will.

I also fully understand how hard it is at first to understand how people can deliberately do hurtful or destructive things to anyone, most of all their own offspring. To help you with this,

Appendix C explains scientifically how people can do this, why they do this, and what kinds of people they are. In contrast to the common tendency to think *Who ARE these rare people?!* sometimes they turn out to be your neighbor next door or perhaps someone living in your own home.

WRONG NOTIONS ABOUT EMPATHY

As pervasive as the concept of empathy is in our culture, the word has only been around for a little more than a century, coined by American psychologist Robert Titchener in 1909. Its etymology can be traced to the German word *Einfühlung* (translation: "in-feeling"), first used in 1873 by German philosopher Robert Vischer, to describe the uniquely human ability to feel emotions artists inject into artwork or literature. The concept subsequently grew to include the human capacity to know and understand another person's emotions, thoughts, and feelings.

As it evolved, the notion of empathy took on positive connotations. Nowadays people conflate empathy with "compassion," a sympathetic concern and pity for other people's sufferings and misfortunes. But it is absolutely possible–and all too common–for someone to have empathy without compassion. If this catches you by surprise, you'll want to read on.

When relationships become strained, partners often say and do unkind things. We tend to attribute their behavior to growing apart, losing their connection, being out of touch with each other's feelings, and lacking empathy for each other. This reasoning lines up well with frequent complaints such as, "My partner never listens to me. I don't feel heard!" Or "My partner doesn't see me anymore. I feel invisible, like I don't exist!" Not feeling seen or heard has seemingly become the ultimate trauma. The possibility that your partner or parent not only sees you, and he or she is targeting you, doesn't exist in this mindset.

Unfortunately, this "no empathy, no connection" viewpoint makes perfect sense to many therapists, who attempt to "fix" the problem by teaching empathy and communication skills to (supposedly) re-establish emotional connection between a couple. This error is particularly common among "attachment-based" therapists, who promote the picture that we primarily suffer because our parents never mapped our minds (i.e., "mentalized" us). For example, one expert described it this way: "Aside from having to deal with unprocessed affect, there is also...the absence of an experience of feeling mentalized, leading to a sense of feeling invisible or absent to others."[1]

Therapists rarely consider the alternative view: Partners do cruel things to each other *because* they have an emotional connection, and they use empathy (the ability to intuit another person's experience) without compassion to accomplish this. These clinicians ignore the fact that partners who do cruel things *closely* track what their mates are feeling. In other words, *when partners' compassion for each other evaporates, their capacity for empathy doesn't disappear.*

It's particularly common to conflate empathy with compassion in countries where attachment theory holds sway, such as the United States and England. Another attachment expert proposed that "...abusing and neglecting a child are incompatible with mentalizing [mind mapping]."[2] In the sunny world of attachment theory, empathy is always presumed prosocial, benevolent, and beneficial; and mind mapping is pictured as an inherently positive interpersonal experience. One claimed, "...mentalizing [mind mapping] is always an imaginative, intuitive, playful activity...."[3]

Nowhere do these therapists account for the traumatic mind mapping that occurs when you realize your antagonists *are* mapping you while they're doing unkind things. They *know full well* the impact they're having on you, and this is precisely what they want. Lack of empathy is always the presumed problem. Their solution? More empathy and the more the better.

Such therapists take a similarly naive view of parents who hurt their kids–parents who starve their children, beat them, chain them in closets, or punish them with scalding water or hot radiators.[4] Or parents who feed their children low dose poisons to gain attention when the doctors can't figure out what's wrong with their kids.[5] Faced with extreme cruelty, therapists from the attachment school claim these parents either don't know what they are doing or they're devoid of empathy. And with a simple wave of this linguistic magic wand, parents who none of us want to admit exist–parents who have "antisocial empathy"–disappear from view.

WHAT IS ANTISOCIAL EMPATHY?

People generally assume empathy is always good (prosocial), and conversely, bad (antisocial) behavior stems from lack of empathy (an inability to know and appreciate other people's emotional states). But while studying the neurobiology research literature on bullying, Machiavellianism, cruelty, and schadenfreude (enjoying other people's misfortune and suffering), I became aware that mind mapping and empathic abilities are sometimes used for antisocial purposes.

For instance, "ringleader bullies" often have particularly good mind-mapping abilities and use their social intelligence (analyzing and interpreting social cues) to select victims, recruit accomplices, and engage in proactive aggression for anticipated personal gains.[6] Mind-mapping abilities are relatively intact for the majority of criminals with antisocial personality disorder and psychopathy, which may have an adaptive function in maintaining a criminal lifestyle. Their key deficits appear to be more related to lack of concern about their impact on potential victims rather than their inability to take a victim's perspective.[7]

Piece by piece, I became aware of a concept I'd never previously considered: *Antisocial empathy*. By popular belief, antisocial empathy is an oxymoron (self-contradictory). Isn't empathy inherently prosocial? According to neurobiologists like Tanya Singer, who avoid the conceptual blinders of many therapists, the answer is, "No, it's not." She notes that empathy is not necessarily linked to prosocial motivation, whereas this is true of compassion. Empathy can be used to find a person's weak spots to make him or her suffer.[8]

You need to understand that antisocial empathy completely changes the picture of reality for many therapists. They suddenly realize they've only been looking at half the picture when they thought they conceptualized the whole landscape. When I discuss antisocial empathy with other therapists, some glare at me, delivering a not-so-subtle message that this viewpoint is wrong, or perhaps even evil. To them, antisocial empathy almost sounds blasphemous, like I'm misusing or debasing their cherished concept.

On the other hand, most therapists I've encountered react positively—if somewhat stunned—to the reality of antisocial empathy. They immediately reinterpret their own difficult cases, and they like the way their new understanding opens up new ways to intervene. Many of them like how this clarifies their own personal histories because they too came from highly troubled homes that "out of connection, no empathy" explanations don't really address.

It's not hard to understand how the mental health industry could embrace a view of empathy that's only half the picture. When we're confronted with extremely bad behavior, especially by parents, our minds want to believe they lack empathy and mind-mapping ability, and they didn't see what they were doing. This makes it easier for us to deal with the situation, but it's simply not true and interferes with coming to grips with difficult truths: Some adults, who do really awful things to kids, do them intentionally and use their capacity for empathy to do it.

For children who feel neglected and abused, the only thing worse than feeling invisible is realizing that you're *not*, that your parents see you and hurt you anyway. When you study their method, timing, accuracy and efficiency of abuse, you realize they're tracking you and using their map of you to make you feel bad. That's when you want to believe you're wrong about what you're seeing. You'd rather feel bad for thinking they deliberately want to hurt you. The alternative involves experiencing traumatic mind mapping.

When your therapist tells you that your abusive mother didn't know what she was doing, or she was doing the best she could, it's meant to be comforting. Unfortunately, this suppresses your functioning and interferes with recovery, because you won't be dealing with an accurate picture of your life. To heal, you need to understand and deal with what really went on. You need an accurate autobiographical memory.

I've had lots of mothers tell me they're disturbed by their own shouting at their kids. It's not just the frequency or volume of their harangues that bothers them. It's the fact they deliberately shout long and loud enough to reduce their kids to tears. These mothers confess to being angry and frustrated enough to *enjoy* seeing and feeling their children's distress. This scenario is more common than you'd ever suspect—even if you grew up with this kind of mother.

As disturbing as it is to hear about these mothers, they aren't the ones we most need to worry about. These mothers are already starting to self-correct. Actually, the best in them is standing up because that's the part of them that's upset about their enjoyment of shrieking. *Only the best in us talks about the worst in us because the worst in us lies about its own existence.* But many sociopaths, con artists, pedophiles, and other manipulative, unsavory characters have empathy too. Their ability to intuit what other people feel allows them to successfully exploit others. What these cunningly deceitful people lack is compassion. Their callous indifference to their negative impact on their victims doesn't mean they're unaware of it. Some of them enjoy it.

(Now is a good time to read Appendix C (Antisocial Empathy), if you want to understand the scientific underpinnings of what we'll discuss here. Keep reading here if you just want self-help. But if you take the time now to read in parallel about how your brain works, you'll understand how the content and case descriptions of Part Two and the methods described in Part Three all fit together. Personally, I think this research is amazing. I've included only fascinating stuff you might want to know. If you get bored, you can simply return here.)

Schadenfreude: Harm-Joy

Earlier I described giving a lecture to 700 people in Germany about my research on couples who fail in treatment where nobody left even though it ran long over schedule. Now picture this: It's Valentine's Day 2014, a year earlier; and I'm speaking about the realities of adult love relationships to about a thousand people packed into the same Urania auditorium in Berlin, Germany. Among other things, I'm talking about antisocial empathy and "normal marital sadism." (Romantic topic for Valentine's Day, isn't it?)

In an effort to explain antisocial empathy, I ask the audience if they're familiar with *schadenfreude*. As I expect, nearly everyone raises their hand. *Schadenfreude* is a German word referring to the enjoyment of other people's misfortune and pain. Roughly translated, it means "harm-joy." There is no equivalent term in English because as a culture, we're not willing to accept that anyone (particularly ourselves) could take pleasure in someone else's pain and suffering.

"Well," I say, "you can't have *schadenfreude* if you don't have empathy. You can't enjoy other people's suffering if you can't detect and experience their feelings. Antisocial empathy lies at the heart of *schadenfreude*. Partners who repeatedly do cruel things to each other have antisocial empathy." Instantly, virtually everyone in the audience understood what I was talking about.

Because German society acknowledges that humans sometimes enjoy other people's misery, it was relatively easy for the audience to accept the reality of antisocial empathy and to recognize it in their own relationships. It's also easier for me to discuss antisocial empathy in Germany because they're not as indoctrinated with attachment theory as they are in the United States. It may take us longer to catch on, but it's important that we do because antisocial empathy provides a more complete picture of human nature than our current one. Without it, you develop erroneous views of marriages and families–but you don't realize this because your picture fits our common bias against recognizing cruelty.

EXAMPLES OF ANTISOCIAL EMPATHY

Antisocial empathy is the enjoyment of other people's pain and suffering. You can't enjoy someone else's sadness, pain and misfortune if you don't have antisocial empathy. In marriage, this shows up in what I call "normal marital sadism."[9] This refers to the emotional mayhem, withholding and psychological torture partners often feel entitled to wreak on each other. This is far more common than reportable domestic violence.

When antisocial empathy shows up in parents towards their children, it blows our minds. We look for excuses or mitigating circumstances to explain this away. As intolerable as it is for us to recognize antisocial empathy between spouses, it's far worse to see it directed at children. Sometimes it's because we empathize with children's stunned minds or because we recoil from parents' enjoyment of instilling misery. But recognizing that parents who behave badly often like what they're doing holds its own particular horror. It goes beyond prodding us to recognize antisocial empathy in ourselves. It's more than dimly-remembered experiences in our own lives that we've yet to accept. This recognition shatters our cherished images of parents innately bonding with their kids and always doing the best they can. It triggers the same terror many children experience with clowns.

Talking to you about antisocial empathy is difficult. The worst in us says *I don't want to hear this!* But the best in us needs to step up and accept its presence in our lives because you can't solve a problem you refuse to believe exists. I'm going to share some examples that may make you feel uncomfortable. Besides furthering your understanding of antisocial empathy, they may remind you of your own traumatic mind-mapping experiences and start your healing. You may also recognize traumatic mind-mapping experiences you've created for other people and be inspired never to repeat them and mitigate their impact (Part Three).

For instance, I knew a man, Bruce, whose father bought him an ice cream cone when he was a kid; but just as he was about to give it to Bruce, his father pulled it back and licked it. "Please don't eat my ice cream, Daddy," Bruce pleaded. "Give it to me. It's mine." But his father just laughed and kept licking the ice cream while Bruce became more distraught with each lick. It wasn't lost on Bruce that his father appeared to enjoy himself at Bruce's expense.

Then there's the mother who made frequent derogatory comments about her daughter's body. When Laurie showed her mother a dress she just bought, Laurie's mother said, "It would look better on you if you lost a couple pounds. You're such a pretty girl, but you really need to lose some weight or no man will ever want you." As you might expect, Laurie was crestfallen and in tears. In response, her mother offered her typical "reassurance." "Stop being over-sensitive," Laurie's mother barked, "some men don't care what a woman looks like, as long as you're good in bed." This would be no less excusable had her father said this, but Laurie's mother's own adolescent weight struggles equipped her to appreciate the impact of her messages. Despite having a trim, attractive figure, Laurie struggled with binging and purging throughout her adolescence.

There's the father who, during a family trip, pulled their car to the side of the road when his eight-year-old son started acting up. Gregory's father yelled at him, ordered him out of the car, and left him standing on the side of an interstate highway in the middle of nowhere without shoes or jacket. Eventually Gregory was picked up by state police, who then located his family. The policeman lectured Gregory's father when he returned Gregory, and Gregory's father acted apologetic and remorseful. But when the state trooper left, Gregory got a beating from his father for "embarrassing" him. When the trip resumed, from time to time Gregory's father came up behind his son and slapped his head, causing Gregory to be wary forever after and never relaxed. In fact, Gregory's father seemed to relish moments when Gregory dropped his guard and focused his attention elsewhere.

I know another man, Pierre, whose mother repeatedly went into her bedroom, locked the door, and screamed like she was dying when he was four-years-old. Pierre would start banging on the door, trying to get in. Panicked, he kicked at the door, while his mother continued her blood-curdling screams. Then suddenly, Pierre's mother opened the door and comforted his sobs as he desperately clung to her. She'd rock Pierre, softly sing to him, and kiss his forehead. Pierre's mother subsequently referred back to these episodes as wonderful times of warmth, closeness and bonding between them.

One mother frequently complained to her kids that raising

them was too much work, instilling pictures in their minds that they were unwanted and unrewarding burdens who ruined her life. To lessen his mother's load, her son, Nelson, made his bed all by himself for the first time. Proudly, he invited his mother to his room and pointed to his bed. "Look what I did!" Nelson said.

But rather than offering praise, Nelson's mother replied angrily, "Well, now that you can make your own damn bed, you can just keep on doing it from now on!"

Nelson was stunned by his mother's response, naively anticipating this would please her. In remapping his mother, Nelson realized it wasn't the work that angered his mother; it was the fact he existed.

I know of a father who repeatedly climbed into his adolescent daughter's bed, saying he wanted to take a nap lying there beside her. When Monica demanded that her father get out of her bed, he refused. When Monica yelled for her mother to intercede on her behalf, Monica's mother climbed into bed on the other side of her. When Monica yelled at them to get out of her room, they laughed at her and refused to leave. To fully understand what's going on, you need to know Monica's father was a guidance counselor and her mother was a lawyer. Both parents were quite able to recognize Monica's distress.

Then there's the mother, whose daughter, Anka, had her first period while on a family trip. Anka was in the bathroom of her parents' suite, upset and crying, unable to insert a tampon for the first time. Anka's father was on the other side of the bathroom door, frantically trying to instruct Anka, while her mother listened in from the couch, calmly reading a magazine.

MIRROR NEURONS AND EMBODIED KNOWLEDGE

In the last twenty years, one of the biggest advances in brain science has been the discovery of "mirror neurons." These highly specialized brain cells, located in your brain's motor cortex, fire when you perform purposeful action or when you witness someone else perform that same action. When you observe another person's doing things, your mirror neurons create a pattern in your mind that's identical to the pattern in the other person's mind, as if you performed that behavior yourself.[10]

This direct, visceral knowledge of someone else's experience is called "embodied knowing." This is an automatic hard-wired brain process that doesn't require deductive logic. The discovery of mirror neurons was cheered by proponents of attachment and emotion-based therapies, who believe we're born to connect and empathize with each other in prosocial ways. To their chagrin, however, mirror neurons and embodied knowing also shed light on the reality of antisocial empathy. This is how, in the moment, people who do hurtful things know the impact they're having, even when their victims try to mask their minds. They can feel it.

Let's revisit the story about Bruce, whose father buys him an ice cream cone but licks it instead of giving it over. It's clear that Bruce's father possesses an embodied knowledge of what Bruce is feeling as he begs his father to stop. He can feel Bruce's visceral yearning to grab the ice cream out of his hand. He becomes gleeful as he senses Bruce's agitation. Even when he relents and relinquishes the ice cream cone after a few licks, he knows Bruce will still feel disappointed and upset. Bruce's father has taken a treat and turned it into a frustration for his son. Then, to cap it off, he insists that Bruce thank him for the ice cream, demanding gratitude and deference he knows Bruce can't possibly feel.

This vignette highlights that, just as our brains are wired for prosocial empathy, we're wired for antisocial empathy too. We're not just emotionally needy children forever searching for love and acceptance. We've got another side to us, and it's much darker and meaner. Most therapists don't realize mirror neurons let people know when they're doing something mean and cruel—*as they're doing it*, if not immediately thereafter. Embodied knowing lets them feel the visceral impact of their behavior on their target. Once you understand this, the cruelty in many people's behavior becomes much clearer.

For example, as uncomfortable as it is, ask yourself *How can pedophiles do what they do?! How can you sexually abuse a kid and live with yourself afterwards?* Most people don't let themselves look at the whole picture. When you live in a mental world where empathy is always prosocial, you want to believe people who exploit children are out of touch with their victims' feelings. But in reality, pedophiles often enjoy inducing inner conflict in children, tempting and seducing them or implanting shock and fear. Some like kids to struggle with their own temptations because it feels good.

Others like seeing terror on kids' faces. Some get sexually excited experiencing children's pain and suffering. All these involve mapping their victims' minds and viscerally feeling their reactions. We want to tell ourselves pedophiles are out of touch with their victims, or they're only interested in their own sexual gratification. But pedophiles commonly have antisocial empathy. That's how they can do something that to a normal person seems insane, including instilling mind-twisting threats to keep kids from reporting them.

I've already explained why I'm telling you terrible stories. I'm trying to help you see things most people deny so you can take care of traumatic mind mapping if you've been on the receiving end. Realizing the important people in your life have antisocial empathy is an important part of resolving traumatic mind mapping.

How You Can Tell Someone Has Antisocial Empathy

This also highlights the importance of identifying people who primarily operate from antisocial empathy. This could be your spouse. Or your parents. Or a sibling. Or your boss. Or a co-worker you thought was your friend. It could even be you, yourself. Recognizing antisocial empathy, wherever you find it, changes your view of your relationships with other people.

To this end, here are some common signs of antisocial empathy so you'll have an easier time detecting it in the future. The most obvious sign is enjoyment of another person's pain, suffering, or unhappiness. Smiling or laughing while other people cry or scream in pain are clear examples. But one variant that often goes undetected is enjoying another person's confusion or inner conflict, including perplexing or stunning other people to the point they're immobilized because they simply don't know how to respond. Rather than grinning, however, your antagonist acts confused and surprised by your confusion. In Part One, I described this kind of person as a mind-twister. Hopefully, you'll quickly recognize when you cross paths with such folks in the future.

Antisocial empathy also shows up in a person's eagerness to punish and deprive. Some parents seize upon any infraction, no matter how small, to take things away from their children, whether it's material possessions or enjoyable experiences. Children of such parents quickly learn not to let their parents know if they enjoy or look forward to something because that's the next thing their parents will take away from them.

Another indication of antisocial empathy is repeated thwarting and withholding, especially when this keeps you in a position of deprivation or discomfort. For example, let's say you have a co-worker who covets the recognition you've been receiving for spearheading development of a new product. You're now tasked with preparing an important report, and she's responsible for a large part of it. You asked her nicely about this a month ago, and she assured you she'd get her part done on time. But every subsequent time you've asked about it, she's given you different excuses for why it isn't done yet, and you're getting nervous she won't finish it on time. You've explained your dilemma to her, and you've even begged her for her stuff. She's keeping you on pins and needles, and it appears to be intentional. It might be overstating things to say she's trying to sabotage your efforts and make you look bad. But your actions, appearance, and statements clearly indicate what you're feeling; and this is at least acceptable, if not enjoyable, to your colleague.

Antisocial empathy can show up in emotional, physical, or sexual arousal. I once knew a woman, Marjory, whose father frequently spanked her when she was a little girl. A lustful smile would come over his face just before he would spank her. It was such a reliable indicator of what was to come, Marjory took to calling it, somewhat ominously "The Smile." When she was young, Marjory realized The Smile reflected the enjoyment her father got from hitting her. By the time she was 11 years old, she recognized his lust. When he spanked Marjory, he would lay her across his lap with her bare buttocks in the air. When she was younger, Marjory thought he did this to embarrass her. But as she got older, she realized her father was sexually aroused by her nakedness and vulnerability.

I met Marjory when she and her husband attended one of the Passionate Marriage® Couples Enrichment Weekends I conducted with my wife, Dr. Ruth Morehouse. When I asked the assembled partners who amongst them enjoyed inflicting normal marital sadism on their partner, everyone raised a hand–except Marjory. This caught everyone's attention, and most people thought Marjory refused to confront herself about enjoying hurting her husband. Many participants nearly fell out of their chairs when Marjory said, "I didn't raise my hand because I didn't think this accurately described me. I become sexually aroused and lubricate when I hurt my partner."

Another sign of antisocial empathy is aggressive mind twisting on top of punitive behavior. As you'll recall, mind twisting manipulates you by someone's implanting a picture of his mind in your mind that's designed to make you do what he wants you to do or feel what he wants you to feel.

For example, let's say you're the Lower Desire Partner in a hostile marriage and your spouse initiates sex. Not only do you decline to have sex, you do it in ways designed to hurt your spouse's feelings and twist her up emotionally. "Well, I *was* interested in having sex with you tonight," you might say. "I was actually going to propose it myself. But now that you're pressuring me, I don't feel like it anymore." The truth is you weren't going to have sex no matter what. But you not only enjoy frustrating and disappointing your spouse, it's even better when you made it look like it's her own fault.

A final sign of antisocial empathy is a notable absence of remorse or regret for hurtful behavior, even if the person expresses guilt, shame, or sorrow. Some folks deliver the classic line, "I'm hurting too!" when they get caught having extramarital affairs. Claiming to be the victim of your own bad behavior *might* convince your partner that you're already being punished and appease her thirst for vengeance. But regretting getting caught doesn't mean you regret hurting your partner or that you couldn't tell this would hurt her. In fact, your "I'm hurting too" move suggests you can feel your partner's anger and you're trying to mollify it.

This short list is a primer for identifying people operating out of antisocial empathy. You're better off avoiding these folks or dealing with them effectively (as I'll describe in Part Three) because interacting with them inflicts traumatic mind mapping if you don't manage the interaction. It often turns out that dealing with these people is unavoidable because they occupy important positions in your life.

Studying Who Fails in Therapy

Around the time I started researching mind mapping, I also started studying clients in my practice who failed to thrive in therapy. Most clients did extremely well, but no therapy is perfect. Sometimes I'd have one couple who wasn't significantly improving. Occasionally, I might have two. But on one "lucky" occasion I actually had three fail-to-thrive cases at the same time, which allowed me to look for commonalities amongst them. I wondered if these cases had something in common, some trait or factor I could identify that would help me turn these cases around.

What I discovered shocked me and ultimately allowed me to develop a new therapy that really helps people from highly troubled backgrounds. Much to my surprise, I found these cases had *two* things in common. I then looked back through similar past cases, and these same two characteristics kept jumping out at me.

The first commonality was all these clients were trying to maintain relationships with poorly functioning parents. In particular these parents functioned so badly that *the only way these relationships could be maintained at their current level of involvement required my clients' own functioning to remain impaired.*

For instance, my fail-to-thrive clients acted like they didn't see what their parents were doing, or my clients went along with things that didn't make sense but acted as if they did. A common example was a spouse who refused to deal with his intrusive, destructive parents, the result of which surfaced within his marriage as arguments with his wife about loyalties and priorities. My fail-to-thrive clients acted like they didn't recognize or understand what their parents were doing. Although they had mind-mapping ability–typically they were hyper-vigilant everywhere else in their lives–they appeared mind-blind to their parents.

In every case, clients understood there was no point asking their parents to change. The current sunny tone in their relationships with their parents only existed as long as they went along with whatever their parents were doing. Things would change drastically if they questioned how their parents treated the adult child or the spouse, or if they challenged their parents' explanations for their past bad behavior. Often these involved shocking episodes that produced traumatic mind mapping in my clients.

By refusing to confront their parents' gross manipulations or unkind treatment of their partners, my clients tried to elicit the best possible version of their parents they thought they could get. They knew from traumatic childhood memories just how punitive, aggressive, or horrible their parents could be. They chose to maintain the superficially friendly rapport even though it only existed on their parents' terms. (If you're familiar with my prior books, you'll recognize these are classic examples of emotional fusion and borrowed functioning [low differentiation].)

The second similarity surfaced when I examined clients' reports of their parents' behaviors. *All these parents were doing terrible things, and I mean **really** horrible.* So much so, I found it upsetting to hear about them, much the same way you probably felt about some cases I've described. My fail-to-thrive clients repeatedly experienced extreme traumatic mind mapping from mapping out what their dysfunctional parents were doing because their parents' actions displayed mind-mapping ability and antisocial empathy. Here are some examples:

- You're ten years old, and you ask your mother to

make some French fries, one of your favorite foods. Your mother is irritated, so she makes a *huge* batch of fries–*and makes you eat every last one of them.* She sits across the table from you, watching you stuff them into your mouth. You're full, but your mother insists you keep eating, which you do until you throw up. When you've finished vomiting, you map your mother. She has a satisfied look on her face. After that, you never ask her to cook French fries ever again.

- You're an adolescent boy, too young for a driver's license, but your father makes you drive him to bars when he's too drunk to drive himself. When you protest, your father describes horrific images of him dying in an accident, and tells you that, if this happens, it will be your fault. When you drive him around, he tells you the two of you are "male-bonding" and you're his pal. He tells you not to tell your mother what the two of you are doing, that this will just be a little secret between the two of you guys.

- You are a six-year-old girl, and your father often plays with your genitals through your underwear while you watch TV together. You feel conflicted because it feels good, but you know what your father is doing is wrong. It felt like you and your father were playing an innocent game until the day your mother walked into the room and your father jerked his hand away from under your pants. In that moment, you realize your father knew all along that what he was doing was wrong; but he did it anyway. This is born out when you are 13 years old, when your father breaks into the bathroom, yanks you out of the shower, shoves you to the floor, pins your arms down, and nuzzles your fledgling breasts with his nose.

- You're a young kid and your mother has a

predilection for hitting you with long wooden spoons from her kitchen. When she's mad, she'll chase you around the house trying to "pop" you with her spoon. Other times she'll threaten to get out her spoon if you don't do what she wants. Sometimes she'll just gesture like she's holding a spoon to give you a warning. She seems to relish the sound of the spoon smacking your skin and the way it makes you jump. Your mother's fondness for hitting you with wooden spoons is so apparent that, when you visit your maternal grandparents, your grandmother hides all the wooden spoons in her kitchen to keep them from your mother.

- You're an 11-year-old girl, and your divorced mother repeatedly asks an 18-year-old neighborhood boy to babysit you while she goes out for the evening. Eventually you let her know the boy touched you inappropriately on several occasions. Your mother acts shocked that you never told her this before and makes you feel like you betrayed her and let her down. She never expresses concern or remorse for what happened to you. You're stunned, confused, and feeling guilty. Your mother proposes that, since you let each other down, you should just forgive each other and move on.

- Your parents frequently have long, violent, drunken fights, followed by loud, aggressive sex. Because this all occurs in public areas of your house, like the living room or the dining room, there's no way you can possibly ignore what's going on. But without fail, at the breakfast table the next morning they act like nothing happened. You are completely baffled, but when you ask them about this, they act like they have no idea what you're talking about.

- When you are 18, your mother takes you on a month-long cross-country vacation with her lover and makes you promise not to tell your father.

As soon as you return home, your father grills you about what happened on the trip. He says he suspects your mother is having an affair, and if they get a divorce it will be your fault. Your mother takes you aside afterwards and asks that, if you decide to tell your father about her lover, you tell her first so she can tell your father.

- You're a 17-year-old girl, and your father locks you out of the house whenever you miss your curfew by 15 minutes. Sometimes he makes you sit outside in the cold for hours. When he finally lets you in the house, he occasionally throws you down a flight of stairs. As you're checking yourself for broken bones, he says you're a piece of filth that doesn't deserve to live in his house. You get the distinct impression he thinks you're having sex with lots of guys. At first you assume his face is flushed with anger—until you realize he's sexually aroused. He's enjoying manhandling you and making you feel worthless.

- Lots of parents have extramarital affairs, but your father has a second family in another town. It turns out you have half-siblings you didn't know about. When you discover this, you confront your mother about not telling you. It's pretty clear she must have known since she made excuses for your father when he often didn't come home. But your mother denies this, insisting she had no more idea about another woman in your father's life than you did. She probably wouldn't have done anything if left to her own devices, but your asking her about this provokes a big family confrontation. Your father decides he wants to stay married to your mother and declares he wants nothing further to do with this other woman or his children with her. He masks his limited investment in these kids by acting like he's making a huge sacrifice in "giving them up" to save his marriage.

These examples are just the tip of the iceberg. Virtually all of the parents of my fail-to-thrive clients were ruthless self-centered intrusive people who felt entitled to do whatever they wanted. Some believed that whatever they did was OK, while others clearly knew what they were doing was wrong and did it anyway. Most lacked the ability to invest themselves emotionally in another human being. Many actively exploited or emotionally tortured their children. In total, all these parents were doing things best described in a single word: Disgusting. These are the parents nobody wants to believe exist. These parents led me to investigate the science of disgust.

DISGUSTING PARENTING

When someone's behavior or thinking is bad enough, it creates such extreme traumatic mind mapping that it triggers automatic primary emotions in another person's brain. The most commonly evoked primary emotions are shock, anger, contempt and disgust. Therapists have focused on clients' anger for decades, but disgust isn't generally part of their lexicon. It's one thing to say you are angry at your parents. It's quite another thing to say they disgust you.

Disgust is a powerful, involuntary, hard-wired brain reaction involving revulsion to offensive, distasteful, or unpleasant substances, situations, and people. Quite often it's a reaction to seeing, smelling, or tasting something physical that's putrid, like rotten food or a decaying body. But as humans evolved and we developed a capacity for morality, our disgust reaction evolved right along with it. Now it can be triggered by disgusting *acts and thoughts–yours or other people's*–that violate universal standards of moral decency, like sexually exploiting children, holding people captive or deliberately starving, beating, or killing them. Your own disgust reaction keeps you from doing antisocial things, and people who have impaired disgust reactions get themselves into lots of trouble.

Research shows that a part of your brain called the *anterior insula* lights up when you experience moral disgust, the same way it reacts to moldy foods and roadkill.[11] It makes no difference whether you actually taste, smell, see, or hear something disgusting, or you imagine or watch someone else experiencing this.[12] The anterior insula of your brain lights up, giving you an overpowering visceral urge—requiring no forethought or decision—to distance yourself from the offending object or person.

Brain studies of disgust are revealing absolutely incredible secrets about the neurobiological underpinnings of our own self-awareness and the awareness of bodily experiences and mental states of others. While the topic of disgust, by its very nature, makes you want to shy away from it, you won't believe how much studying disgust has taught us about how the brain works, creating new opportunities for treating traumatic mind mapping. (Appendix C contains a summary of relevant fascinating research on disgust.)

I began studying disgust when I discovered the parents of my fail-to-thrive clients were doing disgusting things. The scientific study of disgust is actually quite new. The American Psychological Association published one of the few books on this topic just several years ago.[13] I soon discovered disgust isn't just a label people attach to things they find objectionable. Just like mind mapping, disgust is a survival mechanism. It's designed to keep you away from things that could sicken or harm you.

So when I refer to disgusting thoughts and behavior, please understand this is brain talk rather than pejorative labeling. I'm referring to how the brain operates. I'm speaking the brain's language.

Many of my clients witnessed parental behaviors that were so severe they triggered involuntary disgust reactions in their own brains. During treatment, visualizing their stories often triggered disgust reactions in me as well. I wouldn't be surprised if some of my examples affected you the same way. After seeing how prevalent this was, I coined the term *"disgusting parenting"* to describe parents who do such reprehensible things they trigger involuntary disgust reactions in their children's brains.[14]

Putting the words "disgusting" and "parents" together is taboo in our culture because we like to believe (a) parents are always well intended, (b) they want the best for their kids, and (c) they always do the best they can. However, the reality of disgusting parenting blows these myths to smithereens.

In the examples I've given, these parents knew what they were doing. Their actions were voluntary. They clearly understood and enjoyed the impact they had on their children. They had mind-mapping ability and antisocial empathy. That's what makes their behavior particularly disgusting.

When your parents repeatedly act this way, you can't avoid the damage it does to your mind and brain. Just as with mind mapping, traumatic mind mapping is very much a bottom-up process. You don't need sophisticated cognitive abilities to figure things out and be hurt by this process. It's not good to have disgust reactions repeatedly triggered in your brain. Children are particularly impacted deeply because, besides being severely emotionally impacted, these powerful disgust reactions are occurring while their rapidly developing brains are particularly susceptible.

When your parents repeatedly do disgusting things and you realize they not only have mind-mapping ability, they're enjoying your frustration and unhappiness, your mind turns to mush. You get "spaghetti brain." You can't think straight. You walk around in a fog. You're forgetful and clumsy. You frequently fall apart emotionally. You're often anxious, agitated, easily distracted, and depressed. You have difficulty concentrating at school or work, and you're at risk of being misdiagnosed as having ADHD and being put on medication. I'll describe the interpersonal neurobiological impacts in more detail in the next chapter.

THERAPISTS, TEACHERS AND DOCTORS OVERLOOK DISGUSTING PARENTING

You may be the child of these kinds of parents. If you are, you're part of a larger group than anyone wants to accept, made up of innumerable young children and adult children being abused and mistreated in disgusting ways. Like most people, you probably want to believe your parents are simply in turmoil, or "out of touch with themselves" or that they don't understand the impact of their behavior. You may think of this as cutting them some slack, but what you're actually doing is allowing yourself to avoid the problem and not doing something about it.

As a culture, we cherish the notion that parents always do the best they can. That's what I try to do as a parent, and that's what my parents did when they were raising me. But if your parents did the best they could while you were growing up, you won the lottery because that's not always the case. Good parents are often far from perfect, and children can handle their parents' flaws. What they can't handle knowing is their parents are doing much worse than the best they can and that they're doing it on purpose.

In the 1960s and 1970s therapists commonly assured their clients that, despite evidence to the contrary, their mothers and fathers were always parenting to the best of their abilities. In fact, I once watched an archival tape of famed psychiatrist and family therapist Murray Bowen working with a family. I'm a great admirer of Bowen's groundbreaking concept of differentiation, and it's become a cornerstone of my clinical work. So when Bowen said, almost in passing, "Parents always do the best they can," I almost fell out of my chair! I subsequently saw a videotape of another renowned family therapy pioneer, Virginia Satir, basically saying the same thing.

This prompted me to embark on an informal research project, in which I asked all the therapists I knew if they thought parents always did the best they could. Half the therapists immediately agreed, and the other half looked at me suspiciously, like I was asking a trick question. My clients who have disgusting parents often say their prior therapists assured them that their parents did the best they could.

This is particularly insidious because the impacts of repeated disgust reactions can be reversed and corrected through treatment. Therapists who tell clients this rhetoric interfere with their full recovery, which doesn't begin until the person recognizes their parents have mind-mapping ability and antisocial empathy. Only then can they understand what their parents were doing, that they did it anyway, and in many cases, enjoyed doing it.

Refusing to recognize cruelty is a serious, widespread problem. We have difficulty accepting that people deliberately do cruel things to each other. We like to presume people have basic decency; but, unfortunately, we can't take basic decency for granted.

Yes, there are many really decent people in the world. But we prefer to deny what we secretly know and refuse to acknowledge: Being cruel can feel *good*. This is why people do it, and what makes their thoughts and actions particularly disgusting is when they do seem to feel good about what they have done.

As far as I'm concerned, this is a national health crisis. As long as mental health professionals, educators, and policy makers refuse to recognize disgusting parenting, more children will struggle with the impacts of what I've described here. However, these affects aren't limited to interactions between parents and children. They occur whenever people do disgusting things in daily life.

So please put what you've learned about the brain's automatic disgust reaction to good use. Recognizing you're having this reaction helps you deal with people who are doing terrible things, whether they're your parents, siblings, friends, co-workers, or your mate. But let this knowledge also sensitize you to your impact on the people around you. If we would all refrain from doing disgusting things, our brains and minds and the world we live in would be much healthier places to be.

10

IMPACTS OF TRAUMATIC MIND MAPPING

Thus far in Part Two, we've explored how traumatic mind mapping, antisocial empathy, and disgust brain reactions occur in daily life. In this chapter we'll look into how these things negatively impact your brain. The next five chapters in Part Three will show you how to turn these deleterious influences around.

The effects of traumatic mind mapping fall into two categories: short-term and long-term impacts. In the moment, people have acute responses to abhorrent behavior, although you might not recognize them in yourself because you're so impacted at the time. Likewise, repeated traumatic mind mapping–especially repetitive disgust reactions–have their own effects. These can be debilitating and difficult to deal with if you don't recognize what's happening. I've identified these impacts in my clients, but I've seen that you can't help yourself if you're unable to see them in yourself. Awareness is the first step to healing.

Short-Term Impacts of Traumatic Mind Mapping

When you observe revolting behavior, it triggers your disgust reaction. Even if it's just a one-shot event, several things usually happen: You stop thinking straight. You get upset and overwhelmed by strong emotions. You don't feel well. And although you're usually not aware of it, your mind mapping collapses, which fools you into thinking your antagonist doesn't know what he's doing.

COGNITIVE IMPAIRMENTS

Moments of traumatic mind mapping give you "spaghetti brain." Your cognitive functioning declines in speed and accuracy. It feels like you can't add two and two together. Your deductive logic and reasoning evaporate. Things that don't make sense get by you. You're distracted by minor details. You lose track of what's important. Even if you manage an intelligent thought, you can't turn it into an actionable plan. One moment you may look smart enough to cure cancer or you're on your way to an important insight, but after traumatic mind mapping it's like you've just had a lobotomy.

A single, vivid event can induce traumatic mind mapping, but less dramatic, frequent troubling events are often more damaging. Day-in-and-day-out repetitions cause the most harm, but they regularly go unnoticed because you accept them as "the way things are." When people think about trauma, they focus on once-in-a-lifetime moments like being caught in a tornado or the toppling of the Twin Towers on 9/11. They overlook disgusting behaviors, which are less vivid than natural disasters or terrorist acts but are no less powerful when they frequently reoccur because (1) the brain wires through repetition, and (2) disgusting behaviors and thoughts involve traumatic mind mapping (Chapter Eight).

For instance, let's return to a case I described last chapter. Tom's parents typically started drinking as soon as they got home from work, and on weekends they were loaded all day. They weren't social drinkers. They drank until they were knee-walking drunk, at which point they got into loud fights that often turned physical. Plates got thrown, and fists got punched through walls. Tom's parents followed each other around the house, screaming vile names and cursing the entire time. On a nightly basis, Tom fled to his room, locked the door and stuck his fingers in his ears. Unfortunately, he could still hear them hurling books and curses at each other. It wasn't any better when the noise shifted to rhythmic thuds and grunts of pleasure.

But what typically happened next really blew Tom's mind. Like clockwork, the next morning his parents would be sitting at the breakfast table together, drinking coffee and reading their newspapers, acting like all was right with the world. When Tom asked about their brawl the prior evening, they'd smile at each other and then turn to Tom with fake puzzled looks that said *What's wrong with you? What's the big deal?* At that moment, Tom's mind and brain got so overloaded with the significance of what he saw that his mind-mapping ability fell apart.

When Tom came to see me, his presenting problem was his inability to keep a love relationship going longer than a year. He became so suspicious and hypervigilant, bordering on paranoia, that no girlfriend wanted to put up with him. Tom didn't trust anyone—except his parents, from whom he frequently sought advice.

You would think, given Tom's history, that he'd tied his distrust and suspiciousness to his experiences with his parents, and he'd never in a million years go to them with his problems. But this is what it looks like when someone's mind-mapping ability collapses. Tom didn't even mention his parents' fights and sexual encounters until I asked him about their marriage.

MIND MAPPING SHUTS DOWN

To the best of my knowledge, there's no published research in scientific journals on how disgust reactions impair your mind-mapping ability. However, I've studied this in my clients for more than a decade and cross-referenced my observations with available neurobiology research. Here's what I've figured out: When your disgust reaction is triggered, it takes control of the parts of your brain involved in mind mapping and disrupts your mind-mapping ability.

Technically, your anterior insula isn't part of your brain's mind-mapping system, but it's wired into parts of the brain that are. When you have a disgust reaction, your anterior insula takes "functional control" over these other parts. In other words, traumatic mind mapping causes your brain's mind-mapping system to collapse. As a result, you become mind-blind to your antagonist. This not only impairs your ability to respond in the moment, it also shapes how you remember that event in the future. You may recall the encounter in vivid detail, but your map of your antagonist's mind is usually missing when you try to recall the incident.

I once stumbled across a noteworthy example of this in a magazine article humorously entitled "Spending to Piss Off My Father." In one section the author wrote: "A few years before he died, my father called and said, 'You know, I was just getting ready to write my monthly check to the life insurance people, and it occurred to me that there is no scenario in which I'm going to benefit from the $225 I am putting into this fund every month. So if you want the 30 grand after I'm dead, you're going to have to start making the payments.'"[1]

This might not seem strong enough to create traumatic mind mapping. However, according to the author who painted a definite picture of affluence, her father wore finely tailored suits, always drove new Cadillacs, belonged to a country club, and played tennis three times a week. She wrote that her father was "so jealous that I got paid to travel—a passion I inherited from him—that he would practically vibrate with rage every time I returned from France or Bolivia on the publisher's dime." She continued, "If he heard me talking about how lucky I was to make a living doing what I loved, he found a way to assure me I would get my comeuppance. 'One of these days you're going to realize you spent your whole life lying in the gutter with someone else's foot on your neck,' was his favorite expression."

Obviously, the author intended to demonstrate good understanding of her relationship with her father. The article was supposedly about her "insights" on why she spent money so freely: She did it just to piss her father off! The author also attributed her father's behavior to his deprived upbringing. She wrote, "My father was a child of the Depression and therefore obsessed with money...."

But, given what you know about mind mapping, would you say the author accurately mapped her father's mind? Is the fact her father grew up in the Depression the best explanation for his behavior? Does "You'll realize you're lying in the gutter with someone's foot on your neck!" sound like an obsession with money? It seems clear the author's mind-mapping system had collapsed, leaving her mind-blind to her father. Had she recognized the gaping hole in her mind-mapping radar, she probably wouldn't have published the article.

I don't personally know this author. But I know that the way she maps her father will greatly shape how she responds to his proposal that she pay his life insurance premiums. It's one thing to tell him, "Dad, you are a child of the Depression, and you've always worried about money." It's something else to say, "Dad, you have always begrudged me most pleasures in my life, even though you have plenty of money for yourself."

You can imagine how these two responses will bring forth different versions of her father, and he's likely to object the most to the latter one. This illustrates what I meant in the last chapter when I said my fail-to-thrive clients' neurobiological impairments maintained the current equilibrium in their relationships with parents who did disgusting things.

REPEATEDLY TRIGGERED PRIMARY EMOTIONS

In the moment it happens, traumatic mind mapping can trigger your "primary" emotions. Although the exact number is endlessly debated, experts say we have five or six primary emotions hard-wired into our brains. The ones most relevant to traumatic mind mapping are disgust, anger, fear, surprise, and sadness. Each one evokes a powerful subjective feeling, coupled with an equally powerful psychophysiological reaction and a dedicated facial expression that transcends culture. People who repeatedly do cruel things become particularly adept at eliciting and recognizing primary emotions of other people by observing their own impact on their victims.

Traumatic mind mapping often triggers the primary

emotion of disgust, but it can also trigger other ones, particularly anger. If this occurs frequently, you can end up with a problem like reflexively crying when you're angry. I'm not referring to people who cry because they're afraid to get angry or those who use their tears to manipulate other people. Some people continually cry when they're angry because their brains are miswired. Their crying is usually superficial and lacks sufficient emotional depth to provide effective relief or resolution. When these folks learn to control their reflexive crying, they usually deal with their anger more directly and go through corrective interpersonal experiences with the difficult people in their lives who rewire their brains and turn their lives around.

IMPAIRED EMOTIONAL FUNCTIONING

Traumatic mind mapping also impairs your emotional functioning. In the moment, your prefrontal cortex ("executive functioning") loses control, and your limbic system ("emotional brain") takes over. You suddenly feel like you're drowning in your emotions. One minute you feel fine, then–bang!–the next thing you know, you feel like crap. Your feelings of self-worth evaporate. So does your ability to tell time. You feel like you'll be stuck in this anxious, depressed state forever without any hope for change. Or, you can suddenly become enraged in the blink of an eye, flipped-out angry and completely unable to control your temper.

I call this shift in brain function an *acute regression*. This refers to a loss of emotional self-regulation and decreased cognitive functioning. (I'm not referring to an "age-regression," where you return to an earlier time in your life under hypnosis or sodium pentothal).

My clients often describe being in an acute regression as "the quicksand reaction." You feel like your emotional stability has suddenly given way beneath you, and you're sinking and drowning in your feelings. You could be a successful business executive, or a brilliant scholar, or a world-famous fashion model, or a veteran politician who's continuously reelected by his constituents. When you're having an acute regression, you're like a magnificent skyscraper that's built on a swamp.

Your "top end" abilities may be fantastic. But when you're regressed, the worst in you is in charge. Your knowledge or native talents become virtually useless.

During an acute regression you typically feel overwhelmed and devastated. You might exhibit sudden outbursts of anger, uncontrollable sobbing, profound sadness, or self-destructive behavior. You might also experience anxiety, panic attacks or profound dysphoric moods. These are some of the most pervasive problems in modern society, which explains why Prozac is one of the most widely prescribed (and over-prescribed) drugs year after year.

Acute regressions are triggered by high anxiety and profound meaning, and traumatic mind mapping overloads you with both. Meaning is tangible to the brain. I can hit you with a stick and I can hit you with meaning, and both can powerfully impact your brain. If you get overwhelmed by enough meaning—good or bad—your brain starts to regress.

For instance, watch when someone wins a prestigious award like an Emmy or an Oscar. This is so full of meaning, some recipients start babbling like idiots, thanking everyone from their third-grade teacher to their dog's personal trainer. It's also why families frequently experience ugly emotional arguments at births, weddings, and funerals. The meaning of these events is just too much for poorly differentiated people to handle.

Repeated traumatic mind mapping predisposes you to having acute regressions in the future. It takes less stress for you to have "quicksand reactions." You sink further down when you have them, and it's more difficult for you to pull yourself out of them. If these traumatic episodes are frequent and severe enough, you can also have difficulty regulating your own body. As we discussed in Chapter Eight, your healthy HPA stress response can become overwhelmed, and the link between your brain and your heart no longer reacts like it's supposed to.

What can you do when you start to feel regressed? This topic deserves a book of its own, which I plan to publish in the future. But what's the single most important suggestion I can give here? It is this: Simply recognizing that you are regressed helps tremendously. Becoming aware enough to say to yourself

"I'm regressed!" will improve your functioning. This is the all-important first step in pulling up your functioning.

You may be down for hours, perhaps even days. But at some point you'll come back up, and you'll feel much better when you do. Establishing a more solid sense of self (becoming more differentiated) also helps because the stronger your sense of self, the more anxiety and meaning you can handle before they trigger acute regressions.

MIND MAPPING FAILS TO COLLAPSE

Mind mapping shuts down for most people when they experience traumatic mind mapping, but a small group have the opposite problem: Their mind mapping *doesn't* shut down. These people retain a full picture of their antagonist's mind. They also end up looking like they're emotionally dead. They have no response in situations that would elicit strong reactions from most people, particularly situations involving cruelty. It's like they have an emotional dead spot. They're like a bell without a clapper.

For example, one client, Susan, told me about going to a drive-in restaurant with her father when she was 12 years old. Susan's father parked the passenger side of the car beside an Econoline van with no windows in the back. He pointed to the van and said, "Don't ever get in a van like that with a boy because, if you do, he's going to make you have sex with him." Susan told me without hesitation she knew, in that moment long ago, her father was imagining having sex with her in the back of that van.

This would have been difficult for most people to recall. It's more likely they'd remember the event without this part of the experience–if they remembered the event at all. In contrast, Susan remembered being acutely anxious and wanting to escape. That's when she realized her father parked so close to the van, that she was unable to open her door, blocking her exit. Instead, Susan stared out her window and acted like she wasn't listening as her father continued with his impromptu "sex education lesson." This was one of many experiences Susan related where her father did things that, in her words, were "disgusting."

Like other people whose mind mapping didn't collapse from traumatic mind mapping, Susan seemed emotionally dead. She wasn't able to have satisfying adult love relationships, although she certainly had lots of sex. Her marriage was breaking up because she wanted to attend group sex gatherings without her husband; and she wanted him to give her the go-ahead, which he didn't. Although Susan denied wanting to participate, claiming she only wanted to observe, Susan's husband knew she engaged in indiscriminate sex. Sex was her primary way of relating to the world and everyone in it.

While at first glance you might think Susan was hedonistic, she actually suffered from anhedonia. She was unable to experience pleasure from activities most people find enjoyable, like sex with her husband, or solitary activities, or non-sexual social interactions. Researchers theorize anhedonia may result from a breakdown in the brain's reward system, involving the neurotransmitter dopamine.[2] My clinical experience suggests this can happen from repeated traumatic mind mapping.

LONG-TERM IMPACTS OF REPEATED TRAUMATIC MIND MAPPING

Traumatic mind-mapping events that upset your functioning in the moment can also leave a lasting imprint on your brain and mind, especially if they happen repeatedly. They become iconic "red hot" memories that forever stand out on your autobiographical memory, even if you don't fully understand them.

I want you to visualize our next example: You're 12 years old and sitting in your dad's car, waiting for him to step out of the house and drive you to your friend's house. You're bored, so you open the glove box and a handful of condoms fall out. You just learned about condoms in your sex education class, so this seems odd to you. You wonder *What is my dad doing with condoms? Why would he keep them in his car? Does this explain why my parents sleep in separate bedrooms?* The full impact of this discovery might not register in the moment, but years later it hits you like a freight train when your parents get divorced.

As far as traumatic mind mapping goes, this example is pretty mild. But think back to our earlier example in Chapter

Eight, of Roger, the six-year-old boy whose father smiled at him when he caught his father having sex with their maid. This incident produced more extreme traumatic mind mapping because the illicit sex was explicitly visualized rather than implied, and Roger's father involved Roger in the experience by smiling at him. Roger walked away with a severe case of "spaghetti brain." This moment stuck with him for most of his life, until he came for treatment.

The human brain can withstand a surprising amount of trauma, especially if these are isolated events. If it couldn't, our species would be extinct or far less adaptable and inventive. But seeing your father having sex—and smiling at you as he's doing it—is too much for most kids to handle, not to mention that the sex act involves someone other than your mother. But as harmful as one-shot overwhelming, traumatic events can be, frequent-but-less-extreme traumatic mind-mapping experiences can be more destructive. It's not good to have powerful automatic primary emotions repeatedly evoked in your brain, whether it's disgust, rage, fear, surprise, or sadness.

This is particularly true of disgust reactions that cause your mind-mapping system to collapse. Although it's not anatomically accurate, I often explain this to clients using an analogy of living in an old house with plumbing that lacks a surge arrester. If you open a faucet and then suddenly shut it off, the pipes emit a loud banging noise from "water hammer." This picture helps you visualize what happens to your brain when your mind-mapping system repeatedly shuts down.

In all, I've identified almost two dozen long-term impairments caused by repeated traumatic mind mapping. (See Appendix B.) As I describe a few of them here, you may wonder how many traumatic experiences it takes to create these deficits. It's hard to quantify, but recent research suggests it may not take too many.

A recent research study involving 62 parents of young children attending a medical clinic looked at their own frequency of adverse childhood experiences like verbal, physical, and sexual abuse, family criminal behavior, substance abuse, violence, mental health problems, and parents divorcing, all of which involve traumatic mind mapping. Results indicated that the more parents reported having childhood traumatic mind-mapping experiences, the more likely their own children had similar experiences by age five. *Forty-five percent of parents had four or more traumatic childhood experiences, and their own children were six times more likely to already show signs of social or emotional problems.*[3]

How do these results fit with my prior statement that the human brain is amazingly resilient to trauma? I think it means that it doesn't take too much trauma to negatively impact your brain, but most of us are able to function at some level despite this. In other words, lots of people have unrecognized impairments from traumatic mind mapping that limit their functioning, which I'll describe next. Some of these can sound pretty bad so, as you're reading, remember that we'll cover how to solve these problems starting in the next chapter.

STEADY-STATE REGRESSIONS

We've already discussed how traumatic mind mapping can trigger acute regressions in the moment. It also predisposes you to "steady-state" (long-term) brain regressions, also known as "low-mode" functioning. This is like going through life as if you're an eight-cylinder engine running on only five or six cylinders. Another way to describe this is you're walking and talking in your sleep—you're not really fully awake or alive. Typically, you're depressed. You lack motivation, and you appear a lot dumber than you actually are. You and everyone else think you're displaying your true personality and level of abilities, but you're not. You're actually smarter and more talented than you look and act, but your functioning is suppressed.

You not only can have both acute and steady-state regressions, it's fairly common. This means you have severe episodes where your emotional functioning plummets; but when it comes back up, it doesn't come up to the level of your abilities. You settle into functioning at a somewhat impaired level that you now misinterpret as your 'good days," which are punctuated by intermittent "bad days" of acute regressions. As miserable as this sounds, lots of people live like this throughout their lives. You have no idea you're dealing with a variety of neurobiological problems. You just think this is who you are.

IMPAIRED DISGUST REACTION

It's not good to have your brain's involuntary disgust reaction repeatedly triggered. Repeated exposure to disgusting things can severely impair your disgust reaction. One form of impairment is a *markedly reduced disgust response*. You're not disgusted by things other people reject much sooner and more readily than you. This can include things like living in wretched conditions, engaging in deplorable behavior, or not being disturbed by other people's disgusting actions and thoughts.

For another example, if you discover your spouse is having an affair, you know better than to share this information with your children. You wouldn't go to your son and say, "Your mother's a whore!" Your disgust reaction limits how far you might go in certain situations, even if you're extremely upset. But if your disgust reaction is impaired, you're capable of doing this and a whole lot more.

Some people develop an *eroticized disgust reaction* from repeated traumatic mind mapping, wherein they get turned on by things most people find revolting or repulsive. Pedophiles are an extreme example. Chronic philandering is a more common one. Some folks can't have an orgasm without imagining things most people would find disgusting, like children being abused or people being raped, strangled, or dismembered. Watching pornography (or engaging in sex) involving physical "blood sport" and/or emotional abuse particularly appeals to people with an eroticized disgust reaction.

The hallmark of an eroticized disgust reaction is its severity and chronicity. In other words, I'm not referring to occasional "rape fantasies" of being overpowered in pleasant ways, or common "partner replacement fantasies." If you have an eroticized disgust response, you rely on disgusting imagery while masturbating or having sex with a partner.

People who have an eroticized disgust reaction are more likely to indulge in inappropriate sexual relationships with children. As I said in Part One, the vast majority of sexual abuse occurs through mind mapping. So for example, parents with an eroticized disgust reaction want their children to know they're having sexual fantasies about them. Think back to our prior example of Alexa (Chapter Eight), whose father asked her 16-year-old boyfriend if Alexa were any good in bed. At that point, Alexa's father wanted Alexa to know he was sexually interested in her. It shouldn't surprise you by this time that Alexa developed an eroticized disgust reaction too.

It's possible to completely lose your disgust reaction. I never imaged this was possible until I met a remarkable young woman who suffered from this condition. When Kathy first came to see me, it was a learning experience for both of us. She had two problems I had never seen before. One was Kathy had absolutely no disgust reaction. The other was she had true multiple personality disorder, the very first client I've seen that I'd stake my reputation on that diagnosis. Kathy had three personalities: One was Kathy herself. Another was "Sex Woman," who engaged in promiscuous sex. And the third was a seemingly innocent little girl who was vicious underneath.

Kathy had read all of my books and had attended several of my workshops for therapists–although she wasn't a therapist. Kathy was determined to have therapy with me, and I am eternally grateful for her perseverance. Kathy taught me more about disgust reactions than anyone else I'd treated, and she inspired me with her courage. Her upsetting story powerfully illustrates the negative impacts of disgusting traumatic mind mapping, as well as the unstoppable power of the human spirit.

Kathy had a difficult childhood. Her mother constantly twisted reality to the point Kathy thought she herself was the crazy one. Kathy's father had a sadistic arousal pattern: He became sexually aroused and derived great pleasure from physically beating Kathy and her two sisters. He made inappropriate remarks to Kathy and watched pornography with Kathy's sisters. Kathy suspected he might have had sex with her sister.

Kathy's father liked to pick up a cucumber in the produce section of the grocery store and say "This is just like my penis, Kathy. You'd like it." Or he'd say "You need to get screwed.

That's all you're really good for." You'll understand the extent of traumatic mind mapping this involved when I tell you that both of Kathy's parents were mental health professionals.

When Kathy came to see me as a client, she knew I'd helped people with troublesome orgasm-trigger fantasies for over 30 years.[4] As our first session ended, Kathy revealed she had this difficulty and volunteered information about her masturbation fantasies. As far as I'm concerned, making herself do this probably saved Kathy's life. From the outset I knew Kathy was unique. She had no disgust reaction.

Kathy could only have an orgasm if she visualized two different pictures. One was a woman being dismembered by her partner in the midst of sex. The other was a screaming baby being raped in every orifice. If Kathy envisioned either scenario, she climaxed within seconds. Kathy also frequently picked up guys in bars; and while giving them oral sex, she'd imagined doing this with her father. This is how I realized Kathy was the first person I'd known who had no disgust reaction at all.

I had never encountered anyone with such a complete disability. I'd been studying suppressed and eroticized disgust reactions in clients for some time; but until I met Kathy, I'd never considered that someone's disgust reaction could be completely obliterated. I've since learned this is common among first responders, emergency room doctors and soldiers in protracted combat. But I had no idea children could lose their disgust reaction from being around their parents.

My best understanding of this is that Kathy had 'neural network dissociation," meaning her anterior insula had stopped triggering her innate disgust reaction. We set about restoring this using the methods I'll describe in subsequent chapters. Because of Kathy's remarkable courage and fortitude, I can tell you it's possible to restore a completely obliterated disgust reaction–even in someone who has multiple personality disorder because we were able to resolve that problem too. My experience with Kathy taught me that no one–no matter what kind of terrible emotional shape he or she is in–should ever give up on himself/herself.

My initial meetings with Kathy consisted of daily three-hour therapy sessions for four consecutive days, which is typical for people who fly in from around the world for my Intensive Therapy Program. We followed up with monthly three-hour sessions via teleconference interspersed with exchanges of written correspondence.[5] Gradually, Kathy's sexual fantasies morphed into healthier imagery in the most amazing ways; and watching this happen was heartbreaking, fascinating, and thrilling. Four months later, Kathy's disgust reaction had been restored. She no longer could stand the troublesome orgasm-trigger fantasies she was previously dependent upon.

Treating Kathy's multiple personality disorder took a little longer, but by conventional therapy standards it was incredibly fast. After nine months of therapy, her multiple personalities rarely emerged. And after 14 months Kathy's brain had changed sufficiently that she was unable to deliberately elicit them when, during a tough period with her boyfriend, she wanted to fall apart. Today, Kathy's psychological resilience has withstood the tests of time and difficult circumstances. Relapsing into multiple personalities again isn't even a concern.

Kathy and I have continued our monthly teleconference sessions for almost four years, during which time she's turned into a delightful, healthy, vibrant, talented, and beautiful young woman. When she first came to see me, Kathy wore so many face piercings I found it difficult to look it her without having a disgust reaction. She also gave off a "kooky" vibe that made people think she was strange or crazy.

Now Kathy displays a single, modest piercing, and her countenance is soft and appealing. She's a pleasure to behold. She no longer has acute, emotional crashes; and she's become strong enough to deal with her destructive parents. And now that she's not living in a steady-state regression, the precision and sophistication of Kathy's cognitive functioning has dramatically improved. She's actually smart, creative, witty, and capable of deeper emotional sentiment.

Like many young women, Kathy now has the normal healthy struggles of handling herself in a primary love relationship and balancing her personal life with her productive career. She is, by every relevant criterion, hugely successful, and a testament to the power of human resilience. I'm deeply touched, not just by Kathy's progress, but by the fact that she now believes she is a good person. She has become someone who brings joy to others, and she's begun to experience peace in her own mind.

I hope Kathy's story inspires you to persevere, regardless of your difficulties and unfortunate circumstances, to straighten out your mind and brain from the effects of traumatic mind mapping. (I suggest you peruse the fascinating research on disgust in Appendix C). At the very least you'll understand how the brain creates moral disgust. But there's a good chance you'll also realize something about powerful experiences in your own life in the process.

CRUEL MENTAL "VOICE"

Another long-term consequence of repeated traumatic mind mapping is a hostile voice inside your head that encourages you to feel bad about yourself and do self-destructive things. Typically, the voice will be critical and demeaning while occasionally presenting itself as a collaborator or a friend. Because it feeds on your insecurities, the voice is as helpful as a rubber crutch.

Many of my clients suffer from this ailment, although they don't fit into current psychiatric diagnostic nomenclature (e.g., DSM-5; ICD-10). They're not psychotic, and they're not "hearing voices" in the way people with schizophrenia do. They have a single, cruel, mental voice stuck in their head, which often talks to them throughout the day. You probably haven't heard much about this condition because sufferers don't talk about this out of fear everyone else will assume they're crazy.

Most people who have a mental "voice" don't know they're part of an international community of "voice hearers" who want no part of traditional psychiatrists who try to give them anti-psychotic drugs. Voice hearers even have their own organization.[6] According to one research study of a large, random sample, one out of ten people has a voice in his head that talks to him.[7] Now that I know how prevalent this is, I routinely ask new clients if they have one.

By studying my clients' "mental voice" and working

backward from what I know about their lives, I've concluded that repeated traumatic mind mapping can wire your parents' thoughts right into your head, which eventually show up as a cruel mental voice that talks to you. Here's an example of how I came to this conclusion.

One of my clients, Patricia, had a cruel, mental voice that constantly berated her in a very specific way. It would say, "You're pond scum, Patricia—nothing but pond scum! You're such a waste! Why don't you kill yourself?!"

As Patricia and I began to focus on this voice, she mentioned this to her brother for the first time. When she did, her brother exclaimed, "Mom used to say you were pond scum all the time!" Patricia had no recollection of her mother's ever having said this, demonstrating a common reaction to traumatic mind mapping. She wouldn't have put this together if her brother hadn't clued her in.

Patricia had another problem, which I had never previously encountered: She couldn't remember what she looked like. Patricia recognized herself when she looked in a mirror, but she couldn't hold onto a mental picture of her stature, general appearance, and facial expression.

Although this was new to me, Patricia and I developed a way for her to overcome this: Each time Patricia stopped at a red light while driving, she looked at herself in her rear view mirror. When she put on or took off her makeup, she took time to observe herself and study her face. This repetition, plus the things you'll learn in Part Three, allowed Patricia to change her brain/mind, remove her mental "voice," and remember who she was.

My clinical experience indicates you can remove a mental "voice." This differs from the advice of the Voice Hearers Association, which proposes your best bet is to attempt an uneasy peace and coexist with it forever.[8]

HARD-WIRED THOUGHT PATTERNS

When you map someone's mind, you're replicating that person's thought pattern in your brain. There's really no way to stop this. Unfortunately, repeated traumatic mind mapping in childhood—particularly disgusting experiences—encode your antagonists' thoughts, feelings and behavior in your mind and brain. Over time, your brain becomes predisposed to fall into these neural patterns of thoughts and feelings, creating the tendency to think, feel, and react as they do.

Because mind mapping is so dominant and automatic, it's hard for children to stop mapping their parents' minds. This repetition likely facilitates "myelination" of these neural pathways, which makes your brain think faster and more easily in these ways than in other ways, creating dominant patterns of feeling and thinking. [9]

For instance, if your father habitually lied when you were growing up, there's a good chance you'll do the same because you wired his thought patterns into your own brain by repeatedly mapping his mind. This is how you can end up thinking or acting just like your parents, even though you swore you'd never do that. This is so reliable. I'm often able to describe for my clients what their parents were like–before they've said a word about them–simply by mapping my clients as they respond to me!

I've also identified fixed-thought patterns like a "batting away" response in people who experienced frequent emotional abuse during childhood. Whatever you say to them, good or bad, usually triggers a thought and speech pattern designed to deflect what you're saying. Even when they don't want to be deceptive or combative, they come across this way without realizing it and then feel unfairly criticized because this doesn't line up with their feelings.

This can be a difficult mental box to escape from because introspection doesn't help you recognize this glitch in your thinking. Instead, you need a collaborative alliance with someone you trust, to help you recognize your brain's deflecting-response pattern, which is particularly hard for these folks to do.

Another way repeated traumatic mind mapping probably affects your brain is "DNA methylation." Epigenetics is the cutting-edge science of how personal experiences influence the expression (or non-emergence) of genetic predispositions in your DNA. Methylation modifies how DNA functions. For example, research indicates some people carry a genetic predisposition for antisocial behavior; but they have no greater incidence of sociopathy if they don't experience abuse during childhood.[10]

DNA methylation is one of the most intensely studied epigenetic modifications, which include inactivation of tumor-suppressor genes[11], and increased incidence of cardiovascular disease.[12] I won't be surprised when future research links repeated traumatic mind mapping to increased incidence of health problems, including cancers and heart conditions. The good news is DNA methylation is reversible through personal experience, which means it may be treated through proper therapies.

HOLES IN YOUR MIND MAPPING RADAR

In Chapter Eight I told you that traumatic mind mapping can also create holes in your mind-mapping radar. This often shows up as hypervigilance toward the general population and narrow-band mind blindness toward your antagonist. You're distrustful of everybody except the people who truly warrant suspicion. You fail to pick up cues that should snap you out of your stupor and make you say *Oh, my God! What is going on?*

Let's say, for instance, your father is a ne'er-do-well. He's always borrowing money from people and never paying them back. He thinks anyone who works a steady job is a sucker. He's always talking about his next "big score," which, according to him, is just around the corner. His grand schemes always fall apart. And whenever this happens, he gets really upset and takes his frustrations out on his family. If this is your father, as an adult you're going to be hypervigilant, always on the lookout for people who might try to take advantage of you. At the same time, you might be completely blind to your father. This odd combination of general hypervigilance plus narrow-band blindness is the hallmark of having a hole in your mind mapping radar, and it's very common.

Mind-mapping radar gaps often show up as errors in judgment when dealing with parents who do disgusting things. I've seen clients continually fund their ne'er-do-well parents' get-rich schemes and repeated requests for money, long after it is obvious my clients are being exploited. Similar situations occur when adult children continue to accept money from rich, disgusting parents in exchange for allowing their parents to control their lives. Both are guaranteed to create conflict in your marriage, as your spouse maps out that you're blind to what your parents are doing. When dealing with highly dysfunctional parents, people display shocking drops in accurate thinking, deductive logic, memory retrieval, and overall intellectual firepower.

This impaired functioning isn't reserved for the grown children of dysfunctional parents. Some people experience this in their marriage. Let's say your husband has affairs with your girlfriends and doesn't bother to hide what he's doing. You're likely to have a disgust reaction, and you're going to hate him whether you like it or not. Many wives get mad and ask for a divorce. But some walk around in a fog talking about how wonderful their husband is or feeling guilty for being suspicious. Their husbands don't need to hide their dalliances because over time they figure out that their wives have blind spots in their mind-mapping radar.

It's sobering to discover blind spots in your mind-mapping ability. Seeing your emotional blindness is difficult and humbling–particularly if you routinely mask your mind and pride yourself on being hypervigilant. But my fail-to-thrive clients, who ultimately got better, did so because they were finally able to realize they were blind to their parents.

The problem is that, when you have blind spots in your mind mapping, it's not like you don't see your antagonists at all or have absolutely no memory of them. You still map some things about them, which makes it harder for you to recognize that you're not seeing the whole picture. The things you see may have entirely different meanings, and you're still missing other important aspects.

This ailment is more common than you'd ever imagine. That's because parents who do disgusting things are invisible to the general public. Nobody wants to believe they exist–particularly their own children. This results in children of invisible parents suffering invisible maladies like holes in their mind-mapping ability. Recognizing this problem is the first step in resolving it.

AUTOBIOGRAPHICAL MEMORY GAPS AND DISTORTIONS

When traumatic mind mapping collapses your mind-mapping ability, this often leads to gaps and distortions in your autobiographical memory. Some people say they remember nothing from their childhood. Others don't recall major traumatic events. But whenever we recall traumatic memories that have stuck in our minds, there's usually something missing that completely escapes our attention: Our mind-mapping information about our antagonist is missing. As you recount your experience, there's no picture of what's going on in the other person's mind.

As common as this is, it often isn't apparent until you know what to look for. You won't even know it's missing until you examine moments from your past that stand out in your mind. These could be major life events like how your family handled your brother's dying or becoming disabled, or your mother's failing to invite your wife to your sister's wedding shower. Only when you stop and ask yourself *What was my mother thinking when she decided not to invite my wife?!* do you realize that your mental map of your mother's mind has seemingly disappeared.

From what I can discern, memories of traumatic events and associated traumatic mind mapping are actually stored, but the mind-mapping information is not retrieved when the event is recalled. You might be able to remember events that took place, but often you can't access the mind-mapping data that goes with them. Since your mind mapping contains the event's real meaning, you won't recognize the significance of what you're remembering. Because of this, it's actually quite common for people to concoct a family history that's nothing like their actual family.

At least one partner (if not both) in my fail-to-thrive couples typically reported an idyllic childhood. But when we got into the nitty-gritty details of what really transpired, a completely different (disgusting) picture invariably emerged, like a horror-movie version of the original picture they presented. That's how gravely repeated traumatic mind mapping can affect your autobiographical memory. Sometimes you can't recall the awful things you witnessed when you were young, or you've assigned much less virulent meanings to them, or you've constructed a completely false-positive memory. (This doesn't mean everyone who thinks they had a wonderful childhood is deluding themselves.)

Working with my fail-to-thrive couples taught me how little I could depend on their being reliable and accurate historians of their family backgrounds. This actually has enormous ramifications. Therapists usually take clients' self-reported family histories at face value and try to work from this. (Remember, as we discussed in Chapter Seven, most people think we're always the ultimate expert on ourselves because we mistakenly believe in the 'privileged access" of introspection.) I've learned this isn't such a good idea with people who seriously need therapy. Their treatment usually begins with a false picture of their backgrounds and heads off in wrong directions from there.

To help you grasp the scope of this problem and how it surfaces in everyday life, let me describe some recent vignettes from training other therapists. In April, 2015, I conducted the first four-day Crucible® Mind Mapping Therapy Workshop for therapists in Ulm, Germany. I was shocked by how many participants had holes in their respective autobiographical memories. After the first day, a quarter of the attendees reported disturbing realizations that radically changed their perceptions of important aspects of their lives. By the fourth day, virtually everyone reported having had this experience.

Therapists don't usually reveal their own traumatic experiences at professional workshops–particularly German therapists–but most attendees were so deeply affected they felt compelled to speak out. One brave woman, Lisbeth, described hurting her finger playing softball several years ago. Lisbeth thought she'd broken it, so she went to the hospital and had it X-rayed. Just as she suspected, Lisbeth's finger was fractured. But the X-rays also revealed something she hadn't expected. All the other fingers in her hand had previous fractures, and so did all of the fingers on her other hand. And so did her toes, feet, and both of her arms and legs. In all, 25 bones in

Lisbeth's body had previously been broken, although she had no recollection of any of this. Lisbeth knew all this before attending my workshop.

But while listening to stories about traumatic mind mapping, Lisbeth suddenly recalled her mother had been a nurse. This hit Lisbeth like a ton of bricks. If Lisbeth's mother hadn't done this to her, then at the very least, her mother had to know about it but never mentioned this to Lisbeth. This is what it's like to recognize gaps in your autobiographical memory and suddenly realize your mind map of someone is missing. What was Lisbeth's mother thinking and doing about Lisbeth's repeated fractures?

Based on my research with fail-to-thrive clients, what would you predict about Lisbeth's mother and Lisbeth's relationship with her? It shouldn't surprise you that Lisbeth' mother often did disgusting things, or that she maintained a toxic relationship with Lisbeth that suppressed Lisbeth's functioning. As a result of attending my workshop, Lisbeth dedicated herself to resolving her acute and steady-state regressions, having a healthier marriage, and dealing with her mother.

At a similar workshop I conducted in the United States, another therapist, Chloe, reported recalling that she and her mother took a cross-country trip when Chloe was 18 years old. The most memorable part of the trip was the fact that Chloe's mother's lover came along. Chloe's mother expected Chloe to lie to her father about this. When her father asked about the trip upon their return, Chloe found herself in an impossible position.

This was so traumatic Chloe went nearly 40 years without remembering it. It only came back after learning what you're learning here. Chloe wondered *How could my mother do that to me? I was 18, old enough to understand about sex and infidelity. Why would she choose to involve me? How could she do that?*

That's when Chloe realized her mental map of her mother's mind was missing from this memory. It wasn't like Chloe didn't know her mother's mind. Chloe knew all too well how her mother could do this. This is what produced Chloe's traumatic mind mapping and subsequent holes in her autobiographical memory.

Lisbeth and Chloe's examples illustrate what it's like to

restore the missing mind-mapping data in your memory of vignettes. As you can imagine, their new appreciation of their own life stories shocked them. It's not a particularly happy thing, but restoring this information helps you heal traumatic mind mapping, and can dramatically change your life.

ANTISOCIAL EMPATHY

This is a good point to bring up another long-term impact of repeated traumatic mind mapping: developing antisocial empathy—enjoying another person's discomfort and unhappiness. Chronic, traumatic mind mapping gives you a taste for cruelty and teaches you how to do it. The best way to grasp this is to visualize how Lisbeth's and Chloe's actual experiences with their mothers affected them.

For example, Lisbeth's physical abuse and her mother's involvement are only a small part of Lisbeth's picture. You have to add in the home environment—the collection of minds—in which these horrific events are possible. Lisbeth's mother apparently thought Lisbeth's abuse would never surface—and for the most part she was right. This not only tells you how Lisbeth's mother's mind works, it shows you her mother's shrewd assessment of the other minds in the house too. These are the minds Lisbeth is mapping every day.

Envision being Lisbeth and you're growing up being physically tortured. Regardless of who it is—someone in your house is deliberately seriously hurting you and systematically destroying your body. With 25 fractures you are in constant pain. Even if you can't remember this as an adult, you've lived through this; and it's had its toll on you. And even if your mother isn't the one abusing you, your mother knows you're being abused; and at best, she doesn't do whatever it takes to make it stop. Maybe she derived pleasure from your suffering—whether or not she administered it—but don't guess. Stick to what's certain.

Someone in your home enjoyed your pain. And your parents and physicians, who were in a position to know better and do better, turned a blind eye to what was happening to you. In other words, you were growing up around at least one person who had antisocial empathy and he/she/they used it when interacting with you. This is occurring in the midst of people with mind-mapping ability, who also masked their minds. This is the mental world you were growing up in.

Now from this perspective, would you be surprised if you grew up to have a mean streak of antisocial empathy? That you knew how to look like you couldn't hurt a fly, but at home you're often vicious and nasty—and sometimes downright cruel? That at times, you enjoy seeing your partner or your children suffer? What else would you expect? Most likely you'll gravitate toward adult love relationships involving combative or collusive alliances instead of collaborative ones. Your relationships will be prone to disgusting dynamics, like chronic lying and mind-twisting manipulation or maybe repeated philandering or physical and emotional abuse.

If this applies to you, you don't have to remain this way. You need to acknowledge antisocial empathy in yourself because this is the doorway out. This is how Lisbeth embarked on a new life, and you can too. The next five chapters are designed to help you get there

BEING TAKEN HOSTAGE

Of all the damage created by parents who do disgusting things, here's the one I find hardest to accept: If you witness your parents do enough bad things, it can *weld* you to them for life. Rather than running away as fast as you can, *their disgusting behavior actually makes you want to run toward them instead.* They can make you their prisoner. How can this be?

Once again, it has to do with interpersonal neurobiology. Parents' behavior can be so disgusting it actually warps their kids' disgust reaction. Repeated exposure to disgusting parenting impairs this reaction, making it less odious for you to be around them. Either your mind mapping collapses to the point you don't recall your parents' disgusting behavior; or you misinterpret it to be better than it is, so that you'll actually defend your parents if someone else questions their actions.

A normal, healthy disgust reaction forces you to remove yourself from the offending object or person. It makes you want to run away from things that aren't good for you. But if your parents repeatedly do disgusting things, it also instills hatred in you, even if you refuse to acknowledge either of these feelings. Instilling disgust and hatred in a child is like mixing two epoxy resins to produce emotional super-glue. Mixing them together reverses the normal disgust reaction, making a child run *towards* the parents, rather than away from them. Children's inability to acknowledge their hatred and disgust for their parents puts them at their parents' disposal. This is what I find so difficult to accept.

Although I didn't know it at the time, many years ago I worked with a kid in foster care who suffered from this condition. His mother was an alcoholic and a drug addict who turned tricks at home to support her addictions. His father was completely out of the picture. Child protection authorities took the boy away from his mother and put him in foster care, but he kept returning home although his mother didn't want him. This kid witnessed his mother doing all kinds of disgusting things. But instead of running away from her, he kept running back to her.

This boy's obesity and severe acne also made him a constant target of harassment and bullying from other children. But here's what upset me the most: His mother gave him a pink glossy raincoat, which made him an even bigger target. This was like putting a "Kick Me!" sign on the boy's back, but the kid insisted on wearing it everywhere. This made a lasting impression on me, although I didn't understand its significance until now.

These kinds of relationships force you to live in a steady-state regression. You're particularly prone to acute regressions involving severe dysphoria, panic attacks, emotional outbursts and/or raging anger. Your mind-mapping system goes on full-time vacation with regard to your parents, although you're hypervigilant with everyone else. Your ability to hold onto yourself in relationships, i.e., your differentiation, becomes severely impaired. You become emotionally fused with your parents.

Anticipatory traumatic mind mapping plays a critical role in this process. Anticipating your parents' next disgusting move commandeers your attention, and you give them whatever they want to forestall more bad behavior. Some people willingly throw away their marriages and/or turn their children over to their dysfunctional parents in order to maintain relationships with the parents.

THE BENEFITS OF SEEING DARK THINGS

I find this perversion of a healthy survival mechanism particularly upsetting. I'm sure reading about this is unpleasant for you too. So why bother examining things that are, frankly, disgusting to look at?

First and foremost, this opens the door to the help you may need if you've experienced repeated extreme traumatic mind mapping. Starting in the next chapter we'll get into suggestions that can give you what you need.

Second, understanding the impacts of traumatic mind mapping gives you a better idea of what many children face. Hopefully, this will improve their chances of getting help. The general public hasn't really wanted a true picture of many children's experiences. But if we're going to help them deal with the debilitating impacts of traumatic mind mapping, we have to help them accurately understand and articulate their experiences. We can't do that if we ourselves are unwilling to look at this.

Third, I hope this discussion encourages doctors, social workers, teachers, and daycare workers to be more discerning about the physical complaints of children. Children of disgusting parents often suffer repeated stomachaches, nausea, vomiting, eating disorders, bed-wetting, and encopresis (stool holding). Hopefully, school nurses, pediatricians, and mental health professionals will recognize the possible connection between these "gut-related" symptoms, disgust reactions, and traumatic mind mapping. With some luck, we'll be less eager to diagnose children as having ADHD and more willing to consider they might have constant "spaghetti brain" from traumatic mind mapping at home.

Fourth, understanding traumatic mind mapping and moral disgust promotes new approaches to common disorders that haven't been fully understood until recently. For example, obsessive-compulsive disorder (OCD) has traditionally been viewed as an anxiety disorder. When people continually wash their hands or won't touch doorknobs to avoid germs, this has often been interpreted as being displaced anxiety about something else. However, recent research suggests OCD may actually be a displaced disgust reaction, which makes sense given the sufferer's preoccupation with germs, diseases, infection, and filth.[13] This permits treating OCD in entirely new ways, and offers hope to people not helped by conventional therapy.

Fifth, facing traumatic mind mapping head-on reduces complacency with dysfunctional families and marriages. Once my clients realize how damaging traumatic mind mapping is, it becomes a powerful call to action. Couples are inspired to get their act together, rather than continuing their covert warfare. Parents address disgusting behavior in their home, rather than ignoring it and just counting the days until their children go to college.

It's sobering to realize how pervasive traumatic mind mapping is. By some estimates, 16 to 32 percent of people in the United States have been physically or sexually abused; and many more have witnessed disgusting behavior.[14] But imagine life in Third World war-torn countries where woman and children are chattel, schooling is restricted, and human rights are flagrantly ignored. The traumatic mind mapping in daily life, and resulting waste of human potential and loss of productivity, is staggering.

Recovering From "Trump Trauma"

As I write this, people in the United States, leaders of foreign countries, and folks around the world are trying to figure out how President Donald Trump's mind works. Lots of them are upset (if not frightened) by his intemperate statements, impulsive tweets, repeated lies (according to PolitiFact), unsubstantiated accusations, and overly personal reactions.

I have no desire for a polarizing political debate. My focus is what's happening to millions of people who have no idea what is happening to them. Trump's tweets and public statements, full-time cable news coverage, and Internet access, all give unprecedented access to this President's mind, thereby creating a national (if not global) mental health crisis of traumatic mind mapping.

We don't want to accept that someone can become incredibly successful by conventional standards *because* he or she possesses antisocial personal traits like habitual lying, scapegoating, bullying and gaslighting. We don't like to think these same traits (a) could make one more likely to be elected President, and (b) render one categorically unfit to fulfill the responsibilities of office once elected.

Trump's traits are more typical than people want to accept: Narcissism, Machiavellianism, psychopathy and sadism–four unwholesome personality traits known collectively as the Dark Tetrad, are considered normal personality styles by mental health professionals. (Read about the Dark Tetrad in Appendix C). Trump doesn't have to qualify for a personality disorder diagnosis to have all of these traits, including lying and manipulation, creating false narratives, overwhelming preoccupation with receiving praise and recognition, inflated self-evaluation, and enjoyment of other people's suffering. Rather than trying to "normalize" Trump's behavior (by making it more acceptable and less visible), my point is Trump is like many people's parents, which is why a sizable portion of the United States seems mind-blind to him.

When Mr. and Mrs. Kahn, the gold-star parents at the Democratic National Convention, demanded that Trump develop more empathy, they made the common mistake of conflating *empathy* (intuiting another person's emotions and personal experience) with *compassion* (altruistic desire to alleviate another person's suffering). President Trump has plenty of empathy. What he lacks is compassion.

This common misunderstanding of empathy makes a bad situation look better than it is: Misperceiving Trump as lacking empathy pictures him as blind to his impact on others. But people who are sociopathic, Machiavellian, narcissistic and sadistic use empathy to figure out other people's feelings and use this to manipulate and exploit them. Like many people who repeatedly do self-serving or mean-spirited things, Donald Trump has plenty of empathy: antisocial empathy.

Trump's vaulted ability as a dealmaker would not be possible if he were unable to understand other people's feelings and motivations. His vindictiveness utilizes his ability to anticipate the impact of his behavior on other people and detect and enjoy their suffering. Trump understood how Paul Ryan would feel when the President declined to endorse him for Republican Speaker of the House during the primaries.

Reconsider Trump's pre-election comment about Second Amendment Advocates possibly "solving the problem" if Hillary Clinton were elected, or his "blood was coming out of her... everywhere" reference to Megyn Kelly. He was implanting ideas in people's minds without explicitly saying so. Re-tweeting other people's statements while denying this reflects his own thinking. These can only be accomplished by someone with highly developed mind-mapping and mind-masking abilities.

Diagnosing the country. Unfortunately, widespread attempts to map Trump's mind are creating traumatic mind mapping on a global scale, creating a subtle-but-pervasive mental health crisis. People are increasingly traumatized by Trump's off-script comments and choice of words and are very concerned about his temperament, judgment, self-control and suitability to be President. During the primary elections many Republican leaders and 50 GOP national security experts broke ranks and refused to back Trump because his campaign rhetoric and behavior instilled traumatic mind mapping in them (the "Never Trump" movement). After 10 months in office, a former national security advisor and members of his own party question Trump's suitability to be president.

Trump's election traumatized people in many nations, sparking unprecedented public demonstrations around the world. Many people were traumatized, frightened, and/or apoplectic with anger over his unsubstantiated accusations of widespread voter fraud and President Obama's wiretapping his phones, discrediting the media and U.S. intelligence community, dismissing possible Russian intervention in the 2016 presidential election, and escalating nuclear tensions with North Korea.

Trump is triggering widespread, involuntary, brain reactions of moral disgust, whether it's his repeated lying (according to fact-checkers), or his comments on *Access Hollywood* about grabbing women's genitals,[15] or his unabashed sexual interest in his daughter, Ivanka, when they appeared together on TV.[16] It's not good to have repeated disgust reactions triggered in your brain.

Recovering from the 2016 presidential election. Calls to honor the democratic process and support the president replicate the childhood experiences of many people with parents who made up "alternative facts." They are torn between loyalty for the institution of the presidency and terror at the mind of the man who now holds that office, i.e., the "father" of our nation. Messages from the White House to the media to "shut up," demands to accept lies as truths, and intolerance for fact-checking replicate childhood pressures to "go blind." Trump media-surrogates who repeatedly claim he misspoke, didn't say what he meant, or claim his words are being misinterpreted are doing what children of abusive parents do: telling themselves *Dad didn't really mean what he said (or did).*

Understanding how traumatic mind mapping and disgust reactions can be instilled through electronic media and news stories goes a long way towards mitigating this impact. Recognizing traumatic mind mapping in yourself helps you unglue from your television or Internet feed. Taking action generally facilitates recovery from trauma. The Women's March, immigration protests, and world-wide demonstrations are not just messages being sent to President Trump. They are part of a world of traumatized minds trying to repair themselves.

CRUCIBLE NEUROBIOLOGICAL THERAPY

Regardless of the benefits of seeing dark things, it's still hard to discuss the impacts of traumatic mind mapping because the conversation is inherently upsetting. I haven't told you the *really* bad experiences I've heard from clients. Basic decency forbids me to mention them here. I've tried to be respectful of your emotional reactions to what you're reading and have shared only enough to help you understand what I'm writing. I also say this to encourage you to not give up if you have the kinds of problems I've been describing.

The good news is this has allowed me to develop a new psychotherapy, and all the impairments I've described can be treated. After more than a decade of studying and treating traumatic mind mapping, I've condensed what I've learned into a method called Crucible® Neurobiological Therapy (CNT). I've boiled things down enough so that most people can understand and apply it to themselves. I believe this represents the next generation of brain-based therapy for the masses.

CNT allows me to treat people I couldn't help in the past, especially folks from extremely troubled backgrounds. In the following section I'll describe some of the therapy's principal methods and show you how to apply them to your own life. Responsibility for judicious application remains with you. Various aspects of the therapy can evoke powerful emotions. When you consider the difficult relationships we've been discussing, strong response shouldn't surprise you.

CNT doesn't require superior intelligence or advanced education. But to work effectively, it does take stubbornness. You are stubborn, aren't you? I'm a big fan of stubbornness. No good thing in human history has ever been accomplished without it. Most people think of stubbornness as something bad or oppositional because they let the worst in themselves run the show. You have to get the best in you in charge of your stubbornness. And when you do, nothing can stop you.

You need to be stubborn to succeed at resolving traumatic mind mapping because this therapy requires repetition. You have to be resolute, making yourself do things you may not be eager to do at first. You have to be willful, making yourself do things repeatedly because your brain wires and rewires through repetition.

I don't know any magic bullet or one-shot solution that resolves traumatic mind mapping. Your brain—and for that matter your spouse, kids, and co-workers—won't respond to your efforts right away. You can't simply think your way to a new solution. You have to live your way into it. You have to actually do things in the real world, over and over again. If you're stubborn, willful, and obstinate, you've come to the right place. New solutions are within your grasp. All you have to do now is turn the page.

Part Three

Mind Mapping Solutions

11

DETECTING MIND MAPPING IN OTHERS

Our final section is devoted to making your world a better place. Now that you understand traumatic mind mapping, we're going to focus on resolving the problems such experiences create. In this chapter and those that follow, I'll provide you with effective tools to help you deal with really difficult people, reverse the negative impacts of traumatic mind mapping, and live a happier and healthier life.

To start with, you need to learn how to detect the mind-mapping abilities of other people. It's not enough to know intellectually that most people have it. You must be able to *see* their abilities in action, as they're using them in real time. It shouldn't take too much instruction for you to get better at this since your brain is designed to do it. The greater difficulty seems to be accepting that you live in a world where most people have mind-mapping ability, including those you love and care about the most. Once you accept this picture of people, it's much easier to map them.

Identify Other People's Mind-Mapping Abilities

The first step in detecting mind-mapping ability involves simply paying attention to the role mind mapping plays in all your interpersonal experiences. Start observing interactions between people wherever you go, especially when you're out in public. Watch the exchanges between people while waiting at the supermarket checkout counter, riding an elevator, sitting in a classroom, or attending a meeting. Pay close attention to the subtle ways people read each other, looking for cues of sexual interest, aggression, or dominance and authority. Notice how they try to implant false pictures in one another's minds. It's much easier to track people once you realize they are mapping other folks all the time.

Quite often you'll be forced to learn this skill out of necessity in response to someone who's trying to thwart you in some way (your "antagonist"). I often teach this to therapists dealing with obstructionist clients, but it's just as important for the average person to learn to do this too. If people use their mind-mapping abilities against you and you fail to take this into account, this will impede seeing them clearly and dealing with them effectively. If you don't factor in their mind-mapping skills, you make it easier for them to manipulate or hurt you.

I'm going to cover the most reliable ways of detecting people's mind-mapping abilities. I'll describe some typical behaviors to watch for. Someone exhibiting one of them isn't conclusive evidence; but if they display two or more behaviors, the likelihood this person has mind-mapping ability greatly rises.

One way to gauge people's mind-mapping abilities is to observe them while they're tracking other people. People reveal their proclivity and ability to map other people when they're describing someone else's behavior and motivations. It might be as simple as this: You're enjoying a short break in the midst of a very trying business meeting when a co-worker whispers to you, "Did you see the way the boss's assistant looked at you when you spoke up? I think he's threatened by you." Your colleague just demonstrated she's a mind mapper.

That doesn't necessarily mean what she's saying is accurate or trustworthy. She could be wrong about the boss's assistant or possibly setting you up to fight with him. Was she complimenting you on the powerful way you're coming across in the meeting? Or was she warning you because she thinks you deal with office politics naively? Was she trying to implant a false belief in your head that she's your ally? Here's what you know one way or another: Your colleague has mind-mapping ability, and her statement reflects her map of *you*.

As a second example, visualize this: You've just gotten home from a party and you're getting ready for bed when your husband says, "Did you see that woman in the red dress? She was dressed like a real slut, and she was all over that guy. She was obviously looking to hook up with someone." Your husband is acting like he can "read" this woman's mind by the way she dressed and acted. Although he speaks disparagingly of her, for some reason she's attracted his attention. Perhaps your husband is a member of the morality police, or maybe he was wishing Madam Red Dress was coming on to him. It may look like your husband is simply describing what occurred in the room; but if you understand mind mapping, he's showing you what went on in his head.

Not only can you monitor someone's ability to track other people, you can also monitor how open they are about divulging their ability. Watch people's reactions when you point out their mind-mapping ability to them. Some people will sheepishly acknowledge it. Others will deny it to the bitter end. So, for instance, let's say you comment to your husband, "That woman got your attention enough that you're mapping out her intentions. You have mind-mapping ability."

You'll map your hubby one way if he says *Duh! It wasn't hard. Anyone could see what she wanted.* Or *I never really thought about it, but I guess I can do that.* He'll look quite different to you if he says *I don't do that. I don't even know how to do that. Could you explain what you mean?* If you're smart you'll realize your mate just demonstrated mind-masking ability too.

If you pay attention, you'll often see people blow their cover.

A couple came to see me, and the husband (a high-powered business executive) remained adamant he didn't have mind-mapping ability, even after I'd showed him copious evidence that he did. His wife said, "You do have mind-mapping ability. You're always telling me you're the smartest guy in the room, and you can go to a party and within three minutes you have everyone there figured out."

The guy's cover was busted. But being used to bluffing his way out of most things, the husband said, "I'll bet you're thinking my wife blew my cover, Doctor. That doesn't prove I have mind-mapping ability." Busted again by his own statement.

Another sign to watch for is a person's ability to lie, i.e., constructing a false reality and implanting it in someone else's mind. This includes the ability to "control spin," like devising explanations that present events in an overly positive or negative light. People can't manipulate how other folks (mis) interpret events if they don't have mind-mapping ability.

For instance, let's say you're cooking dinner and your child runs into the kitchen and says, "Don't believe what Mary's mother is going to tell you. I didn't break that window. She just wants to get me into trouble." Your kid is demonstrating classic signs of mind-mapping ability: implanting false beliefs and creating misleading explanations. There's no reason, on the basis of this, to assume you're raising a budding sociopath. But you can be sure your child is mapping you all the time, and she knows you're not clairvoyant.

Another sign of mind-mapping ability is asking trapping questions. A good example is *How do you know I'm the one who put that dent in the car?* So is *If you don't have anything to hide, why won't you let me look through your emails?* Is the speaker asking trapping questions to put you on the defensive or to figure out what you know, or is he trying to deconstruct your answer to prove you're wrong? All three tactics require mind mapping and the ability to play several moves ahead.

Repeated, indirect communication using innuendo and implication is another solid indication of mind-mapping ability. People who do this get their message across by implanting ideas in your head without explicitly saying them. They avoid using words that might be held against them. If you try to hold them accountable for their message, they'll say *I never said* that. Or *That may be what you heard, but that's* your *interpretation.*

It may be true they didn't say the exact words you reiterate to them, so it's foolish to quibble about this. But you can be sure someone has mind-mapping ability if they (a) lead you to a conclusion or interpretation they wanted you to arrive at without saying it directly, and (b) create plausible deniability for themselves.

When people are able to repeatedly thwart your efforts to discuss particular topics or block you from confronting them, they can't do this without mind-mapping ability. To see this you have to stop focusing on the topic being discussed and attend to what your antagonist is doing. Focus on the process of the conversation.

People with mind-mapping ability are good at derailing conversations or preventing topics from ever coming up. If you pose a question they don't want to answer or bring up a topic they don't want to discuss, they'll respond with a defensive question like *We can get to that in a moment. What I want to know first, is why do you always assume you're right and I'm wrong?*

If you take the bait and say *I don't always assume I'm right!,* you've blown it. While you're talking, your antagonist doesn't have to, which allows him or her time to figure out which direction to steer the conversation. He or she will also map out that you'll follow where you are led, which also reveals you don't see yourself as solidly positioned in the discussion. *Let's finish what I just asked you about first,* is a much better response.

Detecting Mind-masking Ability

Detecting another person's ability to mask his own mind is just as important as recognizing his ability to map your mind. The fact he has mind-masking ability indicates they also have mind-mapping ability. People who lack mind-mapping ability don't have any reason (or ability) to mask their own minds.

Mind-masking ability can be used for prosocial and antisocial purposes. If you can't mask your mind, you can't pleasantly surprise someone or tell your party hostess an "emergency lie" that her terrible dinner was wonderful. But just because someone demonstrates mind-masking ability, that doesn't prove he or she is doing something nefarious with you. You probably have mind-masking ability, but this doesn't prove you're trying to manipulate everyone you meet.

For many people mind masking is a habitual response. For instance, if your parents are intrusive and controlling or overly eager to punish you, mind masking becomes a way of life. The same holds true if your family members are manipulative, conniving, and constantly lying to one another. Or maybe you just like to get away with things. Mind masking is a critical component of successful deception. People will perceive you as shifty and untrustworthy if you constantly mask your mind.

Realizing someone isn't being transparent triggers one's survival-based mind-mapping radar to search for deception and manipulation. You don't have to get paranoid, but it does pay to be vigilant. Many people benefit from guidelines for recognizing mind masking. A checklist of reliable mind-masking indicators can walk you into important realizations you don't want to accept–like your mate is having an affair or your friend is stealing from you. Or perhaps you're an open book that's too readable by everyone else. Even if you're a master mind-masker, that doesn't mean you can recognize this in others.

A list of mind-masking behaviors comes in handy for you too. As before, a single behavior isn't conclusive in itself, but several (especially when used repeatedly) make this more likely. Common mind-masking behaviors include:

1. Openly refusing to answer questions, choosing to remain silent instead.
2. Generally asking questions that draw the other person out, getting that person to take the lead in a conversation.
3. Asking why you're asking specific questions.
4. Answering questions with questions and refusing to make a statement.
5. Pretending not to understand what you're asking.
6. Changing the topic through tangential thinking.

Steering the conversation toward less relevant issues and digressions.
7. Appearing to answer your question but not really doing so.
8. Always "adding one more thing" before answering your question but never getting around to actually answering it.
9. Making noncommittal, ambiguous, or indefinite statements.
10. Creating room for ambiguity, e.g., *Well, I guess it could be a lot of different things.*
11. Refusing to be pinned down (yes or no, right or wrong). Claiming not to see things as being black or white like you do.

Obviously, this is not an exhaustive list. There are infinite ways someone can mask his or her mind. I once watched a guy (Phillip) deflect his wife (Briana) by pointing out things she couldn't possibly know and using this to fend her off. Briana thought Phillip was having an affair. She compiled a list of facts that convinced her she was correct and started presenting them to Phillip one by one.

Phillip replied, "You don't know the context when I said something. Maybe I was just talking about *my* life. You don't know what I actually said. All you know is what you heard second hand. You can't be sure it's the truth." Very quickly, compelling facts suggesting Phillip was having an affair disappeared from view, replaced by Briana's defending the accuracy of what she knew.

How could Briana fall into this trap? I'll bet you weren't snowed by Phillip's maneuvers. In fact, he probably triggered your deception detector. However, lots of people look really dumb because discovering they're being deceived creates traumatic mind mapping, which shuts down their mind-mapping ability. All Briana had to hear was Phillip say something like *How do you know for sure that's true?* Or *Who are you to tell me what's going through my mind?* Or *How do you know what I'm thinking?!* Briana got traumatic mind mapping from Phillip's mind masking, which explains why she, like many smart women, look deaf, dumb, and blind.

After being sidetracked a few minutes, Briana got her thoughts back together and returned to her talking points. "I spoke to Sally, and she said she saw you with another woman last night. I found lipstick on your shirt. I saw a lewd text on your phone. And you were talking on the phone after I went to bed."

Briana's first three pieces of evidence had Phillip dead to rights, but her fourth was more equivocal. And, much to Briana's dismay, that's the one Phillip zeroed in on. "How does making a phone call automatically mean I'm having an affair?" Phillip said, demonstrating mind-mapping ability by avoiding her stronger points. He told Briana she was being suspicious and paranoid. When Briana tried to return to her other points strongly suggesting his infidelity, Phillip replied, "I can see you're really not interested in what I have to say," and walked out of the room.

On the surface, Phillip had won the argument. But Briana came away with traumatic mind mapping because he hadn't beaten her radar. What's the best way for Briana to respond? It isn't yelling *Well, you're not interested in what I have to say, either!* As long as she wasn't trying to avoid dealing with this, Briana's best move would probably be *We'll finish this later. I need time to think about what just happened here.*

Briana needed time to get through her "spaghetti brain," pull her thinking back together, and get her emotions under control. She needed to consider what Phillip had revealed about himself and decide how far she was willing to go. When Briana demonstrated she was on to him, Phillip continued to mask his mind and remained unrepentant. When I first spoke with Briana, she acted like she couldn't put two facts together to draw a logical conclusion. This, as you may imagine, was just fine with Phillip.

It turned out Phillip was doing all sorts of things that triggered traumatic mind mapping in Briana and suppressed her functioning. Briana exemplified my fail-to-thrive clients whose impairments maintained the status quo in relationships with dysfunctional parents—only in this case the dysfunctional person was a spouse. It may have looked like Briana couldn't see what was right in front of her, but unfortunately, this was happening because she *could*. Briana had traumatic mind mapping, and she wasn't going to be able to deal with Phillip until she had time to turn her mind-mapping ability back on.

DISCOVER DECEPTION AND LYING

Growing international terrorism has dramatically increased public interest in detecting deception. The U.S. Department of Homeland Security as well as major police departments across the country have initiated lots of research on this subject.[1]

R. Edward Geiselman, professor of psychology at the University of California at Los Angeles, also conducted numerous studies that help train thousands of detectives, intelligence officers, and military personnel to identify someone in the act of lying. Geiselman's meta-analysis of more than 60 research studies indicates catching liars isn't easy, and even highly trained experts like experienced crime investigators aren't perfect at detecting deception and lying.[2] However, some indicators have proved to be fairly reliable signs of someone's trying to be deceptive.[3] You'll find these augment the signs of mind-masking ability we just covered.

For instance, deceptive people generally want to say as little as possible when questioned. Geiselman initially thought liars would tell elaborate stories, but his studies with prisoners and college students revealed that liars are more inclined to give bare-bones sketches of an event. They generally don't volunteer information. Occasionally, however, they'll blurt out a justification for what they're saying without any prompting.

If challenged, truthful people will deny they're lying and explain themselves more thoroughly. They provide more specific details of events to buttress their explanation. When pressed, liars generally won't furnish more specifics, although they may adamantly deny their deception and claim to be offended by the accusation. Sometimes deceptive people appear to "open up," but they actually provide misinformation that goes beyond what was asked of them. They're hoping to shift the focus off of them and onto something or someone else.

If you try to steer the conversation back on track, they'll insist on "describing things their own way, not your way."

Liars tend to repeat questions they've been asked before answering them. One might say *So, you want to know why I didn't come home last night? ... I see, well, you know, that's a legitimate question, let me explain....* They're buying time to devise a plausible excuse. As they're speaking, liars carefully monitor the reactions of their listeners, trying to gauge whether or not they are believable. If they detect their story isn't being bought, they'll switch directions and adopt another line of defense.

Liars often speak slowly at the outset, creating time to monitor the reactions and put their story together. Then they'll spew it out faster when they've decided what to say. Deceptive people often think slowing their speech makes them look suspicious, so they speed up their story as soon they can. Honest people, in contrast, aren't bothered by this and don't dramatically alter their speech rate within a single sentence. Liars speak in sentence fragments more frequently than truthful people do—often starting an answer, then immediately backing up or not completing the sentence.

When asked difficult questions, liars are more likely to press their lips together, gesture toward themselves with their hands, play with their hair or engage in other "grooming" behaviors. They look away only briefly, if at al–unless they know they should appear to be concentrating intensely. Truthful people, on the other hand, often make outward gestures with their hands and look away because they're concentrating on the answer.

So, when all is said and done, how hard is it to catch liars in the act? According to Geiselman, it's pretty difficult. Another researcher well versed in identifying deception, Dr. Paul Ekman, concurs. Ekman found that, without any training, the average person's ability to identify when a stranger is lying is no better than a flipping a coin.[4] So Ekman developed a detailed taxonomy of cross-culturally stable facial expressions which, he maintains, cannot be totally inhibited.[5] These "micro expressions" leak out for a fraction of a second. Some research bears out his contention that people can be trained to detect micro expressions,[6] while other research is inconclusive.[7]

There's a big improvement detecting lying in someone with whom you are very familiar because you already know his or her mind. However, this holds true unless you're dealing with *really* good liars, in which case the liar has probably already gotten to you without your being aware of it. These people use your mind map of them in reverse.

First, they implant false pictures of themselves in your mind, which make them look trustworthy, beyond reproach, and having too much character (or not enough nerve) to lie. The specific pictures they implant don't matter; they just have to make it seem improbable, if not impossible, that they would lie right to your face. You're at a distinct disadvantage dealing with such people because you mistakenly think you know their minds when, in fact, you don't. This liability exists until you realize your error and update your map of your adversary. Only then will your advantage of familiarity return.

IMPROVING DECEPTION DETECTION

There are several ways you can better your chances of detecting deception. Because deceptive people try to say as little as possible, Geiselman developed interviewing techniques designed to get them to talk. When deceptive people attempt to cover up their lying, they become more obvious liars. Geiselman attributes this to the fact that, when liars are manufacturing a story and delivering it convincingly, their "cognitive load" is high (the mental effort used in their working memory).

Therefore, Geiselman tries to further increase liars' cognitive load to push them into making obvious errors. Here's how you can accomplish this when dealing with deceptive people:

1. Ask general questions at the start of a conversation to see how they define the situation and structure the content and style of their response. Then shift to open-ended targeted questions that require specific details in the answers.

2. Don't interrupt when people attempt to answer, and don't make assumptions that fill in gaps in their explanation. Allow them to fill out the story on their own.

3. Then ask them to tell their story from a different time perspective. You might say *I want to go back to your apology last night for forgetting my birthday two weeks ago. When did you actually realize you forgot it?* Once you get an answer, you follow up by asking, *What did you do when you realized you forgot it?*

Events that are initially presented as being causally linked often have less connection when viewed this way, creating a need for a more adequate explanation. It's also harder for deceptive people to remember their lies and control their self-presentation while performing complex cognitive tasks such as this.

Spotting Someone Playing Three Moves Ahead

Besides manipulating information (withholding facts, creating falsehoods), deceptive people control the topic and style of conversations. Spotting this requires recognizing common interaction patterns like "playing three moves ahead." Deceptive people tend to be hypervigilant. They not only pay attention to what's happening right in front of them, they extrapolate from this to see what's likely to come their way. Thinking ahead allows them to make small, preemptive moves that steer things the way they want without making this obvious.

Determining people are playing three moves ahead helps you recognize and deal with manipulative people. You have to establish two things in your mind to handle them: One is whether or not these people have mind-mapping ability and know what they're doing. The other is recognizing they are being combative or adversarial rather than collaborative. Recognizing people are playing three moves ahead clarifies these issues (some play five moves ahead!).

Earlier I said it's easier to detect lying in someone you know (vs. a complete stranger) *unless* you're dealing with real good liars who have already implanted false beliefs in your head that they couldn't or wouldn't lie to you. This is an excellent example of someone's playing three moves ahead: It requires advance planning and implementation. In such cases, it becomes particularly important to recognize deception through ongoing patterns of interacting, rather than by detecting lying about facts.

For example, people who play three moves ahead are adept at detecting errors in their own stories by looking at things through your eyes. They map out what you're focusing on and how you're thinking and feeling to anticipate where your mind will likely take you. By doing this, they can see holes and contradictions you might recognize in their story. They then use this to casually offer more lies to plug the gaps and inconsistencies. They may start off with something like *You're probably wondering about (insert the contradiction or inconsistency here)...* and offer explanations that appear to resolve the discrepancy.

People who play three moves ahead also give themselves away by repeatedly derailing topics they want to avoid. They use their mind-mapping ability to figure out when to steer the conversation in a new direction, or stop it entirely, before they're proven wrong. For instance, people who play three moves ahead know when it's time to put you on the defensive through wild accusations. It doesn't matter if they're ultimately proven wrong. That wasn't the issue anyway. They succeeded in diverting the conversation.

To start with, addressing this situation requires getting over the shock that you didn't succeed in discussing topics you wanted to address, a difficult realization that only comes in retrospect. Recognizing what your antagonist is doing can be extremely upsetting, even frightening, especially if you thought he or she was your ally. You won't want to believe it. You'll have to fight your way through "spaghetti brain" to gather your thoughts.

Next, you need a game plan. You need to prove to yourself that your antagonist is really doing what you're starting to suspect. You can detect someone's playing three moves ahead by using a time-tested method: Introduce a topic you want to talk about that you think this person wants to avoid. But this time, you're not trying to have a substantive discussion. (If that occurs, take advantage of it.) You're looking to see if your antagonist changes the topic. If this happens, this suggests he or she is playing three moves ahead.

Now repeat this several times, either by continuing to address the same topic or by addressing other contentious issues. If you return to the same topic later by taking a different tack and your antagonist cuts you off again, this indicates he or she can see exactly where you want to go and is using mind-mapping to thwart you. When people smoothly and skillfully ward off difficult conversations several times, the scientific principle of "parsimony"[8] says they have mind-mapping ability and know what they're doing, regardless of how much they deny this.

Because playing three moves ahead requires deliberate targeted action, you'll have proof they know what they're doing. Prior efforts to plug holes in their own stories become self-incriminating, further revealing their mind-mapping ability and taste for deception. Armed with this knowledge, you'll improve your chances of not getting fooled by their attempts to derail, sabotage, or preempt conversations, explain away their bad behavior, or get you to do what they want.

RULE OUT MIND BLINDNESS

It's upsetting to recognize someone is playing three steps ahead with you. If this person or the outcome of your interaction is very important to you, this often produces traumatic mind mapping. I don't mean you're necessarily traumatized for life. I'm referring to developing "spaghetti brain" in the moment, and being ill-equipped to deal with it. The strategy I've just outlined can do more than help you zero in on people's mind-mapping ability. Sometimes this will pin you against the wall and make you face truths you'd rather avoid.

If those you love deceive you or play mind games, you won't want to believe they're doing it on purpose. You're going to want to protect them in your mind—actually, protect your own mind and keep your anxiety down—by blaming their behavior on something else. You'll seek out alternative explanations your mind can accept. Most commonly, you'll attribute their behavior to "mind blindness," meaning they lack mind-mapping ability and don't understand their impact on you. You may think of this as "giving them the benefit of the doubt," but your mind is seeking refuge in doubting they have mind-mapping ability.

Here's where knowing the signs of mind-blindness is very helpful. This allows you to exercise due diligence in following up on your observations. It also walks your mind into accepting difficult truths you don't want to accept. Just as there are detectable signs of mind-mapping ability, the telltale signs of mind blindness help you determine whether someone actually has this deficit or not. True mind blindness is pretty rare and signs of its existence are fairly obvious.

For example, people who are mind-blind don't understand that misunderstandings are possible. If your antagonists repeatedly suggest they miscommunicated their true feelings or intent or that you misunderstood them, they are not mind-blind.

Another sign of mind blindness is an inability to detect implied meanings in conversations or understand innuendo. When you're mind-blind, you don't accurately pick up on other people's intentions. If your antagonists accurately pick up on implied meanings or demonstrate an ability to read between the lines, they're not mind-blind.

If you're mind-blind, you don't understand that other people can mask their minds and deliberately try to deceive you. You never act suspicious. By the same token, you're also not very good at deceiving others. You're a terrible liar. To deceive someone, you have to be able to mask your mind; but if you're mind-blind, the idea that people can read each other's mental states doesn't even occur to you. If your antagonists are good liars or manipulators, it's a good bet they're not mind-blind.

Other signs of mind-blindness are less dramatic and harder to distinguish from other explanations. For instance, people who are mind-blind can't accurately gauge other people's interest in what they're saying. They just talk and talk. This tendency can be so annoying to others that people who are mind-blind often can't hold steady employment. You have to distinguish this from terminally narcissistic, self-preoccupied people who love to hear themselves talk. It isn't that they can't see that other people don't want to listen, they don't care. Here's how to figure out the difference: If your antagonists know when to talk and when to shut up—even if it's just with friends and co-workers, but not with you—they don't have mind blindness.

A final sign of mind blindness is an inability to accurately anticipate another person's reactions to your behaviors--in other words, not anticipating or seeing your impact on other people. I saved this for last because, unfortunately, it's the one people jump on without considering if any other signs of mind blindness are present.

Repeatedly doing cruel and hurtful things doesn't prove someone's inability to anticipate the negative impact of things they say or do. This may reflect their egocentrism and self-centeredness. Or it may reflect their desire to be deliberately hurtful and destructive or their willingness for other people to be collateral damage in their agendas. Don't jump to conclusions that make you look good-hearted when you're avoiding difficult people and lying to yourself to avoid traumatic mind mapping.

HEARING IS A LARGE PART OF MIND MAPPING

Humans can detect a candle in the dark from five miles away. But hearing is actually a more accurate mind-mapping mechanism.[9] Hearing is our "fastest" sense. Your brain requires a quarter of a second to recognize visual information, but you can recognize a sound five times faster. Your ability to hear differences between sounds is even quicker, allowing you to sense changes in less than a millionth of a second.[10]

How about accuracy? Your ears have your eyes beat here too. You're no doubt familiar with optical illusions that play tricks on your mind, such as a vase turning into two faces. M. C. Escher made a career out of painting optical illusions like hands that appear to be drawing themselves and staircases ascending and descending at the same time. Although you may never have seen a mirage, you know they exist. In contrast, there are no auditory illusions to naturally occurring sounds. Your hearing is less subject to trickery than your vision.

Thus, it's not surprising that hearing is a mainstay of accurate mind mapping, since mind mapping is foremost about survival. Accurately detecting sounds is a key part of your evolution-driven, emergency-response system. Your ancestors' ability to hear a twig snap in the woods at night, while ignoring all the other natural noises, kept them alive. This involves your brain's ability to zero in on small details by sifting through a flood of auditory information while avoiding sensory overload.

According to researchers at Georgetown University Medical Center, the right and left brain hemispheres specialize in different kinds of sounds. Your right hemisphere focuses on slowly changing sounds and excels at tracking syllables and intonation. Your left hemisphere is more attuned to rapidly changing sounds such as consonants.[11]

When you're interacting with dishonest people, you're more likely to hear deception before you see it. Something jumps out at you. Maybe their tone of voice doesn't match their facial expression. Maybe the sound of what they're saying doesn't align with the actual meaning of the words.[12] The same holds true whenever you're mapping another person's mind. For instance, hearing a minor quiver or tremor in someone's voice immediately tells you this person is emotionally upset and flooding with emotions, even if she's attempting to mask her mind. If you're ever confused, trust your ears.

Mind Mapping Isn't Perfect

Like most people, you're probably worried about making mistakes mapping the important people in your life, particularly if these mistakes encourage you to think badly of them. You don't want to accuse them unjustly or treat them unfairly. This is a relevant concern. It's foolish to think you can always accurately discern what other people are feeling or thinking and why they behave the way they do. Once you discover holes in your autobiographical memory and gaps in your mind-mapping radar, how can you trust your memory, your judgment, or your ability to accurately map other people's minds ever again?

Mind-mapping errors fall into two broad categories. In the first, you assume (incorrectly) that you understand other people's thinking and motives in a particular situation without considering other alternatives or testing your assumptions. In the second, you steadfastly presume you're wrong about what you're mapping until proven absolutely right beyond the shadow of a doubt (because you'd rather not deal with it).

Mapping someone's mind involves detecting patterns of thoughts, feelings, and motivations. But you could have pattern recognition problems that prevent you from accurately doing this. You could perceive patterns that don't actually exist or fail to see patterns a prudent person would recognize. You might fail to discern that what you're seeing now differs from what you've seen before.

Attribution bias leads to mind-mapping errors. One attribution bias is a tendency to attribute positive motives to oneself and negative motives to others or vice versa. For example, you may perceive other people as being very competitive when, in fact, they're responding to your competitiveness. Another mind-mapping error involves attributing actions or intent to someone. In one version of this you make unwarranted attributions; in another, you fail to make attributions that are warranted.

For example, let's say you're in a crowded bar and the man sitting next to you spills his beer on you. He may not have done this deliberately or even have been aware that this happened. Or if he did deliberately spill his beer, he may have just been pouring off the head of foam. In other words, there may have been no connection whatsoever between the thoughts in his mind, his behavior, and the outcome. On the other hand (as every four-year-old mind mapper knows) what people do is usually connected with what's in their minds. More specifically, what people do reflects what they want.

So if you put this all together, the guy sitting next to you may or may not know that he spilt his beer or that it landed on you. He may not have wanted this nor feel bad about it. But this doesn't mean you should never make assumptions about what other people are doing, especially if you love them. You can figure this out by paying attention to their preceding and subsequent behaviors.

I'm telling you this to keep you mindful that you could be wrong about what you map out in other people, either by making unwarranted attributions or failing to see what's really there. But sometimes the issue isn't the limitations of mind mapping, it is how manipulative people like to play around with them. If you watch closely, deceptive people often try to implant the false premise that what they say and do are not good indications of what they're feeling, thinking, and wanting. This would be like someone's saying *I misspoke yesterday when I said you're fat, inconsiderate, and lazy.* I meant to say *You've been working hard at losing weight, and this consumes your attention. It didn't come out the right way.*

"Global labeling" (making over-generalizations based on unrepresentative sampling) is another common mind-mapping error. In other words, you take something and blow it way out of proportion. For example, your boss gives a great assignment to your co-worker rather than to you, and you interpret this to mean the boss wants you to quit. Or your co-worker does one nice thing for you, and you mistakenly presume you're friends. Or your friend doesn't sit with you at lunch one day, and you figure she's lost interest in you.

You can also make the opposite error, like failing to recognize the importance of being passed over for a promotion—or telling yourself your husband loves you if he brings you flowers after he repeatedly beats you up.

Given these potential errors, how can you ever be sure your map of someone is correct? I've learned the answer depends on whether it's the best in you or the worst in you asking the question. If it's the worst in you, nothing I say will help you become more discerning because you don't want to believe what you're seeing and hearing (or you won't give up your erroneous notions of this other person). However, if the best in you is questioning your mind-mapping accuracy, I have an answer that can help you move forward.

Healthy self-doubt improves your mind-mapping accuracy. Self-doubt makes you consider alternative possibilities and review known facts. Because of your uncertainty, you have less reason to automatically assume you're jumping to conclusions. But if doubting your accuracy immobilizes you, rendering you unable to come to a conclusion, then you should consider the possibility that questioning your perceptions isn't coming from the best in you. If you persevere without being dogmatic or self-righteous, you're more likely to see where you map is off, and you can make the necessary changes accordingly.

This approach has worked for many of my clients. It's as methodical, pragmatic and scientific as possible. It doesn't make lots of assumptions or jump to unwarranted attributions. It's not perfect, but it doesn't have to be. You're responsible for making a good-faith effort to be as accurate as you can.

Mind mapping doesn't have to be perfect to be useful. Despite these shortcomings, a person's ability to map another person's mind is remarkably good–good enough to allow humans to have incredibly sophisticated interactions, detect deception and recognize cooperation, and maintain long-term relationships. Mind-mapping accuracy gets even better once you learn about these pitfalls because you can take them into account.

WHEN MIND-MAPPING ABILITY IS REVEALED

We've focused a lot on establishing other people's mind-mapping abilities. Why is this so important? I've found this is the hardest part for us to accept when someone does something hurtful, offensive, or disappointing. Some of us are too quick to take offense when we're dealing with strangers. But when we're dealing with important people in our lives, the greater their offense, the more we want to believe they didn't know what they were doing, didn't recognize their impact on us, or didn't want the outcome they created.

For example, think back to my research with my fail-to-thrive clients (Chapter Eight). They were trying to maintain relationships with parents who repeatedly did disgusting things.

They described their parents as self-centered, selfish, shortsighted, stupid, scatter-brained, careless, foolish, immature, bad-tempered, and even mentally ill. But what they couldn't acknowledge was that (1) their parents had mind-mapping ability, (2) they could see their own impact, and (3) either this is what they wanted or it was acceptable to them. These three things greatly contributed to the disgusting nature of their parents' actions and, thus, were crucial to recognize. As you'll see in the next chapter, two important first steps in reversing the negative impacts of traumatic mind mapping are ruling out that your antagonists are mind blind, and confirming they have mind-mapping ability.

Why are we so unwilling to recognize another person's mind-mapping ability? It's because this knowledge completely changes the meanings of difficult interactions you have had with that person. It changes the significance of important events in your life. It thrusts you into coming to terms with aspects of your life that you haven't accurately dealt with heretofore.

Verifying other people's mind-mapping abilities makes their bad behavior less acceptable and forgivable because of the all-important issues of intent, responsibility, and accountability. We're more lenient with people who commit unintentional or unknowing, harmful acts or who create unwanted outcomes. We deal more harshly with those who deliberately and repeatedly do destructive things. We want to see them as nicer people than they actually are because this forestalls our anger and our impulse to break off our relationship with them.

We're also loath to recognize how flawed our assessments of other people might have been. But this pales in comparison to our anticipatory traumatic mind mapping: We know they won't take kindly to being seen more negatively, making them more likely to engage in further bad behavior, including daring us to say they knew what they were doing and/or wanted the outcome they created. This is where most adult children cave in when dealing with parents who do disgusting things.

If you've experienced lots of traumatic mind mapping, you have reason to believe your picture of your life and the people who populate it may be seriously inaccurate. *If you're like most people, your autobiographical memory will be **inaccurate** enough to keep your anxiety down, but **accurate** enough to keep your lie detector from going off.* If you start by presuming you have perfect mind-mapping ability, you'll probably never find your errors because you're looking for subtle but highly meaningful distortions.

But here's the good news: Don't get too upset when you discover gaps in your mind-mapping radar and distortions in your autobiographical memories because they're pure gold. These gaps and distortions are never random. By studying the inaccuracies in your prior pictures of your life and those of other people, you can understand yourself better.

People sometimes feel chagrin about their prior distortions. But this is a small price to pay for a new lease on life. Reversing the impacts of traumatic mind mapping will vastly improve your functioning and your appreciation of your own life. This is where we're headed in the next chapter.

12

Reversing Traumatic Mind Mapping

Have you ever encountered people who are so rude, self-centered, hostile, or misguided that you are left speechless? How about being so shocked that someone is doing cruel things to you (or someone you care about) that it renders you unable to control your emotions or think straight? Ever been terrorized by someone's hostile intent or behavior although he or she never physically touched you? Or emotionally crushed when you realize someone you trusted has betrayed you?

If you've had any of these experiences—the examples are endless—you've probably experienced traumatic mind mapping, particularly if the people in question were important to you. You looked into your antagonist's mind, and what you saw was so upsetting you got "spaghetti brain." Your cognitive sharpness and efficiency dropped. In the moment, you couldn't organize your thoughts into the snappy response you might have come up with otherwise. Your emotions began to overwhelm you and became difficult to control. Your sense of perspective was lost, as was your sense of time. You were lost in a terrible moment that threatened to go on forever.

People get "spaghetti brain" frequently in the course of everyday life. Watch what happens to you when people you care about get really angry, belligerent, or threatening. Or they raise their voices, cry hysterically, or scream. Or suddenly storm out of the room. Or start slapping or hitting. Or threaten divorce.

These highly charged behaviors don't even have to be directed at you. All you have to do is witness this and map out the minds of the people involved, and you'll have traumatic mind mapping. If this goes on all the time in your home when you're growing up, this has long-term impacts on your brain.

As traumatic experiences go, these aren't the worst by a long shot. But now consider these scenes, giving yourself time to visualize the mind mapping that goes on in each one.

- You discover your long-term trusted employee has been syphoning off your business clients and now he's opened a rival business.

- Your husband has been having extramarital affairs and transferring money from your joint account to his private account in the Cayman Islands. He tells you he's leaving, and you're flat broke.

- A good friend commits suicide not long after meeting you for dinner, and you had no idea she was upset.

- During adolescence you are physically and sexually abused by your father. Between times he likes to leer at you, but your mother never seems to notice. Later in life, you have intermittent contact with your parents for the benefit of your child. You think it's important for kids to have grandparents. Then you discover your child has been sexually abused. It turns out that your child's abuser is your father. You confront your father, and he denies this.

Obviously, these traumatic mind-mapping events differ in magnitude. If any of them actually happened to you, each one is bad enough to give you "spaghetti brain" in the moment and may continue to disorganize your thoughts and feelings whenever you subsequently try to make sense of these experiences by mapping out your antagonist. Where these vignettes differ, however, is in the severity and repetitiveness of traumatic mind mapping. Unless addressed, repeated severe traumatic mind mapping can cause long-term brain impairments going forward.

In this and the following four chapters, I'll give you practical strategies for handling the negative impacts of traumatic mind mapping experiences and restoring your psychological health. These methods are part of Crucible Neurobiological Therapy, and the science underlying this approach is laid out in detail in Appendix D.

One of the best things about the powerful methods I'll share with you is that you can apply them to yourself. I can teach them to you, just as I teach them to my clients. You can learn the basic methodologies and apply them to yourself, at whatever depth and intensity you choose. During therapy sessions I guide my clients through the process, but the real curative impact comes from what they do with their own minds. My role is to keep them on track because their brains usually balk at accurately reprocessing traumatic mind-mapping experiences. Making *themselves* do it anyway (stubbornness!) is an important ingredient of this treatment.

This brings up an important issue for my clients that you will face as well: When you start to get "spaghetti brain" as you reprocess prior traumatic mind mapping, you have to persevere and fight for clarity. But both the worst in you and the best in you may wonder if this is good for you to do. Many clients, understandably, want to avoid looking at terrible things in their lives. I tell them I've never had clients become psychotic, severely depressed, or unable to go about their daily responsibilities from doing this. I've never had to hospitalize anyone. But I leave them to struggle with themselves about deciding whether they're going to do this or not.

I'm not saying my clients aren't saddened by what they discover. Very often they're dealing with *terribly* sad things. Their pictures of their lives get worse, but they get better–and the benefits they derive far outweigh what they have to go through to get there. All things considered, clients don't regret applying what I'm about to teach you. But I don't promise my clients this outcome, just like I won't promise you.

If you're dealing with severe and/or repeated traumatic mind mapping, your brain isn't going to want to do the very things that could help you. At least initially, you'll have to make yourself repeatedly do things you don't want to do. On the other hand, you may wonder *What if this is my intuition telling me that I shouldn't do this? Won't revisiting traumatic experiences just re-traumatize me and make things worse?* All I can tell you is I haven't seen this happen with any of my clients. But that doesn't mean *you* should do this. Just like my clients, you need to take responsibility for yourself and decide what you're going to do. Making this decision is an important step in taking control of your brain and mind.

VISUALIZATION AND TARGETED MIND MAPPING

The first step involves realizing you've experienced traumatic mind mapping in a specific event or interaction with a specific person. You can't solve a problem that you haven't acknowledged to exist. Simply recognizing this moves you down the road to better mental health and freedom from your past.

The next step in resolving traumatic mind mapping, as mentioned in the last chapter, involves assessing your antagonists' mind-mapping abilities, so you can develop an accurate map of their minds. It's imperative that you make yourself decide whether the person or persons involved have mind-mapping and mind-masking abilities or not.

This positions you to use the methods I've developed to reverse the effects of traumatic mind mapping. I've condensed years of research and clinical development into a one-sentence description of a four-part process that will help you: *Resolving traumatic mind mapping involves (1) mapping out the mind (2) of the person who exists (3) in the hole (4) in your autobiographical memory.*

I suggest you repeat this sentence to yourself several times and memorize it. I don't think something this profound can be said more concisely, and I hope by the end of this book you agree.

Let's take a moment to unpack what this means: Previously I've said traumatic mind mapping causes your mind mapping ability to collapse, creating a memory retrieval error when you try to recall the experience. In some cases you won't remember the event at all, but more commonly here's what's missing: your mind-mapping information about the person or people involved, meaning your map of your antagonist's mind during this event. This is the "hole" in your autobiographical memory. The person who exists in the hole in your autobiographical memory is your antagonist. This is the mind you need to map out.

In other words, to reverse the effects of traumatic mind mapping, make yourself map the mind of your antagonist who caused the trauma and created the gap in your autobiographical memory. This step-by-step process involves re-visualizing the prior traumatic event and taking a long, hard look at it. You have to map out your antagonist's mind in slow motion as you watch the scene transpire. This requires determination, willpower and good old-fashioned, jaw-setting, head-banging stubbornness.

Let me show you how stubbornness can create remarkable positive brain changes. In *The Brain That Changes Itself* author Norman Doidge documents how brain-damaged children who couldn't tell time by looking at a clock were subsequently able to tell time *extremely* well after 30 days of forced one-hour practice.[1] I say "forced" because the kids didn't want to do this at first. The activity highlighted their deficits. They failed repeatedly. They were frustrated. They felt inadequate. They didn't believe they would succeed. But perseverance brought benefits they never imagined: After a month of daily practice they could tell the time, day, month, and year on a five-handed clock! Through repetition, their brains made new connections that circumvented their brain damage and allowed them to succeed.

Some of my clients struggle to remember any event from their childhood. Many have told me they don't remember *anything*. Invariably this turns out not to be true. They might not recall much, but at least they can recall *something*. Start there, using a method I'm about to describe. Work with whatever you can remember involving the person in question.

This usually triggers a cascade of memories, much more quickly than people expect.

When people don't want to look at their past, they think it's going to surface in difficulty remembering events. It can indeed show up this way, but it's not the predominant form. Most of my clients recall plenty of upsetting events; but when they visualize those events, they have difficulty mapping their antagonists' minds.

If you have this common pattern, it shows up in several ways: When you remember the traumatic event, you don't remember your antagonist's motivation, feelings, and thoughts towards you. Then, when you try to revisualize the event, you have difficulty getting into your antagonist's head. In difficult cases where clients persevered for hours to finally "break through," some had difficulty remembering what they mapped out an hour later.

As long as you don't give up, there's every reason to believe you will succeed. Virtually all my clients succeed if they keep trying.

SEE THE SETTING

To set this into motion, select one of the "hot" images in your memory in which you were shocked by your antagonist. It's important that you *see* the event, rather than intellectualizing or speculating about it. You must *watch* things unfolding. "Watching the picture in your mind" is a right-brain visualization process, whereas thinking about it and "putting two and two together" is a left-brain, deductive-logic activity.

The solution must be *visual*. It won't work if you approach this using your left brain, even if you come up with lots of interesting ideas and insights, which is exactly what your mind will try to do. All the left-brain activities you can think of—journaling, intellectualizing, hypothesizing and telling yourself self-affirming statements—won't solve your problem.

This process is similar to the one used by the brain-damaged children who learned to tell time in Doidge's book. They spent an hour each day staring at a picture of a clock on a computer screen. In your case the visual image you're forcing your brain to track is in your own mind as you watch a troubling scene unfold and make yourself map out your antagonist's mind.

Start off by seeing the room or location where the interaction took place. For instance, can you see the color of the walls, or the floor coverings? Where are the windows and doors located? How are you and your antagonist positioned in the room, relative to each other? How far apart are you? Are you sitting or standing? What are you wearing? Do the same for your antagonist. This gets the initial picture established in your mind.

It's important that your mental picture resemble a movie rather than a still photograph. You're visualizing moments of meeting in the interaction unfolding between you and your antagonist. I understand this is the part you want to avoid, but it's also the part you most need to look at. If you revisualize the examples I described at the outset of this chapter, every single one involved a traumatic moment of meeting.

Here's how I walk my clients through this process: Let's say you're a business executive consulting me because you got fired a year ago, seemingly out of the blue. Ever since, you haven't been able to pull yourself together. I would have you visualize and describe your meetings with the business owner, with particular attention to the termination interview.

Or let's say you caught your husband having an affair and the discovery severely traumatized you. In this case, we would go through your memories of your sexual interactions and failed initiations with your husband, especially visualizing the moment when you confronted him with your evidence.

Perhaps you were traumatized by a friend, a sibling, a boss, or a co-worker who betrayed your trust and then blew you off when you brought it up. I would have you visualize and describe the moments this occurred.

When I'm working with adult children of highly dysfunctional parents, I often start with the most difficult images they can remember. Where they have large gaps in their autobiographical memories, I have them choose a childhood event they've never fully understood. The most important thing is—whatever vignette they settle on—it must involve *visual* images. It can't be an abstract idea or theory like *I feel like my mother never loved me.* It has to be a remembered event that, in this case, made you realize your mother didn't love you.[2]

WATCH THE MOVIE UNFOLD

Let me give you an example of what this looks like in practice: Rachel and Stephen came to me for their sexual and marital problems. Rachel was highly sexually responsive, but she was also extremely passive. Stephen was fed up having to carry their sexual relationship, especially since he understood that Rachel liked sex. If Stephen didn't initiate, they'd go months without sex. When Stephen asked Rachel why she never instigated it herself, she would say she didn't know. She loved Stephen, enjoyed sex with him, and considered him a kind and gentle lover. But unless he proposed it, Rachel never seemed to think about sex.

Lower Desire Partners often report they don't know what's going on because they won't face the truth. In some cases they don't love (or even like) their partners. Sometimes they don't find partners sexually attractive anymore, or they want a divorce. They usually set off my deception detector because they're masking their minds. But Rachel seemed different. She truly seemed mystified and worried about the fact that sex never entered her mind without prompting. Rachel said she had never masturbated in her life, and it never even occurred to her to try. She didn't know if she could bring herself to orgasm, but she had no difficulty on her first attempt. There was no evidence of Rachel's being sexually inhibited.

Rachel vaguely remembered touching herself at some point during her childhood and her mother's admonishing her that this was wrong. Rachel's mother, I discovered, was a nasty alcoholic who liked to say Rachel was her smart daughter and Rachel's sister was the pretty one. Her frequent advice to Rachel was, "Find a smart, wealthy, homely guy to marry because a good-looking guy will want a woman like your sister." Rachel's adolescence was full of constant messages from her mother that she wasn't sexy or attractive.

According to Rachel, her father was the more reasonable parent, someone she could talk to when she was depressed. Rachel's father was her ally against her mother, although he never stood up for Rachel when her mother made unflattering comments. He'd only come around afterward to tell Rachel that he didn't agree with what her mother said.

I asked Rachel about her father's attitudes toward women, women's sexuality, and her sexuality in particular. Rachel said she couldn't recall her father's ever having said anything about these topics. I replied it was impossible for him to have no opinion about these three things, one way or another. Rachel agreed she ought to know her father's attitudes about women and female sexuality, but she came up with a complete blank. The topic of his attitudes towards Rachel's sexuality seemingly escaped Rachel's attention.

Notice Rachel was using a left-brain approach requiring abstract intellectual conclusions. I suggested Rachel use a different method to get around her mental block. "Describe a vignette from your childhood. It can be anything you like. It just needs to have something to do with your father and the topic of sex. You're not going to tell me the topics of sex and your father have never come together in your entire life, are you?"

Rachel thought for a few minutes. "This is the only thing that comes to mind. When I was 14, my father took me scuba diving, which was his passion. He's been diving all over the world. We went to the Crystal River Springs, a well known cave diving area in Florida where we lived. As we were suiting up, we met a woman my father seemed to know; and he suggested she join us because she didn't have a dive partner. Although nothing explicit was said, I came away thinking there was a distinct sexual vibe between my father and this woman. When we got home, my mother was very angry."

"Why was she angry?" I asked.

"I don't remember," Rachel replied.

"Are you telling me you watched your father exchange sexual vibes with a woman, which caused your mother to be very angry when you got home, but you paid no attention to why your mother was angry?"

"I honestly don't remember."

"Let's try something." I said. "Can you visualize the scene when you got home? Where are you when your mother gets angry at you?"

"My father and I have just walked into the living room. I'm standing next to my Dad. Mom is sitting on the couch folding clothes. She looks up at both of us with scorn. First, she unloads on my father, and then she lets me have it."

Before I let clients delve into the meaning of what they're seeing, I encourage them to make their picture as vivid as possible. I'll progressively ask questions (one at a time) like *Are you inside or outside? If you're inside, which room are you in? What color are the floor and walls? What type of furniture is in the room? Where are the doors and windows located?* When I asked her, Rachel wasn't sure about the color of the walls, but she remembered the bay window looking out over the front lawn.

"Tell me about the living room. What does the floor look like?"

"It's got carpeting."

"What color is the carpeting?"

"Kind of reddish."

"What color is the couch your mother is sitting on?"

"It's red. My mother liked things to match."

Rachel seemed surprised by the precision of her memories. As she described the living room, Rachel closed her eyes, focusing on the picture in her mind. Rachel spontaneously engaged her right brain. If she hadn't done this, I would have suggested closing her eyes and visualizing the setting.

This is exactly what you want to do. You want your process to be *I can see...*, as opposed to *I'm thinking about...* or *I wonder if...*, both of which use your left brain. It's important to recall as many details of the scene as possible. This will make your mental picture as vivid as possible. You'll probably be amazed by how much you can remember. Clients frequently see seemingly insignificant details they're surprised they remember--like the pattern of the drapes, or sagginess of the couch cushions or objects on a coffee table.

I also visualize the scenes my clients describe. As I'm looking at the vignette, I realize what I can see and what I can't. This helps me recognize missing details needing to be fleshed out. I'll ask questions like *Where are you located in the picture? Where is your antagonist? Who else is in the picture? How far apart are you? What is everyone doing? Where is your antagonist looking?*

Oftentimes, clients' pictures become so vivid and tangible to me that I'll experience traumatic mind mapping too.

You'll be amazed at how much you remember about the setting of the event compared to the poverty of mind-mapping information you have about your antagonist. This discrepancy became apparent when I asked Rachel what her mother was thinking when she chastised Rachel and her father. Rachel said she didn't have a clue; she'd never even thought about it.

"In your visual image of the scene, your mother is looking right at you, and you're looking at her, right?"

"Correct."

"How far apart are you?"

"Five or six feet"

"Fine. Look at the picture. Tell me what your mother is telling you."

"I should know better. I'm no better than him. I've betrayed her."

"Good. Now look at your mother's face in the picture. She's looking right at you. Tell me what she's thinking about you."

"I'm not sure."

"Take your time. Keep looking at your mother's face in the picture."

"I can't tell. This doesn't seem to be working."

"I suggest you keep at this. What is your mother thinking about you?"

"It was a long time ago!"

"Look at the picture!"

Rachel closed her eyes to focus on her picture. After a long pause she said softly, "My mother hates me."

"Why does she hate you? Look at your mother's face. Map her mind. What is she thinking?"

Rachel glanced down, but she wasn't looking at the floor. She was staring at the picture in her mind. After another long pause she said, "My mother thinks my dad is having an affair, and she thinks I'm his accomplice."

"Do you think it would matter to a 14-year-old girl if her father was having an affair?" I asked.

"I'm sure it would."

"Did your father have affairs with other women?"

"I've never thought about it." Rachel seemed shocked by her own answer.

"Do you think most 14-year-old girls would wonder if their father was having an affair if they met a young woman their father seems to know, who exchanges sexual vibes with him, and then their mother yells at them afterward and apparently thinks they're an accomplice to his infidelity?"

"I'm pretty sure they would."

"These facts don't prove your father was having an affair, but they certainly suggest this possibility. It's noteworthy that you've never let yourself think about things most 14-year-old girls would be preoccupied with." Rachel nodded in agreement.

I said, "I can't see your father's reaction when your mother says you betrayed her and you're no better than your father. Where is he looking? What's the expression on his face?"

"He's looking at my mother. He has a disgusted expression. He's angry at her for embarrassing him in front of me. He never says so...but I think he's happy he got to spend time with this other woman and that my mother is furious about this."

"What makes you say this?"

"He's telling my mother she's a controlling bitch and he doesn't deserve this kind of treatment."

"Does your father ever look at you?"

"Not until he walks out of the room a minute later. On the way out, Dad gives me this look that says *Your mother's a lunatic. She doesn't know what the hell she's talking about. Ignore her. You and I are buddies.*"

I concluded my session with Rachel and Stephen by saying, "When we started, Rachel, you said you didn't have the vaguest idea about your father's attitude towards women, female sexuality, and your sexuality. I think you've got the beginnings of an answer."

The look on Rachel's face said it all: This vivid childhood picture was shaking loose remembrances that filled in other gaps in her autobiographical memory.

USE A FIRST-PERSON VIEW

After you've developed a well-defined picture of the incident you've selected, in your mind's eye, forward the action to a point where you establish face-to-face contact with your antagonist. This usually isn't difficult because many traumatic moments of meeting involve eye contact with your antagonist.

If someone is beating you, insulting you, or lying to your face, you're going to track that person's eyes. Even if you don't make eye contact during the encounter, there will be subsequent moments when your eyes will meet. For example, if your brother sexually abuses you late one night, there will be a moment the next time you are together, perhaps sitting at the dinner table, when your eyes meet.

Focus your attention on these moments. You'll find a wealth of mind-mapping information embedded in this visualization. Eye tracking is a fundamental component of mind mapping.

So, for instance, think back to the examples I gave you in Chapters Nine and Eleven. There was the boy, Bruce, whose father ate his ice cream cone while Bruce begged and pleaded. Bruce's father was looking at Bruce and laughing as he did it. He licked the cone several times before pretending to hand it to Bruce, and then pulled it back at the last second to take one last, gleeful, big bite.

As hurtful as this behavior was, it was Bruce's father's triumphant look of satisfaction that gave Bruce traumatic mind mapping. Bruce's father wasn't just happy about getting to eat ice cream. He was enjoying Bruce's distress. To reverse the traumatic mind mapping, Bruce needed to visualize this again and digest its meaning. This involved Bruce's struggling through massive "spaghetti brain" until he could see what his father was doing and not have his own mind go to pieces.

Remember Gregory, whose father got angry at him in the middle of a family road trip, and left him standing on the side of a highway? Gregory's father grabbed him and shook him, and told him he was a worthless piece of crap. But it was the look he gave Gregory just before he drove away that caused the most trauma. Gregory's father was enjoying himself. He felt all-powerful, like a king in his castle who could do whatever he wanted. He wanted Gregory quaking in his boots, scared out of his mind, under his authority.

In both of these examples, the eye contact was mutual and prolonged. However, the vignette I gave you of the 13-year-old girl, whose father broke down the bathroom door, threw her to the floor, and nuzzled her breasts, was different. The girl never made eye contact with her father until he stopped and lifted himself to his feet. Only then did their eyes meet, and she could see how much he enjoyed what he'd just done, how little he cared how she felt about it, and his lack of concern if his wife found out.

EYE CONTACT ISN'T ALWAYS NECESSARY

It's beneficial to focus on moments where you and your antagonist make eye contact. But you don't need eye-to-eye contact to experience traumatic mind mapping. In fact, let's say you actually never made eye contact with your antagonist while the traumatic moments of meeting occurred.

In Chapter Eight I briefly mentioned another client who unwittingly gave me a great example of this. Peter told me that, when he was growing up, his father was gone all the time and never spent much time with him. Peter's father never came to his high school football games. This wasn't so surprising in itself, except for the fact that Peter was the captain of the team, and he received a full scholarship to play college football. Peter was going to be the first person in his family to have a college education.

"Your father never came to *any* of your football games in high school?" I asked incredulously.

As I delved deeper, I discovered this wasn't the problem. The real issue was the one time Peter's father *did* come to see him play. Among Peter's many roles for his team, he was the main punt returner. As he received a punt during the one game his father attended, Peter momentarily took his eyes off the ball to look at his father, causing him to fumble it. What was Peter's father's response? He popped out of his seat, ran down to the sidelines, and screamed at Peter, "You stink! You're an embarrassment to your team! You should quit right now!" And that's exactly what Peter did the next day, ruining his chance of going to college.

Peter stared at his shoes while his father berated him. He couldn't bring himself to look his father in the eye. However, Peter didn't have to do that to know what his father felt about him at that moment. Peter had mapped his father's mind for many prior years, and he knew his father felt that Peter had let him down and embarrassed him.

But there was more going on that Peter couldn't bring himself to look at: Even if Peter had been staring into his father's eyes, the message couldn't have been clearer: Peter knew his father felt inadequate because he didn't have a college education. Peter's father didn't wish the best for him, and that was the hardest part to take.

Then there was the mind-twisting part. Although Peter's father felt that Peter had let him down, actually it was the other way around: In truth, Peter's father was embarrassing and disappointing Peter. But whenever Peter recalled this momentous event, the picture he visualized lacked mind-mapping data about his father. He couldn't see his father's motivations.

As so often happens, Peter's resulting memory coincided with his father's version of what took place: Peter had let his father down. Peter had no awareness of the disgust he felt about his father's appalling behavior, his betrayal of Peter's best interests, and his smallness as a human being. Until Peter came to see me, the hole in his autobiographical memory maintained his long-term self-perception that he simply wasn't good enough.

You can imagine Peter's surprise when we revisualized the fateful moment when he dropped the punt and his father screamed at him from the sidelines. "Look at your father standing on the sidelines in front of the bleachers," I said. After Peter gave a graphic description of seeing his father standing in front of the bleachers, red-faced, shaking his fist at him, I asked, "Tell me what's going through your father's mind. What is he thinking and feeling about you?"

Peter couldn't do this at first. After several minutes of fruitless effort, I asked him, "Does the person screaming at you from the sidelines have mind-mapping ability? Does he understand the feelings he's instilling in his son?"

"I don't know."

I pressed Peter. "Can your father see your reactions? Does he say anything else to you?"

"He tells me to stop sniveling and suck it up." In other words, Peter's father saw Peter's reaction.

"What does your father do after he yells at you?"

"He tells me he needs a drink, and walks off the field like he's disgusted with me."

"Look at your father in this picture. Does he think he's helping you somehow? What do you see?"

Peter looked at me in anguish. "I want to say he doesn't know what he was doing, but I'm pretty certain that he did. He tells me to stop feeling sorry for myself. He knows I feel bad. And he keeps doing it anyway!"

"Is he doing this *despite* your reaction, or is this the reaction he wants you to have?"

"He doesn't care what I'm feeling. He doesn't care what the other parents think of him, but he knows that I care. I'm embarrassed by how he's acting."

"Does your father want you to feel embarrassed?" Peter nodded. "Then maybe your father *does* care about what you're feeling—and he wants to make sure you feel it." This took a moment for Peter to wrap his mind around.

Finally, I asked Peter the most difficult question of all. "How does your father feel about your being the first person in your family to go to college and becoming more educated than him?" At that moment, Peter got "spaghetti brain." He looked at me like I suddenly started speaking a foreign language, and he couldn't understand a word I'd said.

So I rephrased the question. "While your father is screaming at you from the sidelines, is he aware that if you quit the team—like he's telling you to do—you won't go to college?"

Peter didn't respond. Minutes passed. Finally, he yelled, "He *knows*! Of course he knows this is my big chance! How could my own father do this to me?!"

"You're acting like you still don't know your father's mind."

With a long, sad sigh, Peter said, "I guess I do. He'd been lecturing me at the dinner table that college would just make me an educated fool. He was jealous."

This was the hardest part for Peter to digest. This was an incredibly productive session although Peter never looked his father in the eye during their confrontation. As you can see, eye contact helps, but it's not absolutely necessary.

Solutions For Resistant Problems

Dissecting traumatic mind mapping in this methodical way often brings up anger, sorrow, and loads of other emotions. This can be painful, but it may be precisely what you need. When you're finally able to see the truth, your brain works better because it no longer has to compartmentalize facts and emotions. You become aware of how your memories actually fit together, which is very empowering.

Clients are often upset by their new realizations, but they don't fall apart or become depressed. Within days of going through this, they usually show marked improvement. This makes them more resistant to "spaghetti brain" when subsequently facing other traumatic mind-mapping experiences.

Since I can't go through this process with you, let me show you what to look for in your own thinking. This is what I knew would surface in Peter's mind: *You're going to approach your antagonist as if he or she doesn't have mind-mapping ability.* You might have this as a conscious thought, or it may be implicit in the way you envision or respond to him or her.

In other words, you probably have your antagonist incorrectly mapped as being too dim-witted, inept and ineffectual, or too devoted to your welfare, or too narcissistically self-preoccupied to be tracking you and deliberately using this to manipulate you. So make yourself revisualize the pictures in your mind, reminding yourself you're most likely watching someone with mind mapping ability.

What could possibly make such difficult realizations worthwhile? Well, think back to Peter's initial complaint, that his father didn't spend time with him. Would you want a deeper relationship with such a man? Would that man want a deeper relationship with you? Peter's map of his father's mind was completely missing, leaving Peter so blind he couldn't see his own confusion.

It shouldn't surprise you that, for most of his life, Peter had difficulty getting appropriately angry when people treated him poorly or that Peter was a bright, talented, guy who felt like he never measured up. Getting his autobiographical memory straightened out allowed Peter to radically change his life.

Wouldn't it be worth enduring some discomfort to emancipate yourself from the strictures of your own mind? Or enable yourself to deal better with destructive relationships? Clients who persist through this are invariably glad they did, and you won't find a better use for your willfulness, obstinance, and bullheadedness.

That's because your mind will look for ways to escape looking into your antagonist's mind. For instance, your mind may focus on what you *don't* know and *can't* be sure of. Peter wanted to say there was no way of proving his father ran down to the sidelines to stop him from going to college. That wasn't the issue. Peter's mind was generating uncertainty rather than looking at what he knew. In your case, you need to refocus on what you *do* know.

It really helps to use simple binomial thinking–the way kids zero in on their parents: Is it this way or that way? For instance, when Peter's father ran down to the sidelines, was he thinking positive or negative thoughts about Peter? They had to be negative, right? By asking yourself this question, you've narrowed the possibilities in half, which is a great first step.

Next question: Could Peter's father see Peter's reaction to what he was doing or not? This is a basic question about his father's mind-mapping ability. Start your process using simply answerable questions. *If you're not sure which way to go, or you get "spaghetti brain," focus on what you can be sure of.*

SHIFT TO A THIRD-PERSON VIEW

Often clients are able to step into their pictures as if they were holograms and rotate them, so they can look at their events from different perspectives. Sometimes this shift in perspective is necessary to be able to see your antagonist's eyes from a different angle or to be able to see his or her face at all.

If you're not getting anywhere visualizing an event from your past, here's a method that can help: *Shift to a third-person view*. In other words, shift from a first-person view of the scene (i.e., looking out your own eyes) to a "fly on the wall" perspective in which you can see both your antagonist and yourself. When your mind mapping shuts down from traumatic mind mapping, the first-person view of your vignette is often blocked. When this occurs, switching to a third-person point of view will usually circumvent the mental block. (In some rare cases it's the other way around: Your third-person view is blocked, but your first-person view will be available.)

Earlier I spoke of establishing eye contact with your antagonist in your mental picture. This changes somewhat using a third-person point of view because you'll have a side view of your antagonist and yourself. You'll be able to observe your antagonist's facial expressions as well as your own. You'll still be able to map your antagonist's mind.

This third-person view also allows you to map out your own mind from the perspective of an external observer rather than through introspection. This is advantageous because it's easier for your brain to map out someone else's mind than your own (Chapter Seven). You'll be able to see your antagonist's reactions to your reactions. It took me years to discover this technique, but as soon as I did, my clients made leaps in progress.

Here's an example of this in action. One client, Tim, told me he grew up on a large farm without much parenting as the youngest of five children. As he described it, Tim was raised more by his older siblings than his parents. When I asked him to tell me a powerful image from his childhood that he didn't really understand, Tim described walking into the living room when he was six years old and discovering his parents violently wrestling on the floor.

I asked Tim to watch this movie in his head and tell me what his mother and father were thinking and feeling. Tim said his parents were blind with rage. I asked him to elaborate because, even when people are so mad they can't think straight, they're still having thoughts. Their minds don't just stop. Tim speculated that his mother felt trapped. When she got pregnant with her first child, she dropped out of college and gave up her career ambitions. She resented being stuck on a farm way out in the country with a bunch of kids, in contrast to Tim's father, whose part-time work as a farm equipment salesman allowed him to come and go as he pleased.

I told Tim that, rather than speculating, he needed to look at the picture in his mind of his parents wrestling on the floor and describe the look on his mother's face. Tim said he couldn't see her face because his view was blocked by his father's back. His father was on top of his mother, holding her wrists with his hands, pinning her to the floor. From where Tim was standing in the picture, Tim couldn't see his father's face either because his father had his back to the living room entrance.

This was the perfect situation to suggest that Tim shift to a "fly on the wall" point of view.[3] I suggested he adopt a visual perspective that allowed him to see his parents' faces, as well as his own, as he stood there watching them wrestle. Doing this allowed him to see and describe what his mother was thinking and feeling in the moment. Tim said, "She's snarling and thinking *I hate you. You've ruined my life. You stole my opportunity, my freedom, and my youth. You haven't done a thing to help raise these damn children. You stuck me with them, and you're always going off on your own.*"

Then I asked Tim what was going through his father's mind as he wrestled with Tim's mother. Without hesitation, Tim said, "*You betrayed me! You're not a good mother. You're not a good wife. You're not a good person. You're a liar and a whore!*"

Next I asked him to tell me what he was thinking and feeling at the time, not by introspecting, but rather, by looking at himself in this scene unfolding in his mind. What did he see looking at his own face from this new vantage point? Tim said he looked shocked, confused, and devastated.

Then I asked, "Are your parents aware of you standing there watching them?" Tim's response was revelatory, although it didn't directly answer my question. He said, "Mom always claimed she 'snuck' me past my dad. He didn't want any more kids when she got pregnant with me. She told me I was extra special to her because she made it happen."

I let the fact that Tim's mother basically forced an unwanted child (him) onto his father sink in. After a moment I said, "Apparently your mom was willing to defy your father and lie to him to get what she wanted. This should give you a better understanding of your mother and also help you understand your relationship with your dad. But what do you make of the fact that your mother 'snuck' you past your father, and then turned around and accused him of sticking her with all the children?"

Tim looked at me with a shocked expression on his face. "I never thought about it this way before, but this is an accurate picture of my parents. My father always acts like I'm a pain in his ass, and you can't trust my mother. She's always manipulating people into doing what she wants, and she blames everything that goes wrong on someone else."

BANG YOUR HEAD AGAINST THE WALL UNTIL THE WALL MOVES

This process isn't terribly sophisticated, but it does take stubbornness. You need to force your brain to revisualize the scene where your traumatic mind mapping occurred. Then you can begin to map your antagonist's mind. Construct a timeline, using your antagonist's actions and statements as important points covering the entire event. Then walk yourself through this timeline, statement-by-statement and action-by-action, mapping your antagonist's mind every step of the way. This is how you force your brain to take another look at the incident(s) that blew your mind.

In all the years I'd been doing this therapy, I'd never encountered someone who was willing to put in the effort that couldn't beat the gaps in his or her autobiographical memory. Sure, I've had clients who didn't really try. But I'm telling you the same things I tell them: Don't give up. Keep trying. There's *always* a way around the strictures of your own mind, but you have to work at this.

Let me give you an example of how this looks in action. It's a pretty good picture of how stubborn you have to be—and how stubborn I can be too. I had a client, Edward, whose life was falling apart when he first came to see me. Edward's wife was ready to leave him because he was passive to a fault. She was particularly fed up with Edward's allowing his father to walk all over him. Edward's father inserted himself into their marriage, giving Edward advice that often overruled what Edward's wife wanted to do. Sometimes this also went against Edward's own judgment, and Edward's father had talked him into a number of bad investments and decisions.

Edward kept telling me he was sure his father meant well. I kept asking him how he could be so sure because I knew this involved mind mapping. Edward was insisting he knew his father's state of mind. Edward just said he was sure. I didn't dwell on this point simply because Edward defended his father every time his wife criticized her father-in-law. I noticed Edward didn't have one bad thing to say about his father. When I asked Edward to tell me something negative about his father, he couldn't do it. I wasn't trying to get him to bad-mouth his father. I was trying to see if Edward could access a less-than-perfect memory of him.

Hours passed without Edward's being able to say anything negative about his father. Fortunately, he had flown in for our Intensive Therapy Program, in which sessions are typically half-day in length. Edward would start talking about him, but then he'd repeatedly veer off and start talking about something else. I could see Edward getting upset as he recognized his impairment. He wasn't fooling around. He really couldn't do it. I started wondering if maybe Edward was going to prove me wrong about there always being a way around or through a mental block.

"This isn't good, is it?" Edward said. "I ought to be able to say at least one negative thing about my dad, right?"

"That's right," I responded. "But you just made another left-brain response, and that's not going to help you. I suggest you keep studying the pictures of your interactions with your father. Take a long, hard look at what's happening. We can keep going over this until you succeed."

Finally, after four and a half hours with countless attempts, and faced with the prospect of leaving my office having failed,

Edward managed to describe how his father belittled him when they disagreed. This opened the floodgates. Suddenly, Edward could recount many of his father's unflattering traits. As it turned out, Edward's father was a sneaky, manipulative guy who had to control everyone else's life and never had a nice word for anyone. Edward managed to resolve the effects of traumatic mind mapping with his father through sheer will and perseverance. We just had to keep working at it until he finally broke through.

It is remarkable how people with traumatic mind mapping often picture themselves having a good relationship with their antagonist, even when their memories actually demonstrate the antagonist's callousness and ill-will. This disconnect can be blamed on the brain's way of constructing a narrative that makes terrible situations more acceptable. It's also amazing how, in contrast to folks' initial difficulties mapping their antagonist's' mind, things suddenly snap together, and they see things clearly once they get past this.

You have to be willing to do this too–to bang your head against the wall until the wall moves, if that's what it takes, and sometimes it does. The floodgates of your autobiographical memories open slowly at first and then with a gush. This process requires stubbornness and willpower from start to finish.

Use Your Left Brain To Corner Your Right Brain

Now is a good time to pick up where we left off with Rachel, the woman who couldn't describe her father's attitudes about women and sex. I want to describe a technique that requires some clarification, given that I've repeatedly stressed the importance of re-visualizing (rather than intellectualizing or introspecting about) traumatic mind-mapping experiences.

I helped Rachel use her left brain, which relies on deductive logic, to close in on her right brain, which sees everything in pictures. This differs from using your left brain to *avoid* looking at your pictures altogether. Using your left brain to force your right brain to see things it doesn't want to recognize corrects a problem called "neural network dissociation," wherein traumatic mind mapping causes your right brain and your left brain to stop communicating with each other. You can do this by confronting yourself with knowledge you've gained from repeated visualizations.

For example, Rachel told me that her father often made suggestive comments to attractive young waitresses whenever her family ate in a restaurant. I asked her to describe how her father actually did this. Rachel said he joked around with them, winked at them, or commented on their physical appearance. He'd often say how pretty they were and leave them a big tip. Rachel said she could tell which ones he was attracted to— typically, the young ones—because he never did any of this with older waitresses.

I asked, "How about your mother? Could she tell when your father was flirting with a waitress? Could she see the pattern?"

"I'm sure she could," Rachel replied. "She gave him disparaging looks whenever he made sexual innuendos, and she said he didn't have to put these girls through college with his tips. I thought she was generally accurate whenever she made these comments."

"So you could see what your father was doing, and so could your mother. And you recognized that your mother saw what he was doing too. How about your father? Could he see that his wife recognized he was hitting on these young waitresses?"

"He never acted like he did."

"I'm not asking how he acted. I'm asking about your father's mind-mapping ability. Did he know that your mother was tracking him and she didn't like what he was doing?"

"He must have. He often told her she was being paranoid and jealous for no reason. Then he'd continue what he was doing. Sometimes he had a sly smile like he enjoyed getting away with this."

"How about you? Could he tell that you also saw what he was doing?"

"I guess so. If I flashed him an incredulous look, he'd say, 'What's the problem with me being nice?'"

I shifted gears and brought our conversation to a head. "How would a prudent person interpret the pictures you're describing?

Perhaps your father isn't doing anything inappropriate with these young women, and that's why he doesn't recognize he's making his wife and daughter uncomfortable? Maybe he doesn't realize he's flirting, or he doesn't recognize when the waitresses are flirting back? Or maybe he's flirting with these waitresses, knowing you and your mother are tracking him. And he's doing it anyway? There are other possibilities. What do you think?"

Rachel took a long time to answer carefully. "I think my father was deliberately flirting with these girls, and he knew my mother and I were watching him. I think he enjoyed denying the obvious and telling my mother she was crazy."

"Was your father asking his daughter to be crazy too? To stay blind to him?" Rachel's answer took me by surprise. She said defensively, "I still don't know for certain that my father was having affairs!" It took me a moment to realize Rachel was suddenly playing three moves ahead. She saw where this was headed.

I replied, "Okay, but that's not what I asked. Will you answer my question?"

After a long pause Rachel said, "I can't think straight!" as tears streamed down her cheeks. "I feel like my mind is falling apart. I don't know what I think anymore. I'm so confused!" The floodgates were opening.

In our next session, Rachel described the time her father planned a wreck-diving trip in the Gulf of Mexico with two of his friends. Rachel and her mother decided to go along. Rachel's father agreed, but as they boarded the boat they discovered an attractive young woman was also going on the trip. This woman seemed as surprised to see Rachel and her mother as they were to see her. When it became clear Rachel's' father had invited the young woman on the trip, Rachel's mother stared daggers at him, but he acted like he had no idea what she was angry about.

Conditions in the Gulf were rough that day, so Rachel and her mother decided not to dive. Rachel's father and his two friends could handle the seven-foot swells and the strong current, but the less-experienced young woman had no business getting in the water. She tired quickly, panicked, and nearly drowned. As dive master for the trip, Rachel's father was responsible for deciding whether she was experienced enough to dive in such conditions. His poor judgment nearly got her killed, but he acted as if he were some kind hero when he got her out of the water.

"I think my dad enjoyed the whole thing," Rachel told me. "I'm starting to realize I've been blind to my father's cruelty. When my mother and I decided to go on the trip, he knew exactly what we were walking into; and he did nothing to prevent it from happening. I think he enjoyed the entire episode. I'm starting to think my father did have extramarital affairs, but I don't think he actually liked women. He sure didn't take care of his girlfriend by letting her get into the water. He also knew she was going to be thrown for a loop when my mother and I showed up—just like we'd be shocked to see this woman—and he let it happen anyway. My father definitely has mind-mapping ability. He knew what he was doing."

As Rachel kept talking she suddenly said, "I'm remembering something else my dad did when I was growing up. When I was 16, he got into a fistfight with one of our neighbors. The man was yelling and pounding on our front door until my father opened it. Then they started swinging at each other. The man got my father in a headlock and kept hitting him in the face. I'm visualizing the scene like you taught me. I can see Dad has a bloody nose. Whenever I've recalled this memory in the past, all I could focus on was the blood. This scared me! But now I'm realizing the man was yelling at my dad. He was screaming, 'You fucked my wife!'"

This treatment doesn't require 20 years of analysis. Traumatic memories that previously befuddled my clients now make sense. At first they feel pain and sadness, but they usually see benefits in improved functioning pretty quickly. This makes it easier for them to accept past traumatic events, even though what they're accepting is more difficult than the picture they've been holding onto. As hard as this can be, clients tell me it's worth it. They often describe it as life-saving.

After this revelation, Rachel's sex life began to change.

She now understood why she appeared devoid of sexual interest when she was actually highly responsive. Her father's unconstrained sexuality made exploring her eroticism and sexual interests feel unsafe as a young girl. There was no room for her to be overtly sexual. Rachel subsequently remembered her father's offering to pay for breast augmentation surgery for her when she was 18. Rachel declined because this felt creepy.

Some people report physical reactions to this treatment that involves mind mapping. They feel like things actually shift inside their heads. Some get headaches. People who have been chronically depressed often describe feeling "lighter." Rachel told me about seeing a little boy skipping down the street one day on her way home from work. She was having one of her frequent bitchy moods, during which she was notoriously bad tempered and lacked a good word for anyone. Her normal reaction would have been to somehow find fault, like criticizing the boy's mother for negligently allowing him to play without supervising him.

But on this day, Rachel said to herself, "That's a really cute kid! He sure looks happy!" At that moment, Rachel was overcome by a feeling that the world was somehow brighter than it had been before. She felt like there was more sunlight shining down upon her. The steady-state regression in which she lived was improving.

Subsequently, Rachel embarked on a path of sexual development that left her husband, Stephen, speechless. Rachel stopped masking her erotic mind. One time while they were lying in bed, Stephen asked her what she was thinking. "Well, quite honestly," Rachel said, "I was thinking about sucking your penis."

Between waves of shocked laughter, Stephen managed to say, "Wait! Stop! You can't do this!"

"Why not?" Rachel asked, feeling a little shocked and scared. Stephen laughed. "What have you done with my wife?!"

Rachel flashed him a wicked smile. "Believe me, *I am* your wife, and *you* had better hold on tight!"

I've seen many clients make remarkable improvements, although I don't take this for granted. It takes lots of hard work. But it's still sort of a miracle to me that people from very troubled backgrounds can make such remarkable recoveries.

So even if you've been repeatedly traumatized and feel like the world is a dark place, there's hope for you too. Now that I've laid out a clear methodology, you can start using this to help yourself. In our next chapter, I'll show you more things you can do to accomplish this.

13

Repairing Your Autobiographical Memory

Do you want an accurate picture of your life? We humans innately want to know who we are. *Who am I?* seems to be humankind's perpetual question, and our life experiences–particularly growing up–constitute our life's story. Having major gaps in your life story is not just an unexamined life, it's a life that lacks depth. It limits the meaning of your life story. The more complacent you are about the gaps in your autobiographical memory, the more likely you are to misunderstand who you really are or just invent yourself out of thin air. Like lots of people, you might want to "fill in the blanks."

On the other hand, many people don't want full pictures of their lives, particularly those from highly troubled backgrounds. If you grew up with constant traumatic mind mapping, you probably don't have any difficulty believing this at all. Your approach to life may be something like *Don't look back. What you don't know or can't remember can't hurt you. You can't do anything about it, anyway. You can't change the past. Put one foot in front of the other and keep moving forward.*

Ironically, if this is you, it's especially important that you develop an accurate autobiographical memory. I've learned that gaps in your autobiographical memory predispose you to holes in your mind-mapping radar, distortions in self-perception, and acute and long-term steady-state regressions that surface as emotion-regulation difficulties such as anxiety attacks, emotional outbursts, uncontrollable anger, and severe dysphoria.

Repairing holes in your autobiographical memory resolves the negative effects of traumatic mind mapping, and likewise, resolving traumatic mind mapping also repairs memory gaps. The methods for achieving these are interchangeable, so you can take care of both problems at the same time. Given what I've seen my clients do, I have every reason to believe you'll benefit from this chapter.

Keep a Person's Mind-mapping Abilities in Mind

I'm going to show you several ways to work with gaps in your autobiographical memory. All these methods hinge on your ability to detect someone's mind-mapping and mind-masking abilities and (antisocial) empathy (Chapter Eleven). You have to keep this in mind as you look back on your life because it changes the meaning of events. This often takes deliberate effort at first because your brain/mind wants to avoid this realization. Clarity about this is essential to correcting the negative impacts of traumatic mind-mapping experiences. You saw multiple examples of this in the previous chapter.

In the next chapter we'll focus on dealing effectively with destructive people–the kind of folks who create traumatic mind mapping for everyone around them. You won't be able to handle these people if you don't detect their mind-mapping and mind-masking abilities. But a bigger problem is not being willing to believe you're dealing with someone who has these skills.

Filling in the gaps in your autobiographical memory isn't just a matter of remembering more events, more mile-markers in your life story. It also involves the accuracy of events you already remember and think you understand. It's not too hard to jog loose more memories and have new things come to mind. The hard part seems to be detecting–*and remembering*–the mind-mapping abilities and antisocial empathy of others because this has everything to do with accurately understanding what happened.

As I've said, many clients come to me saying they don't remember anything about their pasts. However, others arrive with life stories they've polished in years of prior therapy–only to discover their "understanding" is inaccurate! Repairing the gaps in your autobiographical memory isn't the same as reinterpreting your preferred life story in terms of your therapist's preferred conceptualizations, interpretations, and theories. But whether you can't remember a thing about your past (and you might have grown up in a laboratory petri dish), or you're toting around a half-baked notion of who you are and where you come from, you can develop a more accurate autobiographical memory.

I'm often struck by how we overlook important, obvious details of our lives, hidden right out in the open. The meanings of these details are frequently disregarded or misinterpreted because their significance is lost when we approach them as "text"–ideas, theories, and psychobabble. But once you approach them visually and more granularly as vignettes, details of great importance start popping out to you. And the meaning of your stories frequently change. In retrospect, you'll be amazed you didn't recognize this sooner. As you'll see, it's probably easier to piece your life together than you might think. The difficult part is accepting what you learn.

AUTOBIOGRAPHICAL MEMORY GAPS
MAKE THIS DIFFICULT

One of the best ways to highlight this is to return to case examples we've previously discussed. For example, over the last two chapters you've been learning about Rachel, who progressively retrieved autobiographical memories, which explained why she seemed sexually disinterested when in fact she liked sex and was highly sexually responsive. Rachel remembered a vivid picture of her father's fighting with a neighbor who claimed Rachel's father had sex with his wife. It's easy to recall if you re-visualize the neighbor's punching her father in the face and yelling, "You screwed my wife!" This pretty much clarified that

Rachel's father had at least one extramarital affair, and multiple events suggested there were very likely others.

But Rachel exemplifies how you can recall things and still have gaps in your autobiographical memory: Armed with this information, did you let (make?) yourself revisit Rachel's past? Did you realize Rachel's father knew of his own sexual escapades while he was interacting with Rachel–including on their dive trips? Given his deceptions, Rachel's father obviously had mind-mapping and mind-masking abilities, so he could read Rachel.

How did he feel about Rachel's meeting his girlfriends? Or sexualizing his interactions with young waitresses during family meals in restaurants? He knew Rachel was watching him. What did he see in her reactions, and how did he feel about what he saw? How did he feel about acting like he was doing nothing wrong? What was his reaction to lying to Rachel, or watching her reaction when he called Rachel's mother a lunatic?

Why did Rachel's father offer Rachel breast augmentation? Rachel was surprised by his offer because she wasn't feeling particularly flat chested. Was he relating to her sexually? Did Rachel's body not measure up to his standards for women? Or did he want her to have a hot body he really desired? Was Rachel living in a house with a father who had ongoing sexual fantasies about her? How did he feel about Rachel's starting to have sex with boys her age, and how did this influence whom he allowed her to date in high school?

When you let yourself really look at this, you'll realize I'm not talking about remembering a specific event but rather a series of events that may have looked different at the time. If you make yourself map out Rachel's father's mind by answering these questions, you'll understand the world Rachel grew up in. You're not just seeing Rachel's father's dismal character or disgusting behavior, you're seeing Rachel's family milieu growing up.

You'd have to do the same thing with Rachel's mother to get the full picture of what went on. So what was she doing when she said Rachel was no better than her father, and Rachel had betrayed her? Was Rachel's mother blaming her for things she herself was unwilling to address directly? What did she think Rachel's reaction would be? What did she want Rachel to feel?

By forcing yourself to look at what you know more closely, you start to develop a more cohesive picture of Rachel's parents' minds. In doing this, you begin to see a more global picture of Rachel's childhood beyond a mere collection of a few events.

If you step back and look at it, Rachel grew up in a lot of cruelty. You didn't see that when you first heard about Rachel's Crystal Springs dive trip with her father at age 14. That's because, at that point, you didn't know Rachel's father had mind-mapping and mind-masking abilities or that he was having affairs and being sexually provocative but acting like anyone who accused him of this was crazy. You also didn't know that Rachel's mother was furious about this and willing to blame Rachel instead of stepping up and dealing with her own husband.

But knowing this changes your understanding of what was really happening in Rachel's life. You gain a deeper appreciation of what was actually happening in the living room between Rachel and her parents after the Crystal Springs dive trip. You begin to understand the swirling hostility, sexuality, and cruelty permeating her childhood world. Both of Rachel's parents were mind-twisters who were quite willing to exploit Rachel for their own personal benefit. Now you can understand the terrible traumatic mind mapping Rachel experienced in that event—and many others.

Right now your brain may be attempting to avoid traumatic mind mapping from looking at Rachel's life. Imagine if these were *your* parents! Then you *really* wouldn't want to believe they had mind-mapping abilities and used them to do awful things. Generally speaking, the closer your antagonists are to you, the more you'll have difficulty believing—or remembering—(1) they have mind-mapping ability, (2) they can anticipate or see the impacts of their actions, and (3) they likely have antisocial empathy and are enjoying what they're doing. When you have traumatic mind mapping, you can't imagine what your antagonists were thinking at the time or how they felt about your reactions.

You may be telling yourself *These parents were out of their minds! They weren't thinking straight! They must not have understood what they were doing, or the impact they were having!*

If so, your mind is trying to construct a picture of their minds being blank or incoherent or having no organized thoughts or feelings. In other words, you don't want to get into their heads.

Your brain operates like a self-organizing entity that's fully capable of protecting itself from overload. This effect is so powerful you might have experienced this yourself when I asked you to consider what was going through Rachel's parents' minds as they performed despicable acts. You probably felt your own mind recoil. Staying out of your antagonist's mind becomes an attempt to hold the experience at arm's length and minimize its emotional impact. Because of this, your brain is prone to erroneously assume other people lack mind-mapping ability and antisocial empathy.

Perhaps now you can see—and feel—why it's so important and difficult to accurately detect another person's mind-mapping ability. This reality check pushes you to consider explanations and motivations for a person's bad behavior that you would prefer not to think about. When reviewing memories from your own life, you need to continually ask yourself *Do the people in this memory have mind-mapping ability? Were they mapping me while we interacted? How did they feel about my reaction? How were they using their map of me to do what they were doing?*

Now let me show you how I take this a step further, reversing the direction of my analysis: As you increasingly appreciate how dark Rachel's father and mother were, do you begin to wonder about Rachel? It's not hard to see how she could become sexually inhibited, but is that all the impact you'd expect from this? If you're shrewd, you know there's got to be more. So where is it? That's how you should know you don't have a full picture of Rachel! That's how I know to look further than my clients want to show me.

Would you be surprised that Rachel often became emotionally vicious, dropping her alliance and "going for the jugular vein" when confronted? Or shocked that there was more to her sexual style than being inhibited? Rachel grew up in a sexual war zone. She was wired for sexual combat.

So, before you paint a mental picture of Rachel as a

"shrinking violet," add in that she completely masked her violent temper and vengeful behavior to the outside world. Being raised by two parents who perpetrated deceptions all the time made it likely she'd be a perpetrator too because she was wiring their thought patterns into her brain every time she mapped their minds.[1]

Detecting Holes in Your Mind-Mapping Radar

I'll show you how Rachel turned her life around a little later in this chapter. But first I want to give you a different example of recognizing autobiographical memory gaps through visualization. Let's return to the boy, Gregory, whose father left him on a desolate stretch of highway during a family vacation as punishment for misbehaving in the back seat of the car (Chapter Eleven).

When Gregory came to me for help as an adult, I asked him to look at this memory and describe what he saw in as much detail as he could. Gregory told me he was standing in the middle of nowhere, miles from the nearest city, with no shoes or shirt on and no money or phone.

"Let's back up the picture to before you end up standing alone on the side of the road. Can you see the family car?" I asked. "What kind of car is it?"

"It's a blue Chevrolet station wagon. Mom never liked driving it because it was so big."

"Can you see what you're wearing?"

"I'm wearing denim shorts because it's summer. Mom bought them for me for this trip. I'm not wearing a shirt because it's hot in the car."

"How do you get from sitting in the car to standing on the side of the road? Do you go there on your own?"

"Dad's yelling at me to get out of the car, but I don't want to go because I'm scared. He rips open the door, grabs me by my arm, and pulls me out of the car. I'm screaming, 'Daddy, don't do this! Don't make me get out of the car!' I'm doing everything I can to get out of his grasp, but this just makes him madder. As I'm getting out of the car, I trip, which he interprets as further defiance. He digs his fingernails into my arm as he drags me through the gravel on the side of the road."

"What does your father look like? What's the expression on his face?"

"He's red-faced, and his eyes and the veins in his neck are bulging out. He's shaking his fist at me!"

"How does your father feel as he's getting back into the car to drive away? Does your father say anything?"

"He's looking at me with a satisfied smile on his face. He tells me I deserve this, that no mother and father would ever want a kid as bad as me. He also says, 'If anyone stops to pick you up, don't get into their car because they could be bad people who want to hurt you.'"

"Can you see yourself in the scene? Tell me what you look like in the moment."

"I'm crying hysterically, begging him to let me back into the car. I'm promising I'll be good and never misbehave again."

"Would a prudent adult know you were petrified?"

"Are you kidding me? I've got tears and snot running down my face. I'm obviously scared out of my mind!"

"Does your father respond to your appearance in any way?"

"He says I should stop crying and acting pathetic because it's not going to get me any sympathy."

"Does your father say how this episode is going to end?"

"He says maybe they'll come back tomorrow and pick me up."

"How does it actually end?"

"A policeman picks me up, and we get off at the next exit to look for my family. We drive up and down a row of fast food restaurants until I see our car. The cop lectures my father and threatens to throw him in jail for child abuse."

"How does your father react?"

"It's the first time I ever saw him act apologetic. He says he realized what he did was wrong and he was on his way to pick me up. He says he lost his temper, and he won't do it again. But I know he's lying."

"How do you know that?"

"As soon as the cop drives away, he drags me behind the diner and gives me the spanking of my life, just like I expected him to."

"Is there anything else important about this incident? Give yourself a minute to look at the entire image."

"Nothing else jumps out at me. That's all I can see."

Gregory had far more information about his father's mind than he realized. Wouldn't you prefer to envision Gregory's father as simply losing his temper? However, Gregory's father demonstrated mind-mapping ability when he saw Gregory was terrified and told him to stop acting pathetic. His father also recognized that sympathy or compassion might be an appropriate response in this situation. Deliberately implanting pictures in Gregory's mind designed to maximize his anxiety and distress and destroy his feelings of self-worth suggests Gregory's father had antisocial empathy.

The same holds true for Gregory's father when telling him not to get into a stranger's car. Perhaps he was looking out for Gregory, but another possibility was that Gregory's father was tricking his son into refusing help from anyone who might stop and assist him.

In this one event, you can see multiple instances of traumatic mind mapping. It's almost overwhelming to picture this, but imagine going through this when you're eight years old! You may be having a disgust reaction as you realize what Gregory's father actually did to him.

But also notice that you no longer envision this scene as Gregory's standing alone on the side of the road, as Gregory initially described it. As scary as this might seem, it's almost peaceful in comparison to envisioning Gregory's father dragging Gregory out of the car, terrorizing him, and beating him after the police left.

Repair Autobiographical Gaps Through Re-visualization

At this point it's tempting to point out that, given the way Gregory's father's mind works, it is likely Gregory had many other traumatic mind-mapping experiences growing up. But this would draw our attention to additional vignettes, and I want to keep us focused on this one. It's certainly good to consider a broad range of experiences, but often people don't go deep enough into any given experience. One of the hardest autobiographical gaps to detect is where what you *do* remember blinds you to what you *don't.*

So, for instance, as you visualize the scene of Gregory's standing on the side of the road, how many people do you see? Of course you see at least one person–Gregory. You might see two people–Gregory and his father. But do you see anyone else? Are you only seeing it to the extent Gregory described it? Remember, Gregory said this was a *family* vacation. Did you realize someone is missing from this picture? Where was Gregory's mother in all this?

This demonstrates why, when I'm working with clients, I make myself see their vignettes as vividly as they do. This helps me understand their experiences much better. As I visualize these events, I experience feelings appropriate to their situation. If I'm struggling with this but my clients aren't, I know they're not really dealing with their own material.

Doing this also allows me to see what's missing or unclear in my client's descriptions. This allows me to recognize holes in their autobiographical memories. For instance, I knew there had to be more people in the vignette than Gregory and his father. When I asked, Gregory said his mother and sister were also in the car, as if it were an insignificant detail.

In the last chapter I described how you can use your left-brain deductive logic to walk your right brain into visualizing things you don't want to see. Here's another example: Remember that Gregory got in trouble for fighting with his sister in the back seat, which means his mother must have been sitting in the front passenger seat. I asked Gregory to tell me where he and his father were standing in relationship to the car before his father drove away. Gregory said they were both standing on the side of the road by the car's passenger doors, away from traffic.

By visualizing this, I realized Gregory's mother had a close-up view of Gregory's interaction with his father. Armed with this information, I asked Gregory to re-visualize the incident and describe his mother's face and her reactions to his being left on the side of the road.

"I really don't know what she's feeling because I can't see her face," Gregory replied.

"You're standing a few feet from her window. Can't you see her face?"

"I can, but I can't see the expression on her face. I only have a side view. She never looks at me once while this is happening. She just stares straight ahead."

"Does she ever speak up and object?"

"No, she's probably afraid of my dad's anger too. She probably feels terrorized like I do. Scared stiff! I guess she doesn't care enough about me to speak up anyway." Gregory said this like he was telling me the worst thing in the world.

"Well, you're starting off with the two best possibilities you can hope for—that your mother is terrorized, and she simply doesn't care about you—because all the other possibilities are worse." Gregory was stunned by what I'd said and didn't respond.

"The best way to resolve this," I assured him, "is to stop hypothesizing and focus on the picture. Look at your mother's face and tell me what you see."

"I can only see the side of her face."

"That's fine. What do you see?"

"I'm not sure."

"Does your mother look terrified? Even from a side view, it will be obvious if she is."

Gregory let out a long, sad sigh. "No...she doesn't look frightened or even upset.... She looks annoyed this is taking so long.... She almost looks bored."

"What you're reporting doesn't support the interpretation that your mother isn't speaking up because she's terrified of your father's anger. When you were growing up, did your mother ever intervene when your father got really angry?"

Gregory paused for several moments before answering. "I never saw my mother do this, but ten years ago my sister and I spoke about Dad's anger for the first time in our lives. She told me she only saw Mom take him on once."

"How did your sister describe this incident to you?"

"She told me Mom said, 'Paul, you're out of control! Go to our bedroom and calm down! Stay there until you get a grip on yourself!'"

"What did your father do when your mother spoke up?"

"He did as he was told. He went to their bedroom and closed the door."

Obviously, Gregory's mother knew she could handle Gregory's father if she wanted to. This single fact added whole new dimensions to Gregory's understanding of this episode's impact on him. That's why I said Gregory's proposal that his mother was terrorized was the best alternative he might hope for. Unfortunately, Gregory's mother, like his father, had mind-mapping ability and antisocial empathy. She was aware of what was happening to Gregory. By saying nothing she was participating in Gregory's being traumatized by his father. At best she was indifferent to Gregory's suffering. The other alternative was she enjoyed Gregory's terror, just like Gregory's father did.

Now that you understand this, your picture of Gregory's traumatic mind-mapping experience is probably more powerful than before. You can't appreciate the full disgusting nature of what happened until you put Gregory's mother into the picture. Nobody, particularly an eight-year-old boy wants to do this, even when he's a grown man. Your mind only wants to consider the best view possible, which, in this case, means you don't realize Gregory's mother is present.

Gregory's story shows how you can spot holes in your autobiographical memory by visualizing a scene from different perspectives. There's no point in continually reviewing traumatic mental images from the same old perspective. Look around inside the image and observe what you're *not* seeing. If need be, adopt a third-person viewpoint to get a different perspective of the scene.

Here's another quick example to show how you can do this: Previously I told you that I had a client whose father broke down the bathroom door while she was showering at age 13. He dragged her out of the tub, pinned her to the floor, and nuzzled her breast buds with his nose. Whenever she reviewed this in her mind, she only saw the two of them. This changed when I asked why her father wasn't afraid of her mother's reaction.

That's when she realized her mother was "off camera," but very much a part of the picture: Her father apparently thought her mother wouldn't do much about it.

MEALTIME MEMORIES ARE RICH RESOURCES

From time to time, when clients say they can't remember anything in their past, I ask them to describe a typical family dinner when they were growing up. Virtually everyone can do this. Even families who don't get along or who don't spend much time together still eat together occasionally. This doesn't have to target a specific event or involve something bad. It's open and inclusive.

Working with a picture of a remembered meal can be very informative. As you saw in our previous examples, start by describing the setting as if you're actually there in real time. Do you usually eat in a formal dining room, in a breakfast nook, or at the kitchen table? What's the shape of the table? Who is sitting where? How are people interacting? Once you start describing interactions between family members and the atmosphere surrounding them, it's amazing how many other things you start to see that you hadn't before.

Lots of families eat together, but no two families eat together the same way. Even if two families eat turkey for Christmas, it's not the same meal (especially if one family is Jewish). There are always distinguishing features: What's the mood at the table? Are you all relaxed and laughing, or are you grim and tense? Are you looking at each other or staring at your plates? Who directs the conversation? What gets talked about? Who participates? What are the typical interaction patterns? What are people's attitudes towards each other? If everyone is silent, what does the silence mean? How does the meal usually end? Do you linger at the table or are you eager to leave?

Let's return to another case we've previously discussed to see this in action: In Chapter Nine, I told you about Laurie, whose mother made frequent derogatory comments about her body when she was an adolescent. Laurie's mother's comments were always delivered as pseudo-compliments, but the real emphasis was an underlying criticism. "You're such a pretty girl," she'd say, "but you'd look even better if you lost ten pounds."

Since Laurie's mother's criticisms often focused on food, I asked Laurie to show me a picture of a typical family meal when she was growing up. Laurie described sitting at an oval dining table with her sister, mother, and father. Her parents always sat at the heads of the table; and Laurie and her sister, who was four years older, sat opposite each other on the long sides. Laurie's father always sat on Laurie's left, her mother on her right. I asked her, "Please describe the interactions you see occurring at the table during the meal."

Without hesitation, Laurie replied, "I feel like my mother monitors every morsel I put in my mouth. She'll offer me dessert, but she looks disapproving when I eat it. My mother also never stops talking while we're eating. She goes on and on about what happened during her day, who she talked to, and what she purchased. Or, she pontificates about current events. Sometimes it seems like the topic doesn't really matter. My mother just likes to hear herself speak. She also likes to ask me and my sister questions about school. We have to tell her one new thing we learned that day. She says our grades don't accurately reflect how smart we are. She says we're probably the smartest students in the entire school. My father rolls his eyes when she says this; but as usual, he doesn't say much."

"What's happening with you and your sister?"

"We're looking at each other with expressions on our faces that say *Isn't Mom impossible?! Can you believe she keeps going on and on like this?*

"Does your mother see you do this?"

"She doesn't appear to. Wait, no, I take that back. I remember she sometimes gives us admonishing looks before returning to whatever she was babbling about."

"Where's your father in all this?"

"My father says almost nothing. My mother keeps the conversation going all by herself, except when my sister and I answer her questions. There's no point in asking my father anything because all he does is growl two-word answers, or he just stares at you and scowls."

"Does your father ever say anything to you or your sister?"

"Only when we do something wrong."

"Like what?"

"I remember my sister and I spilling glasses of milk a few times. Whenever this happens, my father explodes with anger and starts yelling. He bangs his fists on the table so hard the silverware dances across the table. Sometimes my sister and I have to hide our smirking when this happens."

"You said your father yells. What does he yell?"

"That we're stupid, worthless girls, with no more brains than our moron mother."

"Look at your picture. When your father says that, what's your mother's reaction?"

"She doesn't have a reaction."

"What do you mean she doesn't have a reaction?"

"She doesn't say anything."

"She might not say anything, but that doesn't mean she doesn't have a reaction. Look at her face. Tell me what you see."

"I see stone-cold hatred."

"When your father says you're as dumb as your mother, does he ever glance at your mother to see her reaction?"

"No, he just looks down at his plate, hunched over the table while he continues eating."

"Is he worried about your mother's reaction?"

"Not a bit. He completely blows her off. He makes all the money. She's a full-time stay-at-home mother. He knows she couldn't get a job to support herself. I don't think he even cares if my mother withholds sex. They slept in separate bedrooms for years."

"What's your sister doing in this picture?"

"She's giving my father the finger under the table. She hates him and can't wait to leave the house. She ended up getting pregnant when she was 17. My parents made her get married and keep the baby. My mother almost died of embarrassment."

"And your father's reaction?"

"I think he made my sister get married because he was so pissed at her for having sex."

Notice the tremendous information in Laurie's picture of eating at the dinner table. Everyone in her family clearly has mind-mapping and mind-masking ability. Laurie's self-preoccupied mother needs to be the center of attention while her misogynistic father keeps a low profile before exploding. Laurie's mother and father hate each other and have a nonexistent sex life. Laurie's father has her mother accurately mapped. He knows he can treat her like crap and she's not going to do anything about it. Meanwhile, Laurie's sister is masking how much she hates her father. She gets pregnant and her parents retaliate by making her get married. All this shows up in a description of a typical meal.

Some clients initially balk at visualizing a family mealtime scene. They have an immediate negative reaction. They'll flatly declare, "I don't care what you say, Doc, I'm not doing that!" That's because mealtimes often have a visceral feel–sometimes even a remembered smell or taste–that triggers profound memories. These folks learned to sit at the table and suck in their emotional "feelers" to avoid whatever was going on around them. They often plod through the rest of their lives like this. Visualizing a typical mealtime scene dredges up powerful memories and reactions, which is why they don't want to do this. Ironically, they are the ones who most benefit from forcing themselves to do so.

Cross-Reference Memories of Traumatic Events

Another powerful method for repairing autobiographical holes involves cross-referencing traumatic events. For example, this could involve examining consistencies and dissimilarities between experiences with your antagonist over time. For instance, I had a client, Phillipe, whose sister, Marianne, three years older, terrorized him physically and emotionally when he was a child. This suddenly stopped for no apparent reason when Phillipe was 11. At the time, Phillipe wasn't about to question Marianne about the change, fearing she would start harassing him again. But as a grown man trying to understand his life, I suggested Phillipe ask her about this now.

When Phillipe brought this up, Marianne didn't say anything at first, she just stared at him. Eventually she said,

"When I was 14, Dad started having sex with me. It made me feel worthless and dirty. The only way I could feel like I had any value was by taking care of you." Phillipe learned a lot more than he ever expected by cross-referencing events in his life with those of his sister. A shocking new view of his father suddenly emerged. Phillipe started looking for other major gaps in his autobiographical memory now that he had certain proof they existed. Best of all, this also opened the door to a more meaningful relationship with Marianne.

For another example of the power of cross-referencing data, look no further than earlier in this chapter when we reviewed Rachel's experiences with her father. Remember what happened when we plugged her father's mind-mapping abilities and antisocial empathy into Rachel's pictures. Cross-referencing this with his behavior—taking his girlfriend diving with Rachel's mother and Rachel, for example—allows you to see the darkness of his mind. This lets you appreciate the simultaneous levels of traumatic mind mapping occurring for Rachel in each vignette.

But I have an even better example of the power of cross-referencing information. It actually concerns Laurie, the woman we discussed a minute ago, who described her typical family dinner with her mother babbling inanely, interspersed with hostile comments about Laurie's appearance. Can you see Laurie's father, the misogynistic man with an explosive temper, hunched sullenly over the dinner table? Here's what you need to know: He happens to be the same man I described earlier who broke down the bathroom door and nuzzled his daughter's breasts. This was Laurie.

Putting these two pictures together is like, POW! Suddenly, Laurie's world comes together in your mind in a more meaningful and powerful way. Now you understand why Laurie's father wasn't worried when he broke down the bathroom door, not that this knowledge is easy to live with. It was part of the pre-existing marital and family dynamics in their home. Laurie's father had Laurie's mother thoroughly mapped, and he discerned she wouldn't do anything serious to disrupt the family routine or "bring shame upon the family." He knew he had everyone in the house intimidated by his surly manner and explosive temper. He acted impulsively when he broke into the bathroom, but he was cunning and calculated—and he was right. Laurie's mother pretended it never happened.

So now can you see what led up to Laurie's father's breaking down the bathroom door? To do this you have to cross-reference some of the facts you know: Can you see the buildup of sexual tension in the house, as Laurie's older sister started having sex and ultimately got pregnant? Remember Laurie's comment that her father vindictively made her sister get married because he was angry at her for having sex? Did you envision Laurie's father as a starch-necked moralist? That's the wrong picture. In the family dinner scene, Laurie's mother wasn't the only one focused on Laurie's body.

Given his subsequent sexual abuse of Laurie, it's more likely Laurie's father had sexual impulses towards his older daughter first—impulses he ultimately directed at Laurie. With this much sexual energy floating around the house, wouldn't you think a grown woman like Laurie's mother would notice, especially since she wasn't sleeping with her husband? Did Laurie's father enjoy himself? Did Laurie's mother make excuses for him when this happened? These are the minds Laurie was mapping in the midst of this terrible experience.

I know you'd rather believe Laurie's father's mind just "snapped" one day, that he "went insane," and "he didn't know what he was doing." When your mind seizes onto this picture, the broader context disappears, leaving you with explanations like this that are easier to accept. This is why people adopt a myopic focus on sexual abuse, picturing only the moments of inappropriate physical contact and only envisioning two people. This is less upsetting than recognizing the broader daily milieu of minds which makes such events possible. However, repairing holes in your autobiographical memory requires cross-referencing what you know about the various people involved to assemble the whole sordid picture.

ANALYZE CORRESPONDENCE

We've been seeing how people's life stories take on deeper meaning as they fill in the holes in their autobiographical memories. I want to give you three additional powerful ways to accomplish this which involve different ways of mapping your antagonists' minds. Besides cross-referencing existing information, another effective way to retrieve mind-mapping information is to analyze past or current correspondence with your antagonists.

Let's return a final time to Rachel, the woman we've been discussing since the last chapter. Rachel decided to confront herself about her past by sending an email to her father.

She wrote: "Dad, I'm in therapy trying to sort myself out and understand who I am and where I came from. I have a recollection of you and me going diving at Crystal Springs when I was 14, and meeting a woman you seemed to know. We all dove together because it wasn't safe for her to enter the caves without a buddy. When we got home, Mom was angry at us. Who was this woman? Anything else you can remember about this trip would be useful to me. Thanks in advance for your help."

Rachel's father wrote back: "Hi Rachel, thanks for writing. I haven't heard from you in a while. I do remember the incident you're describing because your mother became irrationally livid, and I was concerned how that would affect you. Let me assure you everything between me and this woman was completely innocent. She was an acquaintance I met through a friend. I wish I could tell you why your mother was so paranoid, but her whole family was like that because her father had several affairs. Other than this, I don't remember much. Hope this helps. (signed) Your loving father. "

When Rachel showed me her father's email, she said this was a dead end because her father couldn't remember any more than she could. However, after reading it myself, I disputed her interpretation. Rachel wasn't dealing with the load of information embedded in her father's reply.

I pointed out he never asked her about her problems or how they might be related to this incident. He blamed Rachel's mother's family-of-origin for her suspicions about him, demonstrating some level of psychological sophistication. But he never touched the topic of Rachel's own suspicions. He related to Rachel as if they were on the same team, and therefore he should get the benefit of the doubt. He then ended the conversation without offering to discuss this further in the future.

Rachel was shocked by the difference in our interpretations of her father's email. She recognized how much she had overlooked. Her eagerness to prematurely close off further discussion with her father jumped out at her.

After due time to reconsider, here's how Rachel responded: "Hi, Dad. Thanks for getting back to me. Since I wrote to you last, I've recalled another dive trip we went on together, this time in the Gulf of Mexico. The water was really rough, and I remember throwing up twice. It was you, me, Mom, your two dive buddies, Sam and Vinny, and a young woman I didn't know who you invited on the trip. She was inexperienced and panicked in the heavy current and big swells. Who was she? When Mom and I showed up for the trip, she seemed surprised to see us. I realize my questions put you in an awkward situation, but I hope you'll bear with me, so I can get healthier."

Rachel's father wrote back. "As always, it's a pleasure to hear from you, Rachel, no matter the circumstances. I'm glad you remember how much I loved diving. I don't get to do it much these days, but I have great memories of us diving together when you were a kid. Your mother never really liked diving, and I think she was jealous whenever you and I went off together on a trip. As for the particular trip you mentioned, how could I forget it? The Coast Guard said that, if I wasn't such a strong, experienced diver, that woman might have drowned. She was a friend of a friend who was interested in diving. I was just being nice when I invited her to come along. Everything was perfectly innocent, but your mother made things uncomfortable for everyone. I didn't want to create a bigger scene, so I put on a game face like nothing was bothering me and tried to make the best of it. Hope this helps."

This time Rachel was able to see more in her father's response. She recognized his attempts to instill pictures that they had good times together, that he'd been a hero and a considerate man, and that her mother was the enemy. Around this time, Rachel remembered her father's fighting with the neighbor who accused him of having sex with his wife. Armed with this information, Rachel's response was more pointed than her previous emails.

"Dad, through our email exchanges I've come to a new understanding of what happened in our family when I was growing up. I now realize you probably had multiple affairs, and maybe Mom wasn't as crazy as you led me to believe. I always thought Mom was my enemy and you were my ally, but looking back I see lots of times I was on your side, but you weren't necessarily on mine. I remember thinking how odd it was for you to offer to pay for breast augmentation surgery for me. That's a strange thing to offer to your daughter, especially when I wasn't flat. When you did that, it made me not want to look attractive in front of you. I'm telling you this so you know I'm waking up, not so you can agree or disagree with me. I can see who I am, and who you and Mom are, much better now. What happens between us now depends on how you respond. I suggest you give this a lot of thought."

Rachel's father responded: "So good to hear from you again, Rachel. I don't know how you've come to these conclusions. All I ever wanted was for you to have a wholesome childhood. I know how important this is because, as you know, I didn't have one. But you can't fault me for not trying because you know I did my best. For whatever mistakes I might have made, I apologize from the bottom of my heart. I hope you'll forgive your aging father during whatever time I've got left."

I hope you've learned enough about mind mapping and antisocial empathy to recognize what Rachel's father is doing. If you think he apologized for anything, you're not tracking him accurately. He's telling Rachel several disappointing things: He's not going to confront himself or deal with Rachel about what he's done. The topics of sexual abuse or extramarital affairs are off the table. He insists on being seen as having done the best he could and operating out of good intent. He acknowledges nothing but expects forgiveness for all shortcomings. He's willing to act pathetic to play on Rachel's guilt.

All this amounts to Rachel's father's saying *Look, kid, my functioning is not going to come up. It's my way or the highway.*

If you want a relationship with me, these are the terms. Here's how I expect you to relate to me, even if you have to lie to yourself to do it. Take it or leave it. This is the classic response of parents who do disgusting things when their children start to wake up.

In the next chapter I'll show you how to handle tough things like this. But before we go there, I want you to have two other powerful ways to repair your autobiographical memories by mapping your antagonists' minds. You've just seen how analyzing written correspondence can give you a better idea of what went on in the past and what's going on now. There are other ways you can slow down the action and study what your antagonists say and do, and how they say and do it.

ANALYZE AUDIO AND VIDEO RECORDINGS

I've also worked with clients who have audio and video recordings from their childhoods. More often than not, these recording include parents, siblings, and even grandparents. You get an actual sample of people's facial expressions, mannerisms, tones of voice, speech patterns, and behavior, in addition to the actual content of the conversations. This allows us to take a fresh look at old issues and use our hearing to decipher what was going on. This also permits me to have independent access to the original data.

With this in mind, let's return to Roger, the boy who walked in on his father having sex with their maid when he was an eight-year-old. Roger actually had old videotape cassette recordings which his father, an amateur photographer, had made. For now, I'll just describe the visual images captured on videotape.

The first video was taken at the beach when Roger was a little boy. In the first scene, Roger's mother is putting suntan lotion on him while he squirms and complains. Then she's trying to feed Roger baby food, which he apparently doesn't want to eat. In the next scene, Roger's father is flexing his muscles, applying suntan lotion to himself, and generally showing off. He complains that Roger's mother isn't holding the camera steady enough. Then you see Roger and his mother playing on the shoreline. In the last scene, as the family packs up to leave, Roger's mother is trying to comb his hair while Roger fights her off.

The next video was from Roger's high school graduation. From behind the camera, Roger's father tells him to move to a spot where the sun will shine so his face isn't in shadows. Roger's mother ignores this request, making Roger stay where he is while she fusses with his hair and gown. Roger tries to fend her off but to no avail. In a stern voice Roger's father tells Roger's mother to get out of the way, but she ignores him and continues fawning over Roger.

The third video was taken at Roger's parents' thirtieth anniversary party. At the outset, Roger's father acts like the event's impresario, but in every other scene Roger's mother clearly is the queen bee. Even when standing together greeting guests, Roger's father seems marginalized while Roger's mother is always center stage. At one point, Roger's father extends his hand to shake hands with a guest, and Roger's mother cuts him off and takes control of the interaction. Roger's father glares at her in anger but says nothing, and Roger's mother appears not to notice.

Do these videos help you dial in what was transpiring when Roger walked in on his father having sex with their maid? There's a good possibility Roger's father thought he was defying his wife's authority and control, and expressing his anger at her. He apparently didn't dare to confront her directly. Rather than demonstrating how powerful he was, Roger's father was a little man who enlisted his son as a co-conspirator. In Roger's recollection of his father's smiling at him, he was being sent the message *This will be our little secret. Just between us guys. This will make us pals. Don't tell your mother. It's us against her.*

Other things came out on the tape that Roger hadn't talked about, like the nature of his relationship with his mother. In the eyes of Roger's mother *and* father, Roger was her property. It finally became clear why he never told her about his father's romp with the maid: This would have drawn him in further with his mother, which he wanted no part of. In the videos Roger seemed to be constantly fighting off her physical intrusions, the same way he fought off her emotional manipulations and preening behaviors as an adult. This helped Roger get a hold of his life-long history of difficult relationships with women.

Now can you see the backdrop to Roger's watching his father

have sex with the maid? From Roger's perspective, his mother is there too, and he's mapping his father's mind as well as hers. He's having multiple levels of traumatic mind mapping. And if he were emotionally close with their maid, this adds another dimension of traumatic mind mapping. Roger had to cross-reference all these levels to have accurate autobiographical memories and resolve the impacts of traumatic mind mapping.

Have Mental Dialogs With Your Antagonist

Another great way to resolve traumatic mind mapping and repair gaps in your autobiographical memory involves having a mental dialog with your antagonist. I'm not referring to having an argument in your mind as you're driving down the road. We all do this, and it doesn't help. I'm encouraging something completely different.

I ask my clients to write the dialog in a Word document, so it looks like the script of a play, and then we carefully study it together. If you want to try this, each character (you and your antagonist) has his or her own voice. It's not a summary. It is move-by-move, statement by statement. Your antagonist says this. You say that. Your antagonist responds (or not). You do or say what? Your antagonist replies how? And so on. General length runs two pages.

Studying mental dialogs turns out to be a great way to pinpoint exactly where your mind mapping is collapsing. It makes no difference whether your antagonist is dead or alive. Death doesn't end the impact of traumatic mind mapping nor remove the opportunity to resolve it. In the next chapter I'll show you how to deal with difficult parents if they are alive. But you can work the process I'll describe here if they are deceased.

Talking to dead people is certainly not the way I was trained to do therapy. But this method means it's never too late to straighten out your brain, and it's never too late to deal with your past. Never.

For instance, think back to the last chapter about Tim, who watched his parents wrestling on the floor as a young boy growing up on a farm. As you remember, I had Tim shift to a third-person view in order to see their faces in his revisualization. This produced powerful information that Tim (like many clients) felt he didn't know how to handle when dealing with his parents. The thought of talking to his parents about this turned his brain to mush.

So I suggested Tim have a mental dialog with his parents about that scene and write it all down so we could look at it together. We could look at what he felt he needed to say and how he thought his parents would respond. I knew this would document Tim's maps of their minds. Here's what Tim wrote:

Tim: "What was going on there? Why were you and Dad fighting like that?"

Mother: "Well, son, that was such a long time ago. I can hardly remember it."

That's it. That's as far as Tim got on his first attempt. Like many clients, as far as Tim was concerned, that was the end of the conversation. But having seen this many times before, I knew he was afraid to confront his mother because he knew what was going to happen if he did. Tim had anticipatory traumatic mind mapping. He was soft-pedaling his questions, and he was too ready to give up for fear of eliciting a negative reaction.

My job was to keep Tim moving forward. "Why aren't you talking straight to your mother?" I asked him.

"She says she can hardly remember what happened."

"Well, if she can *hardly* remember, that means she remembers *some* of it. Why don't you ask your mother what she does remember?"

The following is Tim's second attempt at a mental dialog with his mother:

Tim: "Mom, did you know I saw you wrestling with Dad when I was little?"

Mother: "Why would you want to know that?"

Tim: "I've just always wondered about it."

Mother: "Don't waste your time thinking about that. Why don't you think about happy things instead?"

When Tim showed me this at our next session, he

apologized for not coming up with much. "There's enough here to get started," I assured him. The mother in Tim's head was shielding her mind, deflecting his question with another question (basically *Why do you want to know if I saw you watching us?*) and telling him not to think about this.

So I asked Tim, "Why didn't you say, 'Mother, you're not answering my question. Did you know I was watching you and dad wrestling on the floor?' That's a much more effective response." Tim agreed and said he'd do better next time. And he did. Here's how his third try went:

Tim: "Why were you and Dad wrestling on the kitchen floor? Did you know I was there?"

Mother: "Why do you want to know that?"

Tim: "I'm trying to make sense out of my past."

Mother: "What makes you think this will help you now?"

Tim: "I'm not sure, but I think it might."

Mother: "You can't be absolutely certain, can you?"

Tim: "No."

Mother: "Well, I wish I could help you; but I don't think this is going to accomplish your goal."

Tim: "Will you just answer my question?"

Mother: "I wish I could."

It's amazing how these mental dialogs evolve. As you can see, Tim's mental map of his mother suggests she's an expert mind-masker. She fends off Tim's questions in ways that don't reveal what she's thinking or feeling. She prevents her mind from being mapped. People can't mask their minds like this if they don't have mind-mapping ability.

We've already discussed how people often suppress their functioning in order to maintain relationships with dysfunctional people. You know that, if you show these people you can see them for who they really are, that will be the end of their relationship. We saw this very thing come up in Rachel's email exchanges with her father. Well, this also shows up quite elegantly when you create mental dialogs with your antagonists. For example, here's Tim's fourth try:

Tim: "Why were you and Dad wrestling on the kitchen floor?"

Mother: "What makes you think this happened?"

Tim: "I saw the two of you. I was standing there. Why were you wrestling?"

Mother: "We weren't wrestling, we were talking."

Tim: "When you tell me you were talking, you know the event I'm referring to. Why were you and Dad fighting?"

Mother: "I don't like your tone, young man!"

Tim: (unapologetic) "I don't like you not answering my question. Why were you and Dad fighting?"

Mother: "That's between me and your father!"

Tim: "No it's not. From what I saw, I'm very much involved, as far as the two of you are concerned. "

As you can see, Tim was standing his ground more with his mother. He's also alluding to his mother's deliberately getting pregnant against his father's wishes but leaving room to see if his mother will actually bring this up. You also get a clear picture of Tim's map of his mother's imperious style. The additional information that comes out in these mental dialogs is pretty amazing. But instead of having a mental conversation in his head and leaving it there, Tim decided to write his mother to see if she was different than his map of her.

Here's what he wrote: "Dear Mother, How are you? I hope you're feeling well. I know it's hard to quit smoking when you've done this all your life. I hate to impose on you, but I'd like to ask you about a childhood memory that's stuck in my mind all this time. One time, when I was about five or six, I walked in on you and Dad angrily wrestling on the kitchen floor. I don't even know if you were aware I saw this. Were you? I don't want to pry into your marriage, but I'm really curious to find out what was going on right then."

Here's what Tim's mother wrote back: "Dear Timmy, It's good to hear from you. It's been months. I wondered if I was ever going to hear from you again. Your brothers and sisters only contact me when they want something. I'm glad to hear you're thinking about me, but I didn't expect you'd be thinking about all the horrible things I went through with your father. The time you're referring to is a perfect example. I'd just found out he was intimate with a woman who lived in town. I confronted him, and he lost his temper. I didn't do anything to warrant this sort of response. I was completely terrified and just trying to protect myself. I put up with a lot so you and your siblings had a nice home to grow up in. No, I didn't know you saw your father abusing me, but if you did, why on earth didn't you rescue me? I can't believe you never told me that you saw what your father did to me. This could make a mother seriously doubt that her son really loves her, but I'm sure I'll get over it. It would help if you'd come to visit."

No doubt Tim's mother's guilt-inducing moves aren't lost on you. It had actually only been a month since they'd talked last. You can clearly see the positions she wants to maneuver Tim into and out of. She wants him on the defensive and feeling like he hasn't done enough for her. She wants him to visit and contact her more often. She wants to be seen as the victim. She can be vicious in her pursuit of the moral high ground. She not only has mind-mapping and mind-masking ability but antisocial empathy as well.

As you can see, Tim's mother didn't do any of the mind masking ("Why do you want to know that?") he anticipated. She also wasn't viciously imperious with him, although it comes through towards Tim's father. Instead, Tim's mother wanted him to map her mind—meaning the picture of her mind she wanted to implant. Tim's mother was a class-four mind twister.

This didn't mean Tim's map of her was wrong. Tim's mother's email is a good picture of how her mind works when she feels free to structure the situation however she pleases. The mother in Tim's dialogs was dodging being pinned down on a particular topic she was defensive about. Tim's initial message to his mother was tentative enough to avoid creating that exact situation. In fact, Tim's ability to gauge this reflected the accuracy of his map. When Tim did pursue this with his mother, she reacted exactly as he predicted in his dialogs.

As disappointing as this was for Tim, it left him perfectly positioned to finally deal with his mother. (Rest assured he also had to deal with an equally difficult father.) Resolving traumatic mind mapping often involves more than realizing you can't trust your parents. You may also have to interact with them because the process of dealing with them *effectively* repairs your brain. Exactly what *effectively* entails is our next chapter's focus.

But we needn't delay recognizing that dealing *effectively* with your antagonist always involves several things: not losing your nerve, having the best in you drive your willfulness; and struggling to keep your brain from turning to spaghetti when dealing with a beloved antagonist who has mind-mapping and masking abilities and antisocial empathy.

Re-visualizing past events with your antagonists' mind-mapping abilities included, studying actual conversations, reviewing audio and video recordings, and conducting mental dialogs are all ways to resolve traumatic mind mapping and remove holes in your autobiographical memory. Doing this makes you less susceptible to subsequent traumatic mind mapping. It also helps you make what I call *"gold-standard responses"*–keeping up with your antagonists in real time. This is another big part of responding *effectively*, which I'll tell you more about in our next chapter.

14

DEALING WITH DESTRUCTIVE PEOPLE

I consider this the most important chapter in *Brain Talk*. This is ground zero. Everything we've covered thus far has been preparation for this point. We're going to address how to deal with the most unpleasant, manipulative people in your life–the ones who create destructive interactions that inflict traumatic mind mapping.

I'm going to show you how to handle these fiascoes to resolve their damaging impacts. I'll illustrate how to handle yourself, what you should look for, and what you should do. These methods also help resolve prior traumatic mind mapping you may have previously experienced with these people.

These methods can radically change your relationships with the most difficult people. Even if they don't change, *you* will change, and your growth will significantly alter these relationships.

DEALING WITH PEOPLE WHO DO DISGUSTING THINGS

Let's begin our in-depth example of what I mean. Like many couples I see, Doug and Tina were on the verge of divorce. They were in their early thirties, and Doug complained Tina was more married to her parents than to him. Tina retorted that Doug pushed her to always put him first. Moreover, Tina complained, Doug asked her to do things that would hurt and disappoint her parents, to whom she felt tremendous loyalty.

Doug said their most important marital decisions–like whether or not to have kids–should be made between the two of them. He felt Tina's parents acted like they had a vote in this matter.

Rather than denying her parents were doing this, Tina argued that what they were doing was understandable because they wanted to be grandparents. Tina described her parents as over-protective but well intentioned and thinking they knew what's best for her.

This led to arguments between Doug and Tina over their latest incident. Tina's mother had called Doug to discuss his feelings about having children. During the call she revealed she was privy to the details of prior conversations between her daughter and son-in-law, including things Doug had asked Tina not share with her parents. Doug was furious she had violated her agreement, but he felt obligated to respond to Mother's questions and listen to her dissatisfactions about not having grandchildren.

Doug claimed this was another example of Tina's being more married to her parents than she was to him. Tina replied that she had the right to talk with her parents and that her parents had a right to know what was happening with their child. She said Doug misinterpreted things because he had not grown up with his own biological parents.

BACKGROUND

Tina was a young woman who seemed more adolescent and naive than others her age. I subsequently learned that her parents had made most of the major decisions in her life, which they believed to be absolutely appropriate. She usually spoke with her mother daily, one of many signs of an unresolved emotional fusion with her parents.

Tina lived in fear of disappointing her parents. She knew what they wanted, and she usually managed to give it to them. This suggested Tina had mind-mapping ability. On the other hand, Tina never saw her parents' motivations as anything less than ideal. Tina justified everything they did with rationalizations that her parents were acting out of concern and caring for her, even if they were sometimes misguided. Tina sent me subtle messages not to challenge these interpretations.

Doug had no family of his own. His mother was deceased, and his father had left the family shortly after Doug was born and had not been heard from since. From the age of four, Doug had been raised by his maternal grandparents while his mother lived and worked in another city. She occasionally visited Doug until she'd died when he was eight.

Luckily Doug's grandparents took some interest in him. The little money they left him upon their death allowed him to go to college and fulfill their admonition to get an education and make something of himself. Doug came away believing that his grandparents wished him well, and he just didn't think about his parents much. This also left him hungry for family.

So when Tina's parents talked about his being part of their family, this was music to Doug's ears. For the first several years, Doug took every opportunity to act like a loyal son-in-law, focusing on family unity and harmony and going out of his way to get along. By the time he and Tina came to see me, Doug had started to object.

Our initial sessions involved Doug and Tina's arguing about their latest incursion with Tina's parents. For example, the parents had told Tina she and Doug were invited to dinner, and Tina agreed before checking with Doug. I was struck by Tina's seeming inability to anticipate Doug's negative reaction, which I could see as soon as they started relating their story.

On the one hand, it seemed like Tina had a mental map of Doug's mind. Tina talked about how Doug's mind worked when she wanted to explain away his negative reactions and justify her own behavior. She often brought up his lack of contact with his parents and how this must have affected him. On the other hand, Tina seemed to lack a map of Doug's mind that allowed her to take his likely reactions into account when dealing with her parents.

PARENTS MESSING WITH COUPLE'S DECISION TO HAVE A CHILD

When Tina and Doug showed up for the family dinner, Tina's parents asked Doug when they were going to be grandparents. Doug didn't know what to do. He was furious about being put on the spot and having his marriage discussed at the table. However, he felt obligated to be civil and conciliatory. Doug looked at Tina, hoping she would tell her parents this was not appropriate dinner conversation. However, Tina looked down and refused to meet his gaze, as if to say *They're talking to you. This doesn't involve me.* All Doug could do was manage to say awkwardly, "We're working on it."

As if things couldn't get worse, Tina's mother said, "It would probably work better if you didn't use birth control!" Everyone laughed as if she were making a joke. But Doug was burning inside. Tina's parents were obviously privy to this aspect of their sex life too. "I guess it would," was all Doug said, while Tina continued to look down, avoiding eye contact with anyone.

Tina's father followed up by commenting, "I would have thought you'd have knocked up Tina by now. It's not as if you have an ugly wife. She's got a great body, good breasts, small hips. She's a good looking woman! What are you waiting for?" This was not the first time Doug had observed Tina's father looking his daughter up and down in a salacious way.

Tina blushed and looked uncomfortable but said nothing as she continued to stare at her plate. Doug was so upset by the father's comment he could only quietly repeat, "I'm working on it." It was obvious that both Tina and Doug were under pressure and that any collaborative alliance between them had been lost.

The final blow was when Tina's mother exclaimed, "I know, you told us that," with no apparent recognition of what her husband had just said. The impact of Mother's doubling down on pressuring Doug while she whitewashed Father's comment was too much for Doug. He stood up and said to Tina, "It's time for us to go."

DISCLOSURE OF DEAD BROTHER

Our next session began with Doug's venting about this experience. He'd had enough of trying to be a dutiful son-in-law, and he wasn't holding back anymore. Doug also wanted me to know Tina's family didn't talk about lots of important things. Tina challenged him to name something to support this claim. Doug said, "How about your brother or our wedding for examples."

Tina looked stuck. "I had a brother," she said quietly, "but he died. It happened a long time ago. I was 11. Bobby was 17."

"How did he die?" I asked.

Tina responded like she was talking in a dream, without understanding the meaning of what she was saying. "He died of carbon monoxide poisoning in my father's car. He stuck a hose from the tailpipe into the car and closed up the windows. I didn't see this. My parents said they thought he changed his mind at the last moment and tried to get out of the car, and it actually was an accident and not something he did deliberately."

"How did your parents explain this?"

"They said he probably was upset about not getting the apprenticeship my Dad set up for him. He had trouble in high school and eventually dropped out. My parents did everything they could to help him, but he was just too far gone. My Dad pulled some strings and got him an apprenticeship to become an electrician, but my brother screwed up and blew the opportunity."

"That's not a reason most kids would kill themselves."

"We don't talk about it much. My mom gets real emotional if the topic comes up. She was too emotional to attend the funeral. She said she should have died instead of Bobby because he was wonderful despite his screw-ups. He was her pride and joy. Dad handled everything for the funeral, including a beautiful eulogy. He didn't shed a tear because he needed to be strong for me."

Doug interjected, 'Bobby also had drug problems."

Tina cut him off. "My parents worked hard to have his charges reduced from selling and distributing controlled substances to simple possession."

"Was this in his best interest?" I asked.

"I guess so," Tina replied, "My folks only want the best for their kids. They say this all the time."

THE WEDDING

"Bullshit!" Doug exploded. "Tell him about our wedding!"

Tina explained, "Mother threatened not to come to our wedding if we didn't take her preferences into account. During the planning stage, Mother was outraged that Doug and I started making plans without asking her how she wanted things handled.

Doug interjected, "Take her preferences into account?! She raked us over the coals with things like *Don't you care about my feelings at all?! How do you think I feel when you don't consider I have family and social obligations to fulfill? I won't be embarrassed by your thoughtlessness! I simply won't come if you don't want me!*"

"How did you handle this?" I asked.

Tina and Doug looked sheepish. "We capitulated," Tina replied. "From then on we didn't make a decision without consulting Mother. It was half her party and half ours."

"It was more like two-thirds her party!" Doug corrected. "I didn't like this one bit, but I was new to the family. I wanted her parents to like me, so I went along."

Tina continued. "Then the week before the wedding, Mother threatened not to attend if I didn't lend her sister the necklace their mother gave me before she died. Mother made up her mind she wanted my aunt to wear it to the wedding, and she wasn't going to come if she didn't get her way."

THE BACHELOR PARTY

Doug said to me, "This is a good time to tell you about my 'bachelor party.' Father said he was shocked to discover no one was giving me a bachelor party before the wedding. He said no man should get married without a proper bachelor party. I remember Tina's mother looking on proudly when he offered to take me out for a few beers the night before. I said, 'Thanks,' not feeling like I could decline.

"That night Father took me to a bar with strippers and pole dancers. I thought this was really creepy. I told Tina about this on our honeymoon, and we got into an argument. Tina said I misinterpreted her father's intentions."

Doug turned to Tina. "Here's the part I haven't told you. At one point, a new dancer came on stage and your father waved at her. She waved back in recognition and danced over to us. He put two $100 bills in her g-string and told her, "We're here for this man's bachelor party. Take him into a private room and show him a good time." I think the shocked look on my face made him say to me, 'It can only be a lap dance if that's all you want.'

"I was suddenly frozen in fear. I wasn't sexually aroused, I was uncomfortable as hell. I didn't want to offend your father. But I also worried about what your reaction was going to be. All I managed to say was 'No thanks, I'll pass.'

"That's when things got *really* creepy. Suddenly, I realized your father was scrutinizing me intently. He wasn't just seeing how far I would go, he was setting me up to do it. I was shocked to think how naively I thought he was being fatherly. I never said a word of this to anyone. I didn't want to hurt you. I wanted your parents to like me. It felt like one way or another you were going to be angry at me. I'm sorry I never told you about this before."

MOTHER

Ultimately, here's what I learned from Tina and Doug's descriptions: Father was a prominent divorce attorney. Both he and Mother were heavy social drinkers, although Tina didn't consider them alcoholics. Father wanted Mother to run the home, raise the kids, and put on nice dinners when they entertained at home.

Mother learned not to buck him, especially in public. She had no marketable skills and compensated by talking about the importance of being a good mother—which she didn't appear to be. The picture of being an ideal mother appealed to her, although performing the actual functions did not.

Mother's self-worth seemed to completely revolve around a reflected sense of self, social status, and an unwarranted sense of superiority. She lied about insignificant things, constantly tweaking and embellishing facts to make herself look good. Apparently having grandchildren was a necessary part of her desired self-image.

Mother used Tina as "confidant" when she was growing up. Mother made inappropriate disclosures about her disappointments with Father and with her life in general, eliciting sympathy and compensatory behavior from Tina to "make it up to her." For example, Mother told Tina she had as little sex with Father as possible, and she "overlooked" Father's obvious signs of interest in other women for the sake of her children. She said she didn't confront Father's affairs because he would have left, and she would have been unable to support them. Tina learned she was supposed to make Mother feel like she was the best mother in the world.

By Tina's account, Mother devoted most of her time to golfing and entertaining friends at the country club. Although a maid handled most household responsibilities, Mother rarely attended her kids' school functions or extra-curricular activities. Tina learned to cook from the maid.

Gossiping was another favorite pastime, especially when Mother was drinking. She loved to talk dirt about other people, portraying them as immoral and untrustworthy and saying the only people Tina could trust were her family. The common finale to her disparaging rants was "Our family is better than that!"

Mother tried to control decisions–large and small–made by her two children while they were growing up. Her intrusive questioning was justified in the name of trying to help. When her influence led to bad outcomes for her kids, Mother lectured about the importance of taking personal responsibility rather than doing this herself.

Mother needed Tina to seek her counsel and take her advice because this reinforced Mother's self-image. She enforced this by interfering with Tina's autonomy and inducing doubt about her judgment. From the time Tina had been a little girl, Mother mentioned the importance of Tina's finding a man who could take care of her. When Tina went to college, Mother encouraged her to focus on finding a husband rather than developing a career.

FATHER

When I asked Tina for an example that would help me see her father, Tina said, "Lots of examples I can think of involve sex one way or another. Father finds a way to bring up the topic in any conversation, whether it's a fact, joke, or innuendo. When I was a teenager Father would make comments about my girlfriends' appearance and laugh when I told him to stop. It really upset me."

Tina paused to gather her thoughts. "I was pretty sure my father was having affairs when I was a young adolescent. But when I was 16 years old Mother told me he had sex with two of her best friends. She also thought Father was hitting on his newly divorced female clients."

"My Dad also has his good side," Tina told me brightly. "He can be very generous with money." But what I heard in Tina's description was how Father encouraged her dependency through money. His pattern was to induce her to do more or buy more than she herself could afford. Then, once she had put herself in a financially dependent position, Father would use this to manipulate her.

Tina's wedding was a prime example. After encouraging Tina to make wedding plans without worrying about the cost, Father expressed concern about his ability to pay for it. He also refused to give Tina an idea of what he could afford. This left Tina in an ongoing state of anxiety and insecurity, while Father acted like he was making a huge sacrifice for which Tina was indebted.

There were other instances where Tina's father went out of his way to do upsetting things to her, with the inappropriate bachelor party being a prime example. Moreover, after the wedding Father sent Doug a subscription to Penthouse magazine. Doug felt obligated to accept this "gift" from Father, provoking a fight between Doug and Tina.

Tina told me about the time during adolescence when Father began needling one of her girlfriends who stayed for dinner. After several exchanges the girl was clearly uncomfortable about Father's probing. Then he really hit a nerve. Ostensibly the girl left the table to go to the bathroom, but clearly she'd been on the verge of tears. Bobby finally broke the awkward silence by saying Father had embarrassed her.

Father responded by slapping Bobby's face. Mother told Father Bobby didn't mean anything by it. But that wasn't good enough for Father, who demanded an apology. Brother bowed his head and apologized. Father dismissed him, and Bobby left the table without making eye contact with anyone.

Tina then described a time when Bobby came home two hours past curfew. Father was outraged. He took this as a personal insult and defiance of his authority, and he repeatedly slapped Bobby's face. Tina described her father as being flipped-out angry and out of control.

Tina was ready to move onto another vignette when I pressed for more details. "Did Bobby just stand there while your father slapped him?" I asked. "Or did he put his hands up to protect himself?"

Tina said Bobby had tried to ward off the blows, but Father insisted Bobby put his arms down. When Bobby acquiesced, Father slapped him again. I pointed out that Father was making Bobby participate in his own punishment. Father couldn't do that if he didn't have mind-mapping ability. Father wasn't just monitoring Bobby's arm, he was mapping Bobby's mind. He wanted Bobby to acquiesce to his authority.

I asked what disturbed Tina more: Father's claiming to teach Bobby self-control while he himself was out of control? Or Father's being angry but *not* out of control? Tina's description suggested Father's behavior was calculated, measured, and targeted. The expression on Tina's face said my question turned her brain to spaghetti.

So I suggested Tina revisualize the moment when Father insisted Bobby lower his arms so Father could slap him unopposed. Revisualization helps resolve traumatic mind mapping. After ascertaining Tina had a vivid image of the encounter, I told her to *watch* what her father said and did and use this to map his mind.

Tina closed her eyes and watched the movie in her head. She said, "He's thinking... *How dare you defy me?! Who do you think you are? You little shit! Submit to my authority, or I will break you! You'd better get this straight, kid! You can't do whatever you want! This is* my *house!* I'm *the only one who can do whatever he wants!*" Tina was stunned by the words coming out of her own mouth.

SITUATIONALLY ACCESSIBLE MEMORY (SAM)

Not long thereafter, as Tina was walking into a department store, she passed a father and his unhappy young daughter on their way out. The girl was crying and holding her dolly. The thought crossed Tina's mind *I wonder what she had to do to get that dolly?*

Tina took two more steps and stopped dead in her tracks, causing several people to bump in to her. Seeing the little girl had triggered memories she hadn't thought about in years. Tina remembered when she was age four her father had offered her a doll if she would pull down her panties. Tina was tempted and started to do it, but at the last moment she pulled them up and ran out of the room. Recalling her father's laughter as she left stood out in her mind.

Tina was describing having her "situationally accessible memory" (SAM) triggered by seeing the crying little girl holding her doll. This refers to memory fragments that have not been processed into "verbally accessible memories" (VAM) that can be voluntarily recalled and talked about. Relevant circumstances or situational cues can trigger vivid recall of traumatic memories. (See Appendix C and D for more on situationally accessible and verbally accessible memory and their role in resolving traumatic mind mapping.)

When Tina described her experience to me, I said, "That's a pretty vivid image if you remember your father's laughter. How confident are you that this really happened?"

Tina sobbed, "I'd love to doubt my memory! But every year, until I was 12, my father repeated his request, knowing I'd get furious and storm out of the room. He'd laugh when I got upset. My father likes to push people's buttons."

"Your description suggests your father has mind-mapping ability and antisocial empathy. Someone can't deliberately push people's buttons or enjoy it, if he lacks mind-mapping ability or the ability to intuit other people's emotional experiences."

"What does that mean?" Tina asked.

"It means your father knows what he's doing and that he can see and feel his impact on the people he goes after. This is important if you want to understand your memory of your father tempting you with a doll to expose yourself and laughing when you ran out of the room. I suggest you look at the picture in your memory and tell me why your father is laughing."

Tina paused. "He's laughing because he's enjoying upsetting me.... He's tempting me to do something I sense is wrong.... I'm looking at him, trying to figure out if he really wants me to pull down my panties... Oh my God! He's mapping my mind!"

"Many people would find what you're describing your father doing to be patently disgusting. I don't say 'disgusting' to speak negatively of your father. I'm talking brain science. Disgust is an involuntary brain reaction. If what your father is doing is truly disgusting, you're going to have the impulse to seek his approval, rather than move away or protect yourself from him. That's another thing that happens when parents instill disgust and hatred in a child."

Tina and Doug exchanged looks that said they were starting to understand their situation. They were an all-too-common example of how disgusting parenting interferes with intergenerational differentiation and simultaneously inflicts interpersonal neurobiological problems like "emotional super-glue." They exemplify how parents can take their adult children hostage simply by repeatedly doing disgusting things.

Handling People Who Make Moves on You

I'm sure you want to know what Tina and Doug did with her parents just like you want to know how to handle the difficult people in your own life. To understand what they did—and what you should do—you need to know how to handle interpersonal "moves."

A "move" is an attempt to influence or manipulate someone else. Not all moves are negative. Deliberately misleading your husband about throwing him a surprise party is an example of a positive move. So is my interrupting my clients' attempts to avoid difficult issues. When they make a move, I make a counter-move. If I weren't able to handle their moves or make good moves on them, I wouldn't be a good therapist. I'm also quite willing for clients to map out my moves. I'm eager to discuss this with them because this creates intense moments of meeting between us.

"Emergency lies," big lies, withholding information, and sexual seduction are all examples of interpersonal "moves." Some are more prosocial than others. As parents of headstrong four-year-olds know all too well, good parenting sometimes involves making subtle moves that encourage your kid to make the decisions you want him to make.

Being able to influence the world around you (i.e., making moves) is a sign of social intelligence. Moves are neither inherently good nor bad. Whether you use them for prosocial or antisocial purposes says something about *you*.

What you'll learn here about handling interpersonal moves doesn't just apply to dealing with your spouse, children, parents and friends. It will help you regardless of your position in the world. Bosses make moves on employees to reject collective bargaining or accept wage reductions. Employees make moves on employers by staging work stoppages or walkouts. Co-workers make moves on each other to establish dominance and territory. Children are terrific move-makers, and good parenting requires handling one move after another.

WRITTEN MENTAL DIALOGS

One way I help clients learn to deal with interpersonal moves from their antagonists is through mental dialogs.

To prepare Tina for dealing with her parents, and to see her maps of their minds, I suggested that she write out dialogs with each of them. Her job was to lay out what she would say and what their responses were likely to be. I wanted to see what moves Tina thought her parents were likely to make and how she would deal with them.

This is Tina's first dialog with her mother:

Tina: "Mom, could you talk with me about my wedding? I've been realizing I'm not clear about a few things."

Mother: "Like what?"

Tina: "Like you insisting that I lend Maw Maw's necklace to your sister."

Mother: "You weren't going to wear it with your wedding dress. I knew she would enjoy wearing it for the day."

Tina: "But it was a gift to me from Maw Maw!"

Mother: "You always just think about yourself!"

Tina: "I don't just think of myself."

Mother: "Why are you doing this?"

Tina: "To help myself."

Mother: "Is this really necessary?"

Tina: "I thought so."

Mother: "I don't. You must be nuts."

It isn't necessary to accept Tina's depiction as being accurate about her mother. However, two things are certain: One is this is what Tina anticipated dealing with. Tina's "anticipated mother's" basic style is accusatory, punitive, and demeaning. She grabs the moral high ground and anticipates she's going to prevail.

The other thing that's clear is Tina doesn't respond effectively to moves she anticipates her mother making. Tina hesitantly brought up her topics and quickly was put on the defensive. She lost control of the conversation, and her mother got the last word.

DECIPHER YOUR ANTAGONISTS' MOVES

Showing you how to deal with other people's moves is one of the most powerful things I can teach you. This not only improves your ability to deal with difficult people, it also fends off traumatic mind mapping and repairs the negative impact on you. The same events that potentially create traumatic mind mapping can actually be good for you if you handle them well. The key is making really good (i.e., "gold standard") responses, which I'll describe in a moment.

Here's what you need to know now: You can always tell what people want when they make moves on you because they're doing it for one of two reasons:

- To *move you into a position* that you don't currently occupy and you don't want to be in, or

- To *move you out of a position* you currently occupy that they don't want you to have.

It's that simple. Once you figure out the positions people are trying to maneuver you into or out of, you will know what they want. And if you know what they want, you can predict their behavior. Knowing what people want and what they're likely to do helps you deal much better with the moves they make on you.

Looking at things this way, Tina recognized her mother (and father) were maneuvering her *into:*

- Producing grandchildren
- Positions of emotional dependency
- Being compliant to their wishes
- Going blind to their manipulations
- Prioritizing them over her marriage
- Emotionally supporting Mother
- Tolerating Father's inappropriate sexualizations

Simultaneously, Tina recognized they were maneuvering her *out of:*

- A collaborative primary alliance with Doug
- A productive and happy marriage
- Complaining about their interference
- Autonomous emotional functioning

After reflecting on these insights, Tina's next mental dialog with her mother looked like this:

Tina: "Mom, I want to talk with you about my wedding. Why did you make me lend Maw Maw's necklace to your sister?"

Mother: "Why are you bringing this up?"

Tina: "Maybe I never understood why you'd manipulate me on what's supposed to be my big day."

Mother: "You had a beautiful wedding. Better than many girls get."

Tina: "Could you please answer my question?"

Mother: "What's your point?"

Tina: "You made me feel like I wasn't important to you."

Mother: "You must be crazy. You're not thinking straight!

Do you think your father and I would spend what we did on your wedding if you weren't important to us?!"

Tina: "No."

Mother: "All we did for you turns out to mean nothing to you."

Tina told me her brain froze in the middle of writing this dialog. She couldn't think of how to respond to her mother. As you can see, Tina anticipated dealing with an emotional viper who goes for the jugular when threatened. Her "anticipated mother" is cruel and won't see herself as anything other than doing what's best for others—a combination that makes someone dangerous and destructive. When Mother talks about fulfilling her responsibilities as a mother, you know not to challenge her. This mother has to be right; and when she takes over the conversation, Tina follows her lead. Tina's "anticipated mother" gets the last word and uses it to berate Tina and induce guilt.

Tina started off more direct in this attempt, showing a bit more of herself, but quickly equivocated her position with "maybe." She tried to set some limits with Mother by demanding an answer to her question but never got one. Even if she had, this wouldn't have solved the problem. If you pay close attention, you'll see Tina never showed Mother that she could see her. In fact, Tina's question suggests she doesn't see Mother yet.

CONCRETE STEPS FOR DEALING WITH YOUR ANTAGONISTS

Here are ten steps to help you deal with the difficult people in your life:

1. Accept that your antagonists are going to make moves on you. This doesn't require a jaundiced view, just an understanding of human nature. Stop being incensed that people are trying to manipulate you. By making moves on you, they're demonstrating they have mind-mapping ability. Use this opportunity to map them.

2. Consider that you're probably experiencing

traumatic mind mapping and that your mind mapping has likely collapsed. Your initial maps of your antagonists' minds are probably off in some way. Perhaps you haven't a clue about what's going on in their heads, or you're jumping to conclusions. Anticipatory traumatic mind mapping makes you not want to map them accurately.

3. Someone who makes difficult-to-handle moves probably has good mind-mapping ability and they're using it to track you. It's spooky to watch someone trying to manipulate you. You're not going to want to see it because you're not going to want to deal with it. When your mind mapping collapses, you assume your antagonist doesn't have mind-mapping ability and doesn't know what he's doing.

4. Discover the position your antagonist wants to maneuver you into or out of. For instance, does this person want to make you feel guilty when that's not the case? Or does he want you to stop feeling good about yourself when you're feeling great? What is it he wants you to do or stop doing?

5. Use your antagonist's moves to figure out what he or she wants. Now you can get into the "why" of things. If someone is trying to make you feel inadequate, what does he get out of this? Similarly, if someone's feeding your ego and saying you're the greatest, don't naively think you're finally getting the recognition you deserve. What's in it for him?

6. Respond in ways that show your antagonists you can see what they are doing. This is a crucial part of the process. You have to demonstrate you can see your antagonists as they make moves on you. You accomplish this by how you handle yourself when interacting with them.

SHOW YOUR ANTAGONISTS YOU CAN SEE THEM

It doesn't work to simply say *I can see what you're doing!* How you interact makes it clear you're tracking them in real time. Here are two of the best ways to do this: One is not letting your antagonists walk around your questions. The other is keeping the conversation on track when they attempt to sidetrack it. Besides steering the situation in the direction you want, both moves demonstrate you can see these people.

Showing your antagonists you can see them accomplishes several important things. For one, it changes your interactions with them. In many cases they become more circumspect and less eager to make bold moves. When they realize you can map them but their map of you is off in some way, they often lay low while they remap you.

But the other thing this accomplishes concerns your own brain. From what I've seen, showing your antagonist you can see him (or her) heals the physical and psychological impacts of traumatic mind mapping. Given these beneficial impacts, it's absolutely amazing how people desperately avoid doing this! You won't believe how difficult this is until you try this for yourself.

Demonstrating you can see your nemesis in real time can be learned with practice. The following steps will help you deal with destructive people in ways that don't suppress your functioning:

7. Respond directly to what's happening in the moment, or reference a similar prior problematic interaction. Shift your comments from the topic being discussed to what's happening between you and your antagonist. Saying You just changed the topic and I don't want to go there shows you're watching what he's doing. The same goes for saying You and I had this same problem when we talked about (fill in topic).

8. Comment on your antagonists' moves as they occur. You might say I thought we could have a collaborative discussion, but you keep positioning me as having wronged you in some way. If you say Your attempts to make me feel guilty are not going to work, they're not likely to acknowledge they were doing this. But even if they don't, as long as you are accurate, they'll know you've mapped them.

9. Don't beat around the bush or be indirect, sarcastic, or snarky. Sarcasm, implication, and innuendo are insufficient. These behaviors suggest you're not prepared to talk straight and deal with your antagonist, even if you can see him.

10. Stay on message. Conduct yourself in a businesslike manner. Don't be belligerent. If you're willing to act on what you know, you don't need to bluff or threaten. If need be, do what you have to do.

Despite knowing all this, Tina avoided showing her parents she could see them. Probably the biggest factor was that Tina had anticipatory traumatic mind mapping. She was scared of evoking the Father and Mother she had seen during childhood. For example, Mother would frequently work herself up and get hysterical, and then blame her kids for upsetting her.

Tina didn't want to lose her relationship with her parents. She was afraid they would reject her if she revealed she could see them. This would also force her to confront the kind of people her parents really were. This meant giving up her secret desire that the best in her parents would finally stand up.

After discussing these issues in therapy, Tina's next attempt to deal with her mother looked like this:

Tina: "Mom, I have lots of leftover feelings from my wedding. For a long time I couldn't understand how you could threaten not to attend my wedding if I didn't do whatever you wanted. It started when you threatened not to come if we didn't handle the wedding plans your way."

Mother: "You're talking about the past. This was years ago."

Tina: "I asked you to put together the guest list for Sunday Brunch after the wedding. For the longest time you wouldn't commit, leaving me hanging. I was afraid to go around you. When you finally agreed, you told me you were going to tell everyone you weren't attending the brunch."

Mother: "I thought I should be honest with people. I didn't want to disappoint anyone if they showed up thinking I would be there."

Tina: "Then on my wedding day you extorted me again to do what you wanted about lending Maw Maw's necklace to your sister. You counted on my feelings for you—my wanting you at my wedding—but you apparently didn't care about how I felt."

Mother: "Your father and I love you. Of course we cared about how you felt. We're your parents. How could you think your parents don't care about how you feel? Think about how I feel having my daughter say this to me. I'm your mother and I don't deserve this!"

Tina: "I finally understand."

Mother: "What do you understand? You understand nothing!"

Tina: "I understand how you could extort me on my wedding day with your threat not to attend."

Mother: "What's your point?!"

Tina: "You wanted to see if I would still put your needs first. This is what you've always insisted on. And I'm not going to do it anymore."

Mother: "You've always insisted that your needs come first since you were a little girl. Maybe I made a mistake by putting your needs first, but back then that's what parents thought they were supposed to do."

Tina: "I'm a grown woman now. I need to have my own life. You can't be the most important person in my life anymore."

Mother: "So you're going to shut me out of your life now, is that it?"

Tina: "No, Mom, that isn't what I'm saying."

Mother: "It sure sounds like it."

Tina: "Mom, that's not what I'm saying."

Tina's ability to deal with her mother was clearly improving. Now she was willing to show her mother she could see her. You can also sense her fear that as she got better, her mother would become more aggressive. But did you notice how Tina's "anticipated mother" actually manages to create two false realities? One was that she had always put Tina's needs and wants first. The other was that Tina should feel guilty about this. Moreover, look at how her mother continually tries to put Tina on the defensive despite being called out. This is the mother Tina was afraid to deal with.

In the end Doug finally gets the last word, but she doesn't do much with it. She's talking about what she didn't say rather than delivering her message. The good news is that Tina hasn't repeated her prior mistake of letting her mother control the conversation.

After we reviewed this attempt, Tina picked up where she left off and finished the dialog better. This is how it went:

Mother: "So you're going to shut me out of your life now, is that it?"

Tina: "I'm not going to take the bait, Mother. You're not addressing what I'm saying about my wedding. Do you have anything else you want to say?"

Mother: "There's no point in my talking. You've already made up your mind."

Tina: "This is helping me make up my mind."

Mother: "Are you threatening me, Tina?"

Tina: "No, Mother. I'm just watching and learning."

Mother: "Don't you realize how you make other people feel when you do this?"

Tina: "Both of us have the ability to sense what other people feel."

Hold on to Yourself (Differentiation)

Holding onto yourself through something like this is differentiation-in-action. Differentiation is the ability to hold onto yourself when the important people in your life pressure you to conform. I'm describing how differentiation and corrective interpersonal neurobiology finally come together. This is how human evolution and brain repair occur in real time in relationships.

Mapping the minds of people who do traumatic things—*while keeping your mind from turning to spaghetti*—repairs traumatic mind mapping and exposure to disgusting behavior. Your brain becomes more resilient. Your mind becomes more stable and coherent. You develop a more solid and flexible self. And you become a real adult.

This frequently involves gut-wrenching experiences that stretch four uniquely human abilities. I call these pillars of the human self–the Crucible Four Points of Balance–and here's what they look like in practice:[1]

To start with, Tina's personal integrity was on the line. People who do disgusting things force everyone around them into integrity crises. When you fold in the face of this, it's easier for them to take you captive.

To prevail, Tina had to be clear about what she was doing with her parents and why she was doing it, so she didn't lose her nerve or get sidetracked into tangential issues. *(First Point of Balance: Solid Flexible Self.)* Tina also had to calm herself down, quiet her mind, and not let her anxieties run away with her. *(Second Point of Balance: Quiet Mind and Calm Heart.)* She had to make herself address their moves with proportionate responses without over-reacting to their reactions. *(Third Point of Balance: Grounded Responding.)* Finally, she had to remain resolute when things got heated or nasty but also recognize when something (or someone) was a lost cause. *(Fourth Point of Balance: Meaningful Endurance.)*

DIALOG WITH FATHER

Tina's mental dialogs with her father were markedly different. Her first attempt looked like this:

Tina: "Dad, I need to talk with you about something."

Father: "What about?"

Tina: "Do you remember offering me a dolly when I was real young?"

Father: "What you mean offering? I gave you lots of dolls when you were young."

Tina: "Do you remember offering me a dolly if I would pull my underwear down?"

Father: "Let's see.... No, I don't."

This is the sum total of how Tina could handle herself with her Father when she started. Clearly she had difficulty maintaining her differentiation. Here's her second attempt:

Tina: "Dad, I need your help clarifying some things I'm starting to remember."

Father: "I'm glad to help. You're my daughter."

Tina: "I remember you offering me a dolly when I was four if I would pull down my underpants in front of you."
Father: "I don't remember doing that."
Tina: "I do."
Father: "Are you certain this happened?"
Tina: "I think so."
Father: "I don't know what to tell you. I can't help you. I don't remember anything like this."

In either case, Tina was unable to handle her father's saying he didn't remember. So I suggested that Tina pick up right where she left off. This was her next attempt:

Father: "I don't know what to tell you. I can't help you. I don't remember anything like this."
Tina: "You can help me by trying to remember."
Father: "OK, let me try." (pause) "Nope. I don't remember."
Tina: "Are you sure?"
Father: "Do you think I'm not trying?"
Tina: "I appreciate you trying."
Father: "Let's see....What was the question?"
Tina: "Do you remember offering me a dolly to pull down my underwear when I was a little girl?"
Father: "Don't you think that's the kind of thing I'd remember if I did that?"
Tina: "Yes, I do."
Father: "Then what does it tell you if I don't remember doing that?"
Tina: "I'm not sure."

Make "Gold-Standard" Responses

The "gold-standard response" I previously mentioned involves keeping up with your antagonist and making appropriate moves in real time during conversations. One example of a gold-standard response involves commenting on what people are doing while they're doing it in front of you. Describing the moves people are making proves you can see them. Their brains register that they've been mapped, and your brain registers that they've seen you seeing them. You should strive to do in every interaction with them.

This may sound like a tall order and it is in some ways. But here's what I discovered in my research on my fail-to-thrive clients. Some had no difficulty keeping up with their antagonists on their first attempt. They were as dumbfounded as I was by their success. But some let their antagonists get around them again. This is the emotional super-glue that repeated disgust reactions produces (described in Part Two).

Sometimes one gold-standard response is enough to improve relationships with employees, bosses, co-workers, and friends. But this is often just the beginning when you're dealing with your spouse, children or parents. You usually have to hang in for multiple interactions to get anywhere.

Maybe you can keep up for two or three interactions to start. Maybe you fold in the very first one. The key is sticking with this until you can stop people from getting around you, regardless of how many times it takes. This willful repetition—trying and failing, trying something different and still failing, and trying again and doing better—helps you change your brain.

Several dialogs later this is what Tina wrote:

Tina: "Dad, I have this memory of you offering me a dolly when I was four years old, if I would pull down my underpants in front of you."

Father: "I don't remember that."

Tina: "I do. I remember I started to do it. I knew I wasn't supposed to do something like that. But I wanted the dolly and I didn't want to believe you would ask me to do something bad. I almost did it, but at the last moment I yanked up my pants and ran out the door, very upset."

Father: "I'm sorry, I don't remember that."

Tina: "The part that sticks in my mind was how you laughed when I got upset."

Father: "If I laughed at something like that, I would have been cruel. Are you saying I'm cruel?"

Tina: "I'm not saying you're cruel, Dad. But I have a pretty clear memory of this."

Father: "Why didn't you tell me this sooner? Don't you think I have a right to know you're thinking these kinds of thoughts about me?! And all this is based on a single memory you think you have?!"

Tina: "It's more than one memory."

Father: "What?"

Tina: "That wasn't the only time."

Father: "What are you saying? That I'm is a repeat child molester?"

Tina: "I'm not saying that."

Father: "Then what are you saying?"

Tina: "I'm asking for your help."

Father: "Jesus, Tina, first you insult me and then you ask me for my help?! I guess all daughters expect their fathers to be perfect!"

DON'T LET YOUR ANTAGONIST GET AROUND YOU

When you start making gold-standard responses, you're usually playing catch-up. Your antagonist has already gotten around you and mapped you in the process. Now they know how to do it. You usually only realize this after the fact.

Some people will try to get around you by changing the subject. When this happens, just say, "You're changing the topic and I'm not ready to move on yet. I asked you about (restate question)."

Others will try to get around you by acting deferential. Or they'll cry and say, "I'm so sorry. Please forgive me." If you fail to recognize what's happening, you'll walk away thinking you accomplished something. Your next interaction with them usually reveals you got faked out of your socks.

If necessary, some people will try to overpower you. They become increasingly aggressive or accusatory until you back off. If one method fails, they'll often shift to Plan B or Plan C. It won't matter how they get around you, as long as they do.

Here's the key: Present them with a question that requires a response. Then all you have to do is *prevent them from successfully getting around your question.* Keeping people from switching topics demonstrates you can see them plain as day.

Don't try to pin your antagonist down. You don't have to

make these folks give in. You don't have to pin them down or get them to agree. You just have to hold onto yourself and maintain your position. This is a subtle point. Preventing your antagonist from getting around you is not the same as pinning him down. Visualize this for a moment and you'll see the difference.

Don't tell your antagonists what to do. It's easy to get angry when people make moves on you, which leads to this common mistake. If you tell people what to do, they can defeat you simply by defying you. If you need them to do something, they control the situation. The solution isn't showing your anger, it's demonstrating your determination. That's what tells people *You better take me seriously and deal with me.*

Don't try to make your antagonists confess, apologize, or acknowledge your point. Exercising authority you don't have weakens your position. You can't control other people, but you can control yourself. Rather than trying to bend others to your will, empower yourself instead. Tell them what *you* are going to do. This puts the locus of control within you.

Here is Tina's next dialog:

Tina: "Dad, I'm in the process of straightening out my life. And I have a clear memory of you offering me a dolly if I pulled my underpants down in front of you. I was tempted and then I got upset and ran out the room. You laughed and enjoyed upsetting me"

Father: "I don't remember this."

Tina: "What's important is that I do."

Father: "How do you know this isn't a false memory?"

Tina: "Because you did it at least once a year until I got my period."

Father: "Did what?"

Tina: "Offer me a dolly if I would pull my underwear down for you. Even when I was a becoming a young woman."

Father: "Why would I do such a thing?"

Tina: "That's what I want to know."

Father: "Let's see... Because I wanted to teach you not to do this with strangers? Maybe I was testing you."

Tina: "I was hoping you'd tell my why you repeatedly did this."

Father: "How can I tell you why I did something I don't remember doing?"

Tina: "This much I know—these events happened."

Father: "Why are you telling me this?"

Tina: "Because I need to show you and show myself that I can see you."

Father: (sneering) "What do you know?!"

Tina: "I finally see there's a pattern of your inappropriate sexual behavior running throughout my life. The "bachelor party" you took Doug on. You sending him a subscription to *Penthouse* magazine with naked women. Lots of things."

Father: "They're not totally naked."

Tina: "It wasn't appropriate. Neither were your sexual comments about my girlfriends when I was a teenager."

Father: "You had a few really gorgeous girlfriends."

Tina: "It wasn't just my girlfriends. One time when I was getting dressed to go out you said to me, 'I hope you are going to see some lucky guy. It wouldn't be right not to share.'"

Father: "I always had your best interests at heart."

Tina: "What does that mean?"

(Pause)

Tina: "Dad?"

Father: "Yes?"

Tina: "What does that mean?"

Father: "What does what mean?"

Tina: "You had my best interests at heart."

Father: "Let's see...."

Tina: "You have to think about what it means?!"

Father: "Goddamit! You asked me a question and I'm trying to answer it! What do you want from me?!"

Tina: "You also had sex with Mom's best friends, Janet and Leslie."

Father: "How would *you* know if I did?!"

Tina: "Mom told me when I was 16."

Father: "She had no business telling you that."

Tina: "That's not the issue now. I know who you are."

Some clients spend hours writing a single dialog, figuring

out what they need to say or what their antagonist is likely to say and how best to respond. They're not re-editing their writings, and they're not making things up like they want them to be. They are forcing themselves to consider things they'd rather avoid, and waiting to see what comes next to their minds.

Written mental dialogs are a powerful method for recognizing interpersonal, neurobiological problems and correcting them. Mental blocks can be identified, and traumatic mind mapping can be confronted. Best of all, written mental dialogs are free and anyone can do them.

Repeated dialogs allow you to push through mental blocks in ways that limiting yourself to actual interactions with your antagonist doesn't allow. Repetition is important because you're forcing your brain/mind to go down a neural/mental path it doesn't want to go. You can't get enough repetition if you only focus on real-time interactions.

These written, mental dialogs involve keeping your brain/mind functioning while envisioning your antagonist's saying or doing traumatic things. Repeatedly forcing yourself to keep your mind together and your emotions under control while you think about what to do next accomplishes three things: It (1) repairs the damage of previous traumatic mind mapping, (2) reduces the power of anticipatory traumatic mind mapping, and (3) makes your brain/mind more resilient to subsequent traumatic mind mapping.

Progressive mental dialogs also have a rehearsal or priming effect for dealing with real life situations. They make you pay more attention to what your antagonist is doing, and they improve your ability to respond. They can significantly increase your willingness to deal with difficult, real-life situations. This is important because ultimately mental dialogs are no substitute for direct interactions with your antagonists.

ANOTHER COMMENT BRINGS THINGS TO A HEAD

This "opportunity" presented itself to Doug and Tina during a subsequent dinner at Tina's parents' house. Doug and Tina were discussing plans to attend a Shakespeare Festival when Father said he liked Shakespeare's plays and Old English poetry too, whereupon he recited an inappropriate sexual ditty he attributed to Geoffrey Chaucer, the gist of which was "maidens all have tightly bores and mothers all have gashes."

Doug was shocked speechless. He couldn't believe Tina's father was doing this. Tina blushed and looked down but said nothing. Mother rolled her eyes in exasperation. Father explained they were discussing Elizabethan culture, and he was contributing to the conversation.

Nothing more was said for several seconds. Then Mother started talking about something else. When what just happened was about to disappear, Doug spoke up.

"That's not appropriate!"

"What's not appropriate?" Father acted like he didn't understand what Doug was saying.

"What you just said."

"It's famous Old English poetry."

Doug's face flushed with anger. Tina said, "Let me handle this." Then she turned to Mother and Father and took a moment to collect her thoughts.

"I have tried to be respectful of you as my parents. Even when it involved sticking my head in the sand about what you were doing. Dad, all my life you've made inappropriate sexual comments and behaved in sexually inappropriate ways. I'm not willing to go blind about this anymore."

"Old English poetry was known for being bawdy."

"This is not bawdy. This is disgusting!"

Father paused a moment to take Tina's measure. "That's strong language you're using, young lady. I'd be careful if I were you."

Mother wasn't going to miss this opportunity to put Tina in her place. "Apologize to your father this minute!"

"I'm not apologizing. If anyone needs to apologize, it's Dad! I've been so careful around both of you I hardly have a life!"

"Don't blame your marital problems on us!" Mother insisted. "Your father was making a joke, that's all. You know he has a dirty sense of humor."

Tina sat silently for a moment, deliberating how far to go. "This is more than dirty humor. You didn't think it was very funny when Dad slept with your best friends."

Mother was clearly shocked Tina brought this up. She was breaking the family rules and violating Mother's map of her. Mother and Father felt entitled to poke around in the intimate details of the sex life of their daughter and son-in-law, but Tina wasn't allowed to reference anything derogatory in the family's past.

"Like I said, young lady," Father warned, "You are using strong language. You need to watch what you say before you get yourself in trouble."

"I'm already in trouble," Tina replied. "The only thing I can do is get myself out."

This was a big step for Tina because she really was afraid of Father and Mother. She had anticipatory traumatic mind mapping about how bad they could get. She knew Father was watching this whole thing and ready to explode. But if she were serious about emancipating herself and prioritizing her marriage, she had to stop maintaining the status quo with her parents and say what needed to be said. This required forcing herself to think straight when normally she'd have "spaghetti brain."

"I was walking into a department store the other day and I passed a little girl carrying her dolly and crying. It made me remember that you offered me a dolly if I would pull down my underwear so you could see my body."

Mother shrieked, "What did you say?! Are you insane?!"

An ominous silence filled the room. Father asked non-committally, "How old were you when this supposedly happened?"

"Four."

Another long pause ensued. "Are you accusing me of doing this on the basis of a memory you think you had when you were four?!" Father's tone was incredulous.

Tina looked at her father. "Not just four."

Mother exclaimed, "Oh my God! Are you claiming your father is a repeat sexual abuser! I'm going to have a heart attack!"

"If you're going to take this back," Father growled, "this is your opportunity."

"Not just age four," Tina repeated emphatically. "You did

this every year until I got my period. You'd ask me if I'd pull down my underwear for a dolly. You knew I hated when you did this. After a while I realized you didn't seriously think I would do it. You knew I would get upset. You'd do it deliberately to upset me. You always laughed when you succeeded."

Father laughed. "You know I have a bad habit of pushing people's buttons. I don't mean anything by it."

"This is more than pushing my buttons. You were tempting me to do something wrong. You wanted to look at my body."

Mother cut in. "I already told you, your father has a dirty sense of humor."

"This was not funny," Tina replied.

Mother countered. "He must have been joking. You said he laughed."

Father added, "I was joking now and you took it the wrong way. How do you know you weren't making the same mistake then?"

At this point Tina did something she had never done in front of her parents. She responded to Doug like they were a couple, and she looked to him for help. Before Doug could say something Mother redirected the conversation away from him. "What's your point, Tina?"

Tina paused for a moment. "You both let me down."

Mother exploded. "We let you down?! Do you have any idea how disappointing it is for a mother to hear what you just said?!"

Tina replied, "It's taken a lot for me to get past being totally preoccupied with your reactions and how you would feel when I said this! I almost can't believe I'm saying this now." Tina looked at Doug, who nodded in support.

"I think you need therapy!" Mother yelled. "What kind of family do you think we are?! We're better than that!"

"No mother," Tina said, "we're not better than that. We're exactly like that. Including what you're doing now."

"I'm feeling what any mother would feel. I'm devastated! You've ruined a lovely dinner and sprung this on us! You should have warned us you were coming here to confront us!"

"I didn't come here to confront you tonight. I had no intention of doing this. But Father pushed my limits with his unnecessary sexual comment, and I couldn't just sit here and not say anything one more time."

"Well, you can just leave right now! I don't need to be abused in my own house!" Mother said. "I expect an apology!"

Tina and Doug quickly gathered their coats and departed. Neither said a word for a few seconds once they reached their car. Both were emotionally overwhelmed with what just transpired.

"Wow!" said Tina. "I can't believe I did that!"

"Wow is right!" Doug replied. "You were terrific! You handled yourself so well! And did you notice what *didn't* happen?"

"What's that?"

"Neither your father nor you mother explicitly denied what you said."

Email Interactions

These days people increasingly work out their relationships through email exchanges. My clients have found it helpful to analyze messages from their antagonists and make thoughtful, deliberate responses.

The next day following the dinner, Tina received an email from Mother:

> Dear Tina, I hardly know how to respond to the things you said when you were here. No matter how I look at it, you let me down. If what you're saying is false, what weird pleasure are you getting by torturing me? If what you're saying is true, why didn't you tell me sooner? I'm your mother. Why didn't you come to me? I thought we had the kind of relationship where we could tell each other anything. I would have protected you. Either way I feel betrayed.
>
> I couldn't believe you'd throw your father's infidelity at me at the dinner table. I've learned to overlook his constant sexual references. I don't approve of his humor. Your father is the way he is. I've never been able to change him and believe me I've tried! "Perhaps it's not too late to salvage our relationship. We used to be so close when you were younger. All I've ever wanted was

for you to be happy. I know you know that in your heart.

Love, Mother

Two days later Tina replied:

Dear Mother, this is hard for both of us. You asked me why I never came to you about Dad so I'll tell you. I knew you'd get upset. I knew you'd fall apart and I'd have to take care of you. It was better for me to not say anything and keep it to myself.

In the past I would have apologized simply because you feel let down by me. But I wasn't withholding information from you. I didn't think you could handle it. If anything I was shielding you from it. I knew you didn't want to deal with anything that would tarnish you picture of us being a "superior" family.

If you feel you've been denied the opportunity to deal with what Dad was doing with me, let's not make the same mistake twice. Let's meet soon and discuss this.

Love, Tina

The next day Tina received the following:

"Dear Tina, I'm glad you realize you're not the only one suffering here. I feel betrayed by your father. I don't know whether to believe he did what you say he did or not. I've tried to keep our family standards high. This sort of thing is not supposed to happen. But if I'm honest with myself, I guess he might be capable of doing something like that. The thought that I've been sleeping with a child molester for years makes me sick.

I never realized you thought you needed to take care of me. I guess that's because I was busy taking care of you. I've loved you since you were born and the nurses laid you on my breast. Considering what I had to go through with your father, I think I did a pretty good job. When your brother had his accident, I thought I'd lose my mind. But I knew I had to be strong for you. If I've failed to protect you in some way, whatever my sins, please forgive me.

I don't know that I'm ready to hear what you have to say. It sounds like you're going to unload your grievances on me and I'm not up for that. I can't undo whatever happened as much as I'd like too.

Love, Mother

Two days later Tina replied:

Dear Mother, It's premature to talk about apologies or forgiveness, until we have some mutual understanding about what happened. I'm more interested in getting a full picture of my life than assigning blame. My preference is that we talk face-to-face. I'll meet you wherever you like (except at home). I don't want to talk about this on the phone. I think we should be able to see each other when we talk. Let's get together somewhere and really talk.

Love, Tina

Four days later Mother replied:

Dear Tina, I'm sorry but I'm not feeling up to rehashing the past. That was a long time ago, and I don't see the point. You can't change it. Coping with your accusation towards your father is as much as I can handle.

I can't always be putting you first. I don't see the point of walking into being attacked, and I'm really not up to hearing your problems. You have no idea how your accusations have put me in a terrible position.

Under the circumstances, I have a right and responsibility to take care of myself. I never thought I'd have to protect myself from my own flesh and blood. For now emails are all I can handle.

I can't live a life of negativity and regret, no matter how much this might satisfy you. I've always encouraged you to have a positive outlook on life, even when your father humiliated us in front of the entire community. For the good of the family, I've learned to hold my head high and not look back. I suggest you do the same.

All I've ever wanted was for you to be happy. I find

it hard to believe you could harbor such animosity towards me. You're telling me I'm a failure as a mother. I've wasted my life. I might as well be dead. If you ever have a child you'll understand what I'm telling you.

Love, Mother

Tina sought my input about her fears that her mother might commit suicide. I told Tina I've had one client who's father suicided by repeatedly canceling his kidney dialysis appointments. Eventually he went into renal failure and died. This forced me to consider what stance I would take in the future.

I told Tina that, first and foremost, she had to decide for herself how she was going to respond to her mother's threats to kill herself, and how she would feel if her mother suicided. For my part, anytime someone threatened suicide to keep one of my clients an emotional prisoner, my stance was that my client had a right to save her own life.

Two days later Tina responded this way:

Dear Mother, I'm sad that you can't see this any way other than a personal attack on you. I don't wish you ill, and I don't take pleasure in you being upset. But for me, this is about understanding what's happened in my life and how it's affected me. I want to be able to discuss difficult topics with you. There's a difference.

I could see what you're saying as pressuring me to go back to denying what I know. But I've decided to take a more positive view. This doesn't have to be a terrible thing.

You say you need to take care of yourself. Rather than disputing your position, I agree. I encourage you to take care of yourself, and I will do the same. This doesn't have to be adversarial. When you're ready to talk face to face, let's get together.

Love, Tina

Face-to-face Meetings

Tina didn't hear from her father for a week, during which time she struggled over whether she should contact him. She knew he expected her to come seeking his approval, and she was sorely tempted. But Tina held onto herself and seven days later she received an email from him suggesting they have dinner. They met the following evening, and afterwards Tina wrote down the details of how it went. It started with Father pushing Tina to give in to her mother.

"Your mother is very upset with you. She's having difficulty sleeping and crying a lot. All she can think about is your outburst at the table. I'd like you to apologize to her to calm her down. You know how hysterical she gets."

"I'm not going to do that, Dad."

"Why not? Your mother is really upset."

"Mother is upset because she won't deal with the truth. We're not the superior family she likes to say we are, and she's not the super-mother she likes to think she is."

"There's nothing wrong with having high standards."

For a moment, Tina couldn't think straight. She had to fight through an acute case of "spaghetti brain" and nausea to gather her thoughts and realize what just happened. Then Tina did something for the first time in her life. She took control of a conversation with her father.

KEEP UP WITH YOUR ANTAGONIST IN REAL TIME

Tina's head was spinning. She was outraged and disgusted by her father's response. Questions kept running through her mind: *How can you do this to me? You're my father! How can you let yourself act like this?!*

These questions distracted Tina from dealing with Father until she was willing to answer them for herself, right then and there. *My father's willing, even eager, to do disgusting things. He knows exactly what he's doing. I've seen him act like this my entire life. I'm so disgusted I could puke. And right now, he's mapping me. He expects me to back off and let him off the hook. I can't allow myself to do that.*

Tina looked Father square in the eyes. "The problem isn't high standards! It's that you and Mom don't live up to them!" The room became deafeningly silent, although it felt to Tina like a thunderbolt had landed.

"What are you saying?!" Father's tone was threatening.

"Here we are again, one more time. You're pressuring me to do whatever it takes to stabilize Mom. This time I'm not going to do it. But arguing about what I'm going to do with Mom shifts the conversation away from me and you."

Father looked surprised by Tina's response. "What do you want to talk about?"

"I'm a grown woman. I'm not your little girl anymore. But either way, you have no business making sexual comments in front of me at the dinner table."

"You and Doug were talking about Shakespeare and I contributed an Old English poem."

"You always have a way of making yourself sound innocent, like you're acting appropriately when you're not. You should not have sent Doug a subscription to *Penthouse* magazine, and you shouldn't have taken him to a strip club the night before our wedding."

"It was just male bonding."

The hours spent writing mental dialogs paid off. Tina's thought processes were more resilient. She could think faster and clearer when Father made moves on her. It was easier for her to see what he was doing and she was less affected by it. This made it easier for her to respond, and her responses were more effective.

"No, it was you injecting sex where it doesn't belong. My marriage is more important than you bonding with Doug. He didn't like when you did these things, but he felt obligated to go along to have a relationship with you. "

Father's eyes squinted as he appraised Tina. "Do you expect me to apologize?"

"No...I expect you to carry on like you always do, claiming what you did was appropriate. One way or another, in your mind you always turn out to be right. Mom does too."

"Watch your mouth!"

Tina looked her father in the eye and met his gaze without flinching. "Believe me, I am.... Repeatedly offering me a doll to pull down my underpants was wrong. You knew I'd get upset. You enjoyed torturing me."

"After all this time you accuse me of this?! How come you never said a word about this until now?"

"This was so traumatic I didn't remember it until I passed that little girl crying with her dolly."

"How do you know this isn't a false memory?"

"It would have to be a whole series of false memories of you taunting me about something sexual and then laughing at me. You did the same thing when I asked you to stop telling sexual jokes to my girlfriends when I was a teenager. You enjoyed upsetting me."

"Why would I do that? What kind of person would I be if I did that?"

At this point, Tina had a stunning realization that her father was baiting her. He was smiling a sardonic grin. She had finally laid out the history of his sexual transgressions. If he wanted a better relationship, it was time for him to acknowledge what he had done. But he wasn't. He was using his typical manipulative methods of waylaying conversations.

Tina thought *My dad is smart and cagey. He makes me uncertain about what he's doing. If I challenge him, he puts me on the defensive, or shifts the conversation so he has the high ground.*

This realization dazed Tina because it involved traumatic mind mapping. But rather than let her mind fall apart, Tina fought for clarity. Tina said, "I've been blind to who you are, and it's nearly cost me my marriage. You never admit to doing anything wrong. Your answers are always noncommittal. You always leave yourself an excuse or a way out. You make sure no one ever pins you down."

This was the first time Tina didn't feel obligated to answer Father's questions, and this wasn't lost on him. "So now do you feel better for having confronted your father?" Father asked dismissively.

"I'm learning there's no point in confronting you," Tina replied. "You're going to say or do whatever you like. I'm confronting myself with the reality of who you are. That's what's really important here. I've been welded to you and Mom for most of my life, and I'm setting myself free. I'm also doing this for Bobby." Father recoiled at the mention of Bobby's name.

Tina concluded, "If you're interested in having a relationship with me and my family, then acknowledge what you did and what you continue to do."

"What do you mean '*your* family'?! You and Doug are part of *our* family." Despite often seeming unable to follow the thread of a conversation, Father tracked the implications of what Tina had said.

"I have to start thinking of myself as a woman with my own family and marriage rather than just part of your family."

Father asked menacingly, "Are you threatening to cut me off from my grandchildren?! You're not even pregnant!" This was the bottom line as far as Father was concerned. He had been tracking this all along.

"I can't wait until I'm pregnant to be a good mother. I'd love to have a father who can be trusted with his grandkids." Tina gathered her things to leave.

Father's bold moves indicated he expected to get away with them. He wasn't about to give up the first time Tina spoke up. "Grandparents have rights, you know! Legal rights!"

Tina paused and looked at her father. "Children's welfare and parents' rights come first. You can contact me whenever you're ready for a serious conversation." Tina said her father's face was an impassive mask as she left.

LOOK FOR "NEWS OF A DIFFERENCE"

Notice how many ways Tina showed her father she could see him:

1. Tina described Father's manipulative moves.
2. Tina didn't let Father control or change the topic.
3. Tina stayed on point, demonstrating she wasn't confused by his confabulations.
4. Tina didn't get intimidated when Father was menacing, but she indicated she'd seen this.
5. Tina didn't overreact when Father treated her like a little girl.
6. Tina brought up prior situations that supported her description of him.

Pay attention for subtle signs that your antagonists perceive changes in you. This is not the same as seeking their validation or approval. Recognition of change puts you in a different position, although they may try to act like nothing has changed. This helps heal traumatic mind mapping, increases your sense of self-efficacy, and enhances your resilience to "spaghetti brain."

Tina saw multiple signs of recognition in her father:

- *Emotional outbursts.* When Father saw Tina standing up for herself, he yelled at her, threatened her, and dared her to draw conclusions about his past behavior.
- *Changes in affect, mood, or attitude.* Father stopped acting like he was holding all the cards.
- *Changes in strategy.* Father subsequently stopped baiting Tina and ramping up their interactions.
- *Increased mind masking.* Father kept a lower profile. He wanted to see Tina's next moves, because he wasn't sure how to handle her for the very first time.

DID MOTHER AND FATHER KNOW WHAT THEY WERE DOING?

Perhaps you'd like to believe Mother was so self-centered she lacked any capacity for empathy. But everything indicated she had *antisocial* empathy as well as mind-mapping and mind-masking abilities. Her effectiveness in manipulating Tina indicated she could recognize Tina's emotional experiences. Presenting herself as looking out for Tina's best interests, while deflecting her own culpability, can only be accomplished by someone with these abilities.

This is what allowed Mother to deflect every point Tina made, create an excuse for everything, cut off further discussion, and instill doubt, insecurity and guilt. How else did she know when it was time to interrupt or start saying *It was a long time ago, I don't remember?*

Mother's threats not to attend Tina's wedding further demonstrate antisocial empathy. Mother knew Tina didn't like what she was doing because Mother was using this to push Tina into accommodating her. Mother may have been blind to herself, but she wasn't blind to Tina. What Mother lacked was compassion for Tina.

Likewise, multiple data points indicated Father had mind-mapping and mind-masking ability and antisocial empathy. Father went after people, stunning and immobilizing them with unsettling questions or biting sarcasm and then saying 'just kidding!'" His interpersonal advantage came from doing mind-blowing things you couldn't believe he would do. Then he'd mask his mind, shrouding his behavior in plausible deniability and implanting uncertainty about his motivations.

Father enjoyed poking people where they were emotionally vulnerable and upsetting their emotional equilibrium. You can't do this efficiently or derive pleasure from other people's discomfort if you don't have the ability to intuit their feelings. The same goes for stunning and immobilizing them, like setting Doug up with a stripper at his "bachelor party."

In particular, Father knew how Tina was going to feel, think, and react when he offered her a doll to drop her underwear. If he didn't know the first time, he certainly knew it thereafter. Then there was keeping Tina hanging about paying for her wedding. And baiting her during her more recent attempts to talk to him about her childhood. Father enjoyed feeling Tina get upset; that's why he repeatedly did things like this.

Finally, Father liked to push the limits of sexual propriety, which requires tracking other people's reactions and knowing conventional social boundaries. When challenged, he became threatening, which also involves monitoring another person's emotional reactions and calculating how far to go. The most parsimonious explanation was that Father and Mother had mind-mapping and mind-masking abilities and antisocial empathy. This is what made their behavior disgusting.

Impacts Ripple Through Dysfunctional Families

When showing your antagonists you can see them, there's always a judgment call about how far to take things. You don't necessarily have to take it to the nth degree. On the other hand, dealing with people who do disgusting things often requires extreme measures because they won't have it any other way. At least for some length of time, you have to address their manipulative moves. You don't have to do this forever. Usually the problem is you won't do this at all.

If you want meaningful resolution with your parents before they die, you're going to have to show them you can see them. If you're unwilling to do this, the alternative requires appearing blind to them until they pass on. Adult children who are super-glued to disgusting parents consider this a final act of love. Unfortunately, they subsequently can't get over their parents' deaths and deify them once they're gone.

People typically die the same way they lived. Hollywood-style, last-minute, bedside reconciliations with disgusting parents rarely happen in real life. They don't occur because the parents don't rise to the occasion. Some clients say their parents' last deliberate act on earth was saying or doing something disgusting.

Caring for disgusting parents who are elderly, ill, or about to die can force heartbreaking decisions. The best strategy involves fulfilling your responsibilities to them (as you deem them), without allowing yourself to be taken hostage or abused in the process.

Ask yourself *What does the best in me say to do? How far do I have to go taking care of my parents? How much personal contact does this have to involve? How much do I want? How can I protect myself from traumatic mind mapping? If they start doing disgusting things, how will I deal with it?*

If you have children, finding an acceptable solution with dysfunctional parents is much harder. Agreeing to disagree about how they act around your kids doesn't work. You have both the responsibility and the right to ensure that they don't say or do destructive things to your children. Their willingness to acknowledge their shortcomings and change their behavior becomes far more important if they want unsupervised visits.

Here's my suggestion: Let their functioning determine how much personal access you give them. The shape and depth of your subsequent relationship should depend on how they respond. Apply the same guidelines when deciding how much access to your children to allow them. You don't necessarily have to cut off contact, nor do you necessarily have to give them access. Your children's welfare—and your own—should be the first priority.

In highly dysfunctional families, once you start dealing with your parents, you'll likely end up dealing with dysfunctional siblings who are super-glued to them. If you're looking for an ally, it's likely to be the outcast "black sheep" in the family. But keep in mind you have to be willing to deal with the whole family if need be, standing on your own.

"FALSE MEMORIES"

I don't simply take what my clients say at face value—except for this being the way they perceive or remember things. Or maybe it's the way they want me to see things. But when dealing with marital discord this is all I need. I ask clients to either confront or act according to their own perceptions rather than stipulating to their accuracy.

However, issues of inaccurate recall, confabulated experience or "implanted false memories" have to be considered when dealing with traumatic mind mapping. Given the high stakes for three generations of family members in these situations, the possibility of false memories merits serious consideration.

I studiously avoid implanting false memories by direct suggestion or innuendo. Clients visualizing experiences they already remember tends to minimize this problem. Vividness of memories does not, in itself, prove their accuracy. However the concurrence of major and minute details, cross-referenced against known facts, add some measure of confidence. The same goes for overlap among "anticipated parents" in mental dialogs, emails and letters from parents, and clients' reports of parents' behaviors in phone conversations or face to face. When there exists a broad spectrum of concurrent cross-validating data, derived from multiple sources and spanning different times and circumstances, the scientific principle of parsimony says it's prudent to accept clients' representations of their childhoods as accurate.

Post-Traumatic Growth

Tina subsequently showed definite signs of "brightening." Her countenance was healthier. She was more energetic. And her eyes were bright and twinkling. She seemed mentally sharper, like she had suddenly gained 20 IQ points.

Tina reported that her mind and brain felt different. She said it was like a fog had lifted and she could finally think straight. She felt lighter, like a weight had been lifted from her shoulders. She also described a funny feeling in her head like bubbles in carbonated water. It was clear to me that Tina's long-term regression had subsided.

Doug said he liked whatever was going on in Tina's head. Her emotional functioning was better. She seemed more stable, more like a mature woman. She also had more self-respect and carried herself differently. Doug liked the quiet, self-confident manner surfacing in her interactions with him and other people.

Tina and Doug's marriage blossomed, based on a proven trust. Tina became more interested in sex, and Doug found her more desirable and attractive. All of this contributed to their decision, finally, to have a child together.

I'm describing what is known as *post-traumatic growth*, the valuable positive developments and optimism about the future that can surface after a trauma. Almost half of trauma victims report some sort of positive outcome.[2]

The methods I've described maximize post-traumatic growth by developing greater self-efficacy. Moving beyond trauma to resilience involves seeing greater meaning in your traumatic experiences.[3] Finding the courage to make sense of what happened often reduces PTSD symptoms. Accepting what has occurred and understanding human vulnerability, paradoxically, gives you more strength to face whatever comes next. Your sense of identity reorganizes as you shift to a higher level of psychological functioning, greater cognitive complexity, and better emotional regulation.

Post-traumatic growth invariably changes your view of yourself, your world, and the people who populate it. This change is an overall gain rather than a loss. It allows you to remain resilient and optimistic in the face of disappointment. You feel self-empowered. This is how post-traumatic growth enhances your differentiation. (Appendix D contains more information on post-traumatic growth.)

On the other hand, post-traumatic growth doesn't solve everything. Tina and Doug's resolution with Tina's parents left a lot to be desired. As I anticipated, Mother didn't commit suicide. She was willing to threaten this to pressure Tina; but when it became clear Tina wasn't going to succumb, Mother was much too narcissistic to hurt herself.

Tina and Doug thought long and hard about revealing their decision to have a baby to the prospective grandparents. Doug didn't want to tell them at all. He didn't think they deserved it, and he didn't want to face the barrage of questions he knew would start. Would Mother and Father be in the delivery room? How soon could they come to see the baby? The thought of leaving his child alone with Tina's parents almost made Doug violent.

Tina was solid and stable. She said, "We'll tell them when we agree it's right and not before. As for the delivery room, giving birth won't be anything like our wedding. Aside from the medical staff, it's going to be you and me and our baby. And as for future visits and babysitting, there's no question our baby's welfare always comes first. We'll see how my folks do during supervised visits before we think about leaving our child alone with them."

THE MORAL OF THE STORY

Someone once said the importance of marriage isn't that parents give rise to children but rather that children give rise to parents. It isn't easy to navigate complex issues like children's benefits in seeing their grandparents, your parents' desires to see their grandkids, and your responsibilities to your child, parents, spouse and yourself. But this difficult path can turn out to be a God-send.

Time and again I've watched people from wretched backgrounds break the intergenerational transmission of trauma with their own children. This could not be possible if our brains couldn't recover from the worst we humans can do. Clients like Tina and Doug reinforce my belief that there is nothing more awesome or unstoppable than when the best in us stands up. Refusing to suppress our functioning to maintain dysfunctional relationships is uplifting to the human soul.

15

CREATE POSITIVE MOMENTS OF MEETING

We have arrived at our final chapter. As you've learned about the incredible world of mind mapping, I hope you've gained new appreciation for how we are all connected through our minds and brains, and how we always affect each other for better or worse. In particular, now you know the neurobiological impacts that parents have on their children's brains through disgusting behaviors like sexual, physical, and emotional abuse. In some cases, all it takes is willful self-blindness and self-deception. You recognize the importance of addressing long-avoided personal, marital and family issues that may exist in your household. Putting this into action will make you a more effective parent and a better marital partner.

Up to this point, we've mostly addressed mind mapping with regard to negative experiences. Now I'd like to switch gears and conclude by discussing how to use mind mapping to create *positive* ones. Mind mapping can help you improve many parts of your life, like becoming a more effective leader, enhancing your teaching skills, providing others with encouragement and emotional support, instilling prosocial empathy and compassion, and making your sex life far more intimate.

Deliberately creating positive mind-mapping experiences helps you repair fractured relationships, and reduce the impacts of prior negative experiences you've previously co-created. This not only helps you, it helps other people tremendously, even if they don't understand what you're doing. You can use what you've learned about mind mapping to become a more positive influence in the world, and you don't need anyone's permission, cooperation, or assistance. You don't even need to tell anyone what you're doing. Just go ahead and do it.

Positive Moments in Love Relationships

Really bad arguments and nasty shouting matches are negative moments of meeting that stick in your mind for a long time unless you deal with them effectively. We've spent lots of time discussing this. But another way—certainly the most gratifying way—to fade bad memories involves creating *positive* moments of meeting. And there's probably no better (or more necessary) place to do this than in a committed adult love relationship. There is no greater gift you can give your partner than creating positive moments of meeting with him or her. Lest you think I've suddenly turned Pollyannaish on you, let me show you what I have in mind.

For instance, be more honest with yourself about your limitations, and stop expecting your partner to live within them. A mind that's blind to itself is a dangerous mind, and being self-deluded makes you unsafe to be around. Confronting yourself in front of your partner allows your partner to relax in your presence. It also earns your partner's respect. Romantic relationships improve dramatically when you stop demanding that your partner go blind to who you really are. Be forewarned, pretending to confront yourself when you're really not is guaranteed to backfire. The moments of meeting that will occur thereafter are not going to be positive ones.

One client, Samuel, approached his wife and said, "I don't know how you live with me." Samuel's bluntness shocked her. "What are you talking about?" she replied.

"I'm so incredibly blind. You keep telling me I'm insensitive, and I always deny it. I refuse to see myself. I don't really listen to you when you talk. You constantly have to repeat yourself, and then I tell you you're always being critical. I must drive you crazy."

Samuel's admission was an incredible gift, much better than money, flowers, or candy. It's also a far more effective aphrodisiac. This made Samuel easier to love, and it was all done through mind mapping. Positive moments of meeting like these are powerful enough to steer relationships in entirely new directions—as long as they are followed up with genuine change. However, if Samuel were making a move to keep his wife from leaving him and he didn't shape up in the future, then what seemed a positive moment of meeting becomes traumatic mind mapping in his wife's mind.

How do you come up with these kinds of difficult self-admissions? I'm willing to bet you had several difficult self-realizations in the course of our discussions. It's all about being willing to map your own mind. I've even given you new ways to do a better job of this. Now all you have to do is take the hit about who you really are.

GIVE "THE GIFT OF MIND" TO THOSE YOU LOVE

Mind mapping is a crucial part of buying people good gifts. When you invest little or no thought in a present, you turn a potentially positive experience into a negative one because your recipients map out your mind and see there was no effort or consideration involved. Your gift says *I bought this for you because I had to get you something, but I'm not emotionally invested in it at all.*

People who enjoy giving gifts find half the joy comes from mapping their recipients' minds before selecting what to give them. Use your mind-mapping ability to discern what your partner *really* wants (rather than what you want them to have). There is no better present you can offer someone than this "gift of mind," something that says you were thinking about him or her when you made or bought it. Gifts with forethought say *You live inside me. Even when you're not here, I carry you inside my mind.*

For priceless gifts that cost nothing, put sticky notes with terms of endearment all around your house—in your partner's underwear drawer, in her book on the nightstand, under her pillow, or on the bathroom mirror. Each one says you were thinking about your partner at the time you put them up. Dare I suggest your kitchen is not off limits (as long as it's not full of sexual innuendo when there are kids in the house). Whenever you offer someone the gift of mind, whatever you give will be received like pure gold.[1]

STOP MASKING YOUR MIND

We've established that it's easier for your brain to map other people's minds than it is to map your own mind, which means you're better at mapping your partner than mapping yourself. It's also pretty common for folks to be more comfortable mapping other people than they are letting themselves be mapped. Guess what happens when this comes together? When you and your partner mask your minds from each other, while you're also furiously mapping each other, you end up driving each other crazy. This is a pretty common emotional "squirrel cage" in relationships.

For your relationship to have a secure foundation, you have to let yourself be accurately mapped. This requires opening your mind without screening your thoughts and feelings and letting yourself be known as you truly are, warts and all. You have to self-disclose this without expecting your partner's acceptance and validation because neither they nor you will like everything they see. I call this "self-validated intimacy" and it's the hallmark of mature adult love. (I understand you don't want to reveal yourself until you feel certain of acceptance, but you can't be sure of this until *after you reveal yourself.*)

To pull this off, you have to hold onto your own self-worth and let your partner see the parts of you that you usually hide or deny.[2] Just accept that, more than likely, your partner already has you mapped. Give up your self-comforting delusion that he (or she) hasn't seen you yet. Stop worrying that he or she won't like every part of you. Believe me, your partner's probably long past that.

I joke with my more secretive clients that they play their cards so close to their chest they might as well be tattooed there. They need greater transparency and less denying, deflecting, or misleading, particularly when it's clear they've been accurately mapped by their partner. Refusals to fess up often don't serve you like you may think they do.

RESOLVING EXTRAMARITAL AFFAIRS

Resolving extramarital affairs can be difficult. You can't really forgive someone until you've accurately mapped his or her mind, and you can't do that if that person is mind masking. When adulterers continue to mask their minds, this increases tensions in their relationships. And when you avoid mapping your adulterous partner's mind because you're afraid of what you'll see, you'll probably never be able to truly forgive that person.

Affairs themselves don't usually destroy relationships as much as adulterers' refusals to stop masking their minds. The affair may be long over; but when adulterers continue to hide small details that contradict previous versions of their stories, there's no peace. Even when you don't know the exact content of what's being hidden, you can frequently tell when adulterers are still mind masking. When this is pointed out, adulterers often make it worse by complaining *Why can't we put this behind us? Why can't you let this go?!*

The only real way to recover from an affair requires you and your partner to confront yourselves and open your minds to each other. As painful as this might be at times, this gives you something solid on which to build a different future, one in which true forgiveness is possible. *This is where you can end up because true forgiveness comes at the end of the healing process, not the beginning.* When you try to jump-start the process with forgiveness, you'll soon be dead in the water. That's because you've actually formed a collusive alliance to *not* look too closely at what really happened.

When partners stop masking their minds from each other, only then will they finally find peace with each other. Now that you know about mind mapping and mind masking, you have the tools to directly accomplish this. I've been impressed by the ability of couples to get over past affairs when partners finally talk straight with each other.

If you want to stay with your partner regardless of what he or she does—or regardless of what you've done—then by all means "protect yourself" by masking your mind. Just consider the real possibility that you're not protecting yourself at all. All you're doing is pacing yourself so you can tolerate future affairs, forever.

DEEPER INTIMACY IN AND OUT OF BED

I know some readers will be gravely disappointed if I don't say something about improving sex and intimacy. Someone once described the difference between dating and marriage as exchanging the hurly-burly of the chaise lounge for the deep comfort of the marital bed. What's lost in this analogy is the fact that you can be married and have plenty of hurly-burly too. Knowing each other (and accepting that you're known) provides the basis for being at peace with each other *and* screwing each other's brains out. Or actually, healing each other's brains.

Here are three well-developed ways to use mind mapping with your partner to make a troubled relationship better and a good one great. They all involve interpersonal neurobiology, so they're also good for your brain. I didn't understand the role of mind mapping when I first wrote about them in *Passionate Marriage*. Thanks to modern brain science, now I do. At first, you might think I'm just describing how to get your body involved; but if you pay attention, you'll realize all three activities have everything to do with using your mind to create intense positive moments of meeting. (You'll find the science underlying these "conjoint somatosensory neuroplastic activities" in Appendix D.)

HUGGING 'TILL RELAXED

Don't be fooled by how straightforward Hugging 'till Relaxed may initially seem. It's also incredibly powerful. Stand with both feet on the floor about a foot away from your partner. Get balanced on your own two feet. Now shuffle forward and put your arms around your partner while maintaining your own balance. Refocus your attention on yourself. Calm yourself down. *Way down.* Relax your body. Quiet your mind. Progressively slow your heart rate and deepen your breathing. Do this for at least ten minutes.

(Did you just think to yourself *Ten minutes?! You must be kidding!* That's all the more reason to give this a shot.)

Why is Hugging 'till Relaxed so effective? From what I can tell, doing it produces seven conditions that brain scientists say facilitate positive brain change.[3] Among other things, quieting yourselves while you're in physical contact with each other creates "body learning," which powerfully impacts your brain.

Your brain maps your partner's mind through his or her body. How does your partner's body feel to you–tense and rigid or soft and relaxed? Perhaps he or she was less-than-eager to do this with you. When your partner relaxes, you feel a kinesthetic shift in your partner's body through your body. Your brain immediately picks this up and turns this into mind mapping.

Great, you think to yourself, *my partner is getting into this.* This realization helps you relax further. This is called body-learning, and your brain eats it up. Hugging 'till Relaxed also produces a second factor that facilitates positive brain change: a collaborative alliance between you and your partner.

Now, the first time you try this, you might be surprised at how long it takes you to calm down. Some people never calm down during their first attempt at Hugging 'till Relaxed. This doesn't mean you failed or you're doing it wrong. You're merely seeing how much anxiety you and your partner are carrying around. But as you do it more and more, you'll calm down more quickly, become more relaxed, and achieve a deeper focus. Pretty soon the phone might ring as you're doing it and you'll barely notice it. Going through this process produces another brain-changing condition: sharing a moderate level of anxiety followed by calm. So I wasn't kidding when I said don't get too whacked out if you don't calm down at first. In the long run, it accrues to the process.

Hugging 'till Relaxed cannot only change you and your partner, it can benefit everyone in your house. When you're just starting, practice somewhere in private where you and your partner won't be disturbed–somewhere you can lock the door and enjoy a little privacy, free of distractions. But once you've grown adept, there's tremendous benefit to doing Hugging 'till Relaxed in a public area of your home, like your living room or kitchen. You and your partner will become "anxiety sponges," calming your family's environment. If you and your partner have been arguing (even in private), you'll absorb the anxiety that permeates your house instead of making everyone anxious.

This is an important benefit if your children are hyperactive or diagnosed with attention deficits or learning disorders. If you do what I'm describing, your kids will be mapping you like crazy. This will have a calming effect on them, and they will have a body-learning experience too. There's a trickle-down effect from your brains to your children's brains. This body-oriented positive moment of meeting will make the mirror neurons in their brains produce corresponding patterns almost as if they themselves were doing Hugging 'till Relaxed. It's not uncommon for hyperactive children to enjoy a newfound sense of calm and for children previously diagnosed with learning disabilities to finally be able to think straight.[4]

HEADS ON PILLOWS

Heads on Pillows is another powerful way of using mind mapping to create deeper intimacy in your relationship, to calm things down if you haven't been getting along, and to create a powerful positive moment of meeting. Like Hugging 'till Relaxed, Heads on Pillows is simple on the surface but powerful and deep.

To do it, you and your partner lie down on your sides facing each other. Each of you puts your head on your own pillow, placed just far enough apart so that your partner doesn't look like a Cyclops. Then lie there, soften your gaze, and look *into* each other's eyes until you've calmed down. This is not about staring; it's about opening your mind to each other, and getting more comfortable allowing yourself to be mapped. You'll also be triggering a special part of your brain that lights up when you look into another person's eyes.

These two techniques differ greatly from sitting alone in a room and meditating–and I would argue they're more powerful and meaningful–because you're harnessing interpersonal neurobiology. These are easy ways to create positive moments of meeting between you and your partner. And if you do them consistently, you'll train your brains to be more comfortable in each other's presence.

EYES-OPEN SEX AND ORGASMS

This final method is perhaps the most fun to practice. Gazing deeply into your partner's eyes in the middle of sex is a great way to make an intimate moment even more so. You might have to say "Look at me!" Because many people have sex with their eyes closed, in the dark, or watching the TV in the background! Don't be surprised if your partner says *Why are you looking at me? Am I doing something wrong?* You might respond *I want to do more than trade orgasms with you.* Or *I want more contact when we have sex.* Either way, you're telling your partner *Please map my mind. I want to map your mind. I want to create a positive moment of meeting with you.*"

When you do this, sex becomes more than a physical act. It becomes an expression of your mind and heart. If you have Eyes-Open Sex frequently, and do this for increasing lengths of time, eventually this can become Eyes-Open Orgasms where you reach orgasm looking into your partners eyes. If you really want to change the dynamics of your relationship and change your brains as well, then taking action–not just thinking about it–is the best way to achieve these goals. And without a doubt, these are two of the most heavenly forms of taking action you'll find.

POSITIVE INTERACTIONS WITH YOUR CHILDREN

One of the most important emotional investments you can make in your children is creating positive moments of meeting with them. Everything we just covered about giving mindful gifts and the gift of mind also applies to dealing with your kids. But here I want to approach this as giving your children peace of mind.

GIVE YOUR CHILDREN PERMISSION TO SEE YOU

Giving your children permission to accurately map your mind is on par with giving them a parent they can respect. Your children can't really respect you if they're not allowed to see you as you really are. Unfortunately, parents often demand their kids' total respect while simultaneously not allowing themselves to be accurately mapped. This gives children the distorted message that loving someone means going blind to them. Don't put your kids through this common mental wringer because it encourages them to develop gaps in their mind-mapping radar.

For instance, I knew two parents who had deliberately given their adolescent daughter an inaccurate picture of their own premarital sexual experiences, hoping she would remain a virgin until she got married. Shortly before their daughter left for college, this couple decided to tell her the truth so when it came to making her own decisions about premarital sex, she'd have a more realistic view of her choices. They wanted their daughter to feel okay if she decided to have sex while she was at school. These parents gave their daughter a great gift, showing her that they were more concerned with her welfare than preserving an idealized image of themselves. This is a great example of parents giving their child the gift of peace of mind.

Here's the cosmic joke: As terrifying as it is to let your kids see who you really are, it's one of the surest ways to earn their respect. You stop looking phony and gain credibility in their eyes. One way to do this is letting your children question how your behavior and values line up. Show your children they don't have to go blind to you, by being open to being seen as you really are.

You're probably not proud of your limitations, but your children need to see you can accept not being perfect. This tells your kids that it's okay for them to see your imperfections. When you do this, you'll actually be showing them the best in you because only the best in you takes responsibility for the worst in you. The worst in you lies about its own existence.

FREEDOM TO UNMASK THEIR MINDS

Another incredible gift you can give your children is showing them it's safe for them to open their minds to you. By showing them you can handle being seen and you won't punish them for seeing you, this allows them to feel comfortable enough to show you their minds. Establishing this when they are young builds a foundation of trust so later in life they'll be more willing to open up to you. Saying *You can always come to me and talk about anything* as your kids head off to college may make you think you've done your job as a parent, but it's pointless if you haven't shown them beforehand that it's safe for them to show you their minds.

I consider myself blessed that my grown daughter, Sarah, still talks to Ruth and me about things so personal it sometimes boggles our minds. Obviously we don't know what we don't know. And as parents we've had to accept that Sarah will mask her mind when she wants to. This wasn't always easy when she was an adolescent. But we now know what's going on in her life because she's not afraid to tell us. I can't fully express just how wonderful this is for all of us. Sarah effectively says to us *Take a look. This is my mind. This is who I am. I'd like to know what you think.* As a parent, these are some of the richest experiences you can ever have with your children, and it's all done with mind mapping.

ENTER YOUNG CHILDREN'S MENTAL WORLDS

Tell every pregnant woman you know to start talking about feelings and desires keyed to her six-month-old baby's behavior *and don't stop.* Put her on to the information in Chapter Four about how this facilitates her child's mind-mapping ability three years in the future. *Oh, you're crying! You look unhappy. Do you want a bottle, or do you have a poopy diaper? Let's take a look* is a good start.

If pregnancy is behind you, many things can foster a toddler's mind-mapping abilities, including talking about feelings, anxieties, hopes and concerns; shared pretend play; creating narrative stories; prosocial deception; telling jokes; sharing memories; and negotiating conflicts.[5] The wider variety of interactions, the better.

You can deliberately use mind mapping to create positive moments of shared attention with older children, in which you and your kids focus on or do the same thing. However, this isn't reducible to simply "doing things together." For example, baking cookies with your kids or washing the family car can involve more than lessons on how to do things or getting household tasks done. Beyond the transfer of knowledge that takes place, your brains and minds can be focused on the same thing at the same time for the same purpose, which is good for everyone involved.

For another example, you might enjoy playing chess with your son; but if there's no mind mapping involved, the experience lacks the level of richness it might otherwise have. However, if you smile at him while moving one of your pieces and say *So, now what are you going to do?* this adds whole new elements to the game. You're not just staring at the chess board; you're focused on your son. You want to see what he can come up with. You're also teaching him about friendly competition. In this moment of meeting, you're showing him it's okay to want to win, but you're not being cutthroat. You're getting into his head and allowing him into yours, and this process makes a positive experience even richer.

For another example, some fathers take their daughters fishing and never allow the girls to get into their heads. For other fathers, catching fish is half the fun, the other half is talking with their daughters in ways that invite them to share what's on their minds. You could say *Let's tell each other the best thing and the worst thing about sitting here fishing right now.* Engaging your daughter this way shows her you're willing to open your mind to her and that you're interested in her doing the same.

Playing games with young kids is great opportunity to create positive moments of meeting. That is, provided you play the games "correctly"—meaning from your child's perspective. This often means abandoning your own notion—or the inventor's notion—of how to play the game. Cast away your reservations, inhibitions, and need to follow the rules; and dive into your kid's world headfirst.

When adults play Monopoly, it's important that everyone moves the correct number of spaces and pays the right amount of rent. But little kids are far more interested in fantasy and make-believe, pursuits based in the mind. They're also testing out implanting false beliefs and seeing if they can get away with things. If you're digging a hole in the back yard, young kids don't want to hear you say, "How about we just dig down a couple inches or else we might hit an electrical line or something?" They want to hear, "Let's make sure we don't dig down to the center of the earth! I've heard it's made of molten metals. What would we do with all that molten metal if it started gushing out of ground?"

Invite your children to inhabit a shared mental world with you. They'll realize you're suspending your adult way of being, and you're getting down on their level. Kids absolutely love when you do this. Something as simple as digging a hole or baking a cake becomes an adventure and a positive moment of meeting.

BRAIN-ORIENTED SEX EDUCATION

Although the United States has less sex education available than many industrialized nations, we're doing better at educating teenagers about contraception and sexually transmitted diseases. However, other important aspects of sex education—the mental part—are rarely discussed.

Growing up in America you'd never believe that every canton in Sweden has its own public sexual health center where teenagers routinely go for sex education and contraceptives. Or that the public school curricula in countries like Finland actually teach younger children how to handle *sexual feelings*.

We need to do more than talk about the "body-parts" aspects of sex. Our kids are taught virtually nothing about how sexual experiences impact our brains.

Sex is a brain-wiring moment because its many elements, for better or worse, facilitate brain plasticity. Before, during, and after sex, you have high anxiety: *What if she gets pregnant or I get a disease? What if our parents find out?* You have high meaning: *I'm losing my virginity! This must mean he loves me! What if we break up?!* And you have close proximity: At that moment, your partner is as physically close to you as two people can get. This is why negative sexual experiences stick vividly in your brain.

The fact that sex wires your brain usually gets lost in conversations with children about sex. Typically, we want to say *Don't have sex! Just say 'No!'* even though research demonstrates this approach doesn't work.[6]

But what we really should be saying is *There's a lot more to sex than simply knowing how to do it, or how to prevent getting pregnant or catching diseases. If you're going to do it, then make it a good experience, and if it's going to be a good experience, you need to take care of yourself and your partner.*

There's much more at stake here than your reputation. Bad sex– meaning upsetting, highly anxious or forced sex, or sex when you're drunk out of your mind–is not good for your brain.

Kids can usually figure out how to plug genitals together on their own. What they need help with is not *how* to have sex, but rather, *why* have sex? And why not, and when not to. We have to engage kids in the decision-making process, which involves showing them we realize this is their decision to make.

To help teenagers learn responsible sexual decision-making, adults must stop masking their minds from their children so that kids stop thinking they were luckily conceived on the one or two times their parents copulated. Kids don't talk about sex with parents who appear to have no recent, first-hand personal experience with the topic.

THREE-STEP REPAIR STRATEGY

Once parents understand children's vulnerabilities to traumatic mind mapping, they're often motivated to repair any damage they might have previously inflicted on their own kids. I encourage clients not to waste time and energy feeling guilty because fixing the problem is more important than feeling bad about it. Instead, I suggest they implement an effective three-step strategy I've developed to address this problem. You can use this to help your children resolve prior traumatic mind-mapping experiences they may have had with you.

The first step involves examining a recent interaction with your children that you handled particularly poorly. Losing your temper and yelling at them is a good example. Pick a specific occurrence you can visualize rather than approaching this as an abstract or general concept. Here introspection is a waste of time. Looking at this through the lens of your intent (as you remember or confabulate it) will muddy rather than clarify your vision.

Instead, visualize the last time your problematic behavior occurred, looking at the scene either through your children's eyes as they observed your behavior or from a third-person "fly on the wall perspective." Fight your impulse to look at this scene from your own (first person) perspective, as you'll probably want to counter the emerging picture. As you look at the picture, ask yourself, "What would a prudent person presume about my intent and my feelings toward my kids, given the way I'm acting?" Next, use your maps of your children's minds to decipher how they saw you at that moment and what they took away from the experience. Ask yourself *What pictures am I implanting in my kids' heads about who they are?*

For instance, forget the self-serving notion that, by screaming at your kids, you're just trying to get their attention to get through to them. It's far more likely you're taking out your frustrations on them and trying to make them feel bad. Realizing what your kids have seen is the first step toward rectifying your prior negative impact. Such self-confrontations can be excruciating so stay focused on what you're trying to accomplish.

The second step is easier. Wait until the same issue occurs in an incident between your children and their friends. Young children (particularly girls) can be shockingly vicious to each other so you probably won't have to wait long. You get to look like the world's best parent, as you help your children decipher an upsetting vignette in which, for example, their friends turned on them and no longer acted like allies. This is a great opportunity to teach your kids that people who care about each other sometimes drop their alliances and do mean things to each other. You can tell them *It's no fun when this happens, and good friends strive to keep their conflicts to a minimum, but it does happen, and this doesn't necessarily mean you have to give up your friends.* Learning about relationship difficulties and being able to recognize them are actually good preparation for adult friendships and marriage.

The third step in this process is tougher. It's time to walk the walk and not just talk the talk. Help your kids revisualize some of your troubling prior interactions with them, spending plenty of time on the details to help them see what you did wrong. Remember, kids as young as four can understand when a parent wants something bad to happen to them. Telling your kids you were angry and wanted them to feel bad won't come as a complete surprise to them, but talking openly and honestly with them about this *will* surprise them. Helping them understand such moments sometimes occur between people who love each other not only makes this a teachable moment, it allows your kids to map your mind so they can see how you feel about them right there and then.

Here's the key to making this work: *Regardless of anything else–and I mean **anything else**–your children have to see the best in you stand up.*

Now is the time to give your children the ultimate gift: a parent they can respect. This means getting your act together. If you've mistreated your children, you can't just say you made a mistake and apologize to them. They've got to see you really mean it, by your not making the same mistakes again. If you really have changed and allow them to map your mind, they'll see you are truly different.

Other parents have to deal with different types of problems. For example, maybe you work too much and always put your kids second, expecting them to revolve around your work schedule or absolve you of showing up for their school activities. You may tell yourself you're working hard for your family; but actually, you're probably doing what you prefer and pursuing your own agenda. The fact that your family secondarily benefits from your overwork doesn't really change things.

As unflattering as this picture may be–and there are countless others–allowing your kids to map you accurately is far more important than apologizing to them. This involves more than just "talking to your kids" and finding the right words to express yourself. This all-important third step involves letting your children map your mind. Once you've crossed that bridge, speaking openly and honestly is usually enough to do the job.

The whole issue of apologizing to your children is extremely complex and deserves lots of forethought. If you are going to apologize, don't do it until you've helped them thoroughly understand whatever it is you've done and they've had a chance to think about it. It's also important you do this in ways that are not self-pitying. I've been through this with countless parents, and I can assure you that you won't beat your kids' radar. They're going to map you like never before; and if it even *smells* like you're faking or asking to be let off the hook, they'll spot it. This will reinforce their prior map of you and they'll disregard your efforts, which isn't good for them or you.

Understand that, when you apologize, you're putting your kids on the spot, forcing them to either accept or decline your apology. Most children don't dare reject an apology from a parent. Make it clear that you don't want or need their support at this moment, and they have time to think over how to respond. If they immediately say they forgive you, thank them. But tell them the focus is what they need, not what you need. Suggest their feelings may change after they think about this. This is your turn to be the parent and look after your children, and their time to be kids and have their needs come first.

Reducing your children's anticipatory traumatic mind mapping is far more important than apologizing to them. You do this by changing your behavior. As their anticipatory mind mapping fades, you'll see a difference in how they react to you. This shift won't feel like night and day, the signs will be subtle at first. But when this starts to happen, it will feel just as good to you as it does to them. Epigenetic research suggests you'll have done something that may even provide trickle-down benefits to the brains of your great-grandchildren born after your lifetime.[7]

If your apology contradicts your children's map of you, you'll have their undivided attention and new opportunities to create positive moments of meeting unlike you've had before. But once you talk with your children about your shortcomings and past failings, the countdown clock starts. You need to summon all your strength and get off your arse and handle things differently. You can't keep telling them you know you're falling short. Repeatedly saying, "I'm sorry I let you down," just becomes another form of abusing and exploiting your children. Frustrated kids quickly supply the perfect answer to lame apologies: *If you're really sorry, then **stop!***

POSITIVE MOMENTS WITH FRIENDS

Relationships with good friends can be tricky. As close as you may feel to them, figuring out how far to go with them in certain situations can be difficult. For example, let's say one of your friends likes to be sarcastic. On his good days he's funny, but on his bad days he's downright mean. When he aims a hurtful remark your way, you tell him you'd prefer he didn't speak to you like that. He says he didn't really mean it and offers a trivial apology. You reject his apology and restate that you think he did mean to hurt your feelings. He walks away mad. Did you do the right thing?

For another example, let's say a different friend lost her job two weeks ago, and she's gotten drunk every night since then. Concerned about her behavior, you bring up her excessive drinking, but she doesn't take it well. Suddenly, your friendly intervention turns into an ugly argument.

Regulating yourself during intense interactions with friends is a very important ability. You have to remain calm, say what needs to be said, and know when to stop. If your friend gets upset, the first thing to do is consider whether you went too far or your delivery was off. You have to decide whether to shift the conversation to a less threatening subject or remain focused on that topic. What you decide hinges on your ability to map other people, your map of this particular friend, your investment in the friendship, and your ability to know your own mind.

Maybe you didn't really have your friend's best interests at heart, or maybe your own feelings of superiority came through. Introspecting about your motives and intent won't help. Re-visualizing your behavior from a third-person view makes it easier to see how you actually came across and helps you understand what your friend saw or how your friend may have misunderstood you. This doesn't necessarily mean you end up agreeing with your friend's assessment of what you said. You have to decide this for yourself.

Previously we've seen how destructive people play three moves ahead, but this is also an integral part of maintaining collaborative alliances with friends (and spouses, children, and parents). Playing three moves ahead in heated conversations with friends helps you achieve the positive outcomes you're seeking. You might say to yourself *I see where this is going, given how my friend's mind works. She's going to steer us into an endless argument, and we're going to end up being angry at each other. But I'm not going to let that happen. I'm going to make a move right now to prevent that from occurring.*

To be a good friend, you have to be willing and able to deal with your friend's attempts to make moves on you. Going through this is what real friendship is about. Sometimes the positive moment of meeting only comes in talking through difficult issues.

But being good friends is also about sharing good times and enjoying each other's company. Too often we don't show the important people in our lives how much we love them and how much they mean to us. Open your mind to the people you love. Friendship requires mind mapping, prosocial empathy, and compassion, rather than masking your mind.

CREATING A HEALTHIER WORKPLACE

Ironically, dealing with co-workers can turn out to be an even bigger challenge. Their importance in your life stems from the fact they happen to be employed at the same place as you and you need to get along, rather than stemming from your liking them as people or caring about them. For instance, imagine co-workers who need lessons in basic personal hygiene or judicious use of perfumes and colognes. They seem oblivious to the fact that you and others in the office find it difficult to be within arm's length of them.

You first need to map out whether or not they have mind-mapping ability. Are they truly as unaware of other people's negative reactions as they seem to be? The next question is: Do they want the impact they're having or not? Some people wear perfume as a personal statement but don't feel the right to inflict this on others. Others stain the air with colognes to claim their personal space. Some folks do this through body odor, and asking them to bathe feels to them like your intruding on their privacy and lifestyle.

Workplace success greatly hinges on your ability to accurately map your colleagues. If they lack social skills (and soap and water!) but hunger for approval and acceptance, you can take them aside and offer some friendly advice. However, this won't work with hypersensitive, over-reactive co-workers who consider their scent a personal prerogative and identity.

With these folks, you have to make it clear the issue is about *your* personal space or allergies (or if necessary, workplace bullying and harassment) rather than their body odors or perfumes.

However, telling them you're fine with their private life, you're complaining about *public* health issues, sometimes gets you nowhere. You may be dealing with people who enjoy imposing themselves on hapless co-workers and get a sense of power out of making other people miserable.

Sometimes you have to go as far as showing certain co-workers that you can see what they're doing because they'll continue to harass you if you don't. Once you've demonstrated you can keep up with them, you don't have to beat this to death. You've accomplished what you needed to, so you can let it go. Even if it's necessary to draw in a supervisor to obtain relief, holding onto yourself and maintaining a quiet mind in heated moments at work will allow you to steer the outcome in a more positive direction.

But let's also celebrate important colleagues and co-workers in our lives. Our all-too-important careers couldn't exist without the open-minded inspiration and creativity we experience in the workplace. Science and industry would cease to exist if our mind-mapping abilities suddenly evaporated. Mind mapping influences everyone in the workplace, from CEOs to janitorial staff. When team members start masking their minds from each other, there's trouble ahead. The more co-workers you have, the more you're probably sitting in the midst of a mind-mapping hurricane. Where do you think workplace shootings come from? How else do you think they're going to stop?

Take it upon yourself to be the quiet-minded, cheerful-dispositioned, good-natured person in your office. That doesn't mean you're the one everyone turns to for advice. Nor does it mean you make yourself important by stirring the pot in other people's disputes. Be someone your co-workers want to be with when they're having a *good* time. When you smile at co-workers, you're not just showing them your face. You're showing them your mind. Be a mind that helps other minds feel healthy and function better. Be a hero. Smile! (It's interpersonal neurobiology!)

Reconnecting with Siblings

Let's say you have good relationships with your siblings, but you're looking for ways to deepen those relationships. If you followed the conventional route of focusing on mutual interests or activities, your upbringings would be prime candidates for discussion. Out of all the people in the world, only your siblings share your family experiences. But you might be amazed how many siblings never discuss such things throughout their lifetimes. Vast numbers of siblings have *no contact at all.* On the other hand, for this very reason you might not be surprised at all: You might be one of them.

For instance, let's say you've been distant with your brother or sister for years. You haven't spoken since you argued over the division of your parents' estate because he or she came out with more money than you did. Now that you're older, you've come to accept that the distribution of money is never going to change. You dare to wonder what life might have been like if you and your siblings had gotten along during the intervening years.

How does mind mapping factor into this situation? First, you need to reach out to your siblings and implant a picture of your mind that opens the door to more contact in the future. Contacting your siblings seemingly out of the blue is bound to set off red flags, and their mind-mapping radar will go on high alert. Just calling them on the phone is going to catch them by surprise and make them guarded.

A letter or email gives them time to reflect without putting them on the spot. Talk of "getting back together" would be premature. Suggesting you have a conversation with them by phone or further emails paints a more measured picture of your expectations. Skype makes it dramatically easier for everyone to map each other's mind. Meeting face to face is great if everyone's ready to invest the time and effort. Let your sibling(s) pick their preferred initial contact.

Your siblings are going to have loads of questions about what you're up to: What do you want? Why are you doing this? Why now? What's in it for you? What's in it for them? Don't be vague or coy answering these questions. Helping them accurately map you out will put their minds at ease. This increases their chances of seeing beyond unflattering mental maps they may have of you. Don't mask your mind! This doesn't protect you. Tell them you've given this some thought. Demonstrate some investment on your part.

Now that you have their attention, tell them why you're reaching out. Stop worrying they'll tell you to go fly a kite. In the example I've been describing, you should make it clear you've gotten beyond the old conflicts about your parents' estate. Tell them you've been wondering what your subsequent life would have been like if they had been in it. Let them know you've been wondering what their own lives have been like. Transparency will save you wear and tear in the long run, and ultimately get you more of what you're seeking.

If you and your siblings were raised by disgusting parents, reestablishing contact can be particularly sensitive. The first question to consider is *Are my siblings 'in the fold'?* Meaning, are they foot soldiers for your parents, who do your parents' bidding? Will they try to make you follow the unwritten dysfunctional family rules? Or are they available as independent adults to have a healthy relationship with you? Offering siblings a collusive alliance and pretending you don't see them won't work.

Typically, siblings who were banished from your parents' home or who left voluntarily as soon as possible are your best candidates. Opening your minds to each other by trading childhood vignettes can be incredibly helpful in repairing holes in your autobiographical memory.

Personally, I wouldn't trade my relationship with my brother, Steve, for anything. Steve is one of the most important people in my life. I can't imagine how our relationship could be more honest and open. We reveal ourselves to each other, and we confront each other. We backpack and snowshoe together, work together, and occasionally drink Don Julio 70 tequila together. Our trust and loyalty to each other have withstood differences of opinion and some deep disagreements.

I know my brother backwards and forwards, just like he knows me. I know his mind. He knows mine. I cherish the freedom and mutual respect this gives us with each other. If you have this with your siblings, you have my congratulations.

If you don't, I wish this for you and hope this inspires you to seek this for yourself.

CHERISHING AGING PARENTS

In most cases, your parents will grow old or contract serious illnesses. And when they do, you can give them something so valuable it will make all the presents they ever received look like mere baubles. At this point in their lives, your parents aren't looking for more material possessions to take with them to their graves. The closer people get to the end of their lives, the more important *meaning* becomes to them.

The best way to fill their lives with meaning is to create positive moments of meeting with them, including helping them review their lives through your conversations. If your parents struggle with Alzheimer's disease or presenile dementia, your autobiographical memory is better than theirs. Use your mind-mapping ability to implant pictures in their minds that they can enjoy. Even if they don't remember some events, the fact that *you* do makes it real to them–they know they've had a life in which that happened. Give them the gift that only you can give: Savor deeply satisfying and meaningful moments of meeting with them before they pass on. Here I speak from recent personal experience:.

My parents, Stanley and Rose Schnarch, both passed away in 2014 at the age of 92. They were married for 72 years, and they were inseparable. In the wonderful Denver-based Shalom Park Senior Community where they spent their final years, they were both referred to as "Stan-Rose" because you never saw one without the other. They were such gentle, good-hearted people that strangers went out of their way to be nice to them. My father died after a nine-month battle with cancer; my mother passed on suddenly from a bowel blockage six months later.

With assistance from my wife, Ruth, (and a team of terrific full-time caregivers), my brother Steve and I helped our parents pass on at home. I can say without equivocation that for the three of us, this was one of the most important events of our lives. None of us anticipated my parents' deaths could be positive experiences, but that's exactly what they turned out to be. We miss them dearly, but we weren't burdened with grief when they passed (or subsequently), which surprised me. I didn't even know this was possible. In fact, for a while I wondered if there was something wrong with me.

I'm still coming to terms with my own mortality and I don't have all the answers. But I've concluded that our lack of grief has to do with how we, as a family, handled their passing. We went far beyond stating our love for each other and saying goodbye. We opened our hearts and our minds to each other to the point that, by the end, there was nothing left unsaid.

I was out of the country when Dad's doctors told Steve and my mother about his terminal diagnosis. Mom couldn't bring herself to tell my dad, so this fell to Steve and me. We agreed that Steve would wait until I returned, so we could do it together. I needed to share this burden with Steve, but I also didn't want to miss out on the experience. Both of us viewed this as a responsibility and an honor. Telling your father he's going to die is a unique moment of meeting. You only get to do it once. There are no do-overs. Either the best in everyone stands up in that moment or it doesn't, and this largely determines how the news is received.

I can honestly say it was a great experience for all three of us. I was acutely aware that Steve and I were lending Dad our minds to prepare himself for his own death. Steve moved in with my parents three months before my dad passed away. I moved in two weeks before he died.

During our constant times together, we talked about everything: my dad's life, our lives, death, Mom, growing up together, Dad's being a prisoner of war during World War II, things he really liked, and why he never spent money on himself. My dad kept our spirits high by cracking jokes; and Steve, Ruth, and I went out of our way to implant positive pictures in my parents' minds about our lives with them. We brought up joyful memories and some hard times our family went through. We showed them old photos and pointed out familiar faces. Even if they didn't recall any of the people in the photos, the gesture wasn't lost on them.

We were making them feel good about what they were leaving behind. We were effectively saying to them *You still have a remembered past, and it will continue to live on, even when you're not here. We'll carry it for you and pass it on to our children.*

I know from personal experience how good it feels to say, "Dad, you've been a great father." But this doesn't produce as powerful a moment of meeting as asking, "Dad, do you remember driving up the Taconic Parkway and taking us to the duck pond in Tarrytown, New York?" Even if my dad had said, "No, I'm sorry. I don't remember that at all," that would have been okay because the important thing is that *I* did. As long as I remembered, then the event still existed because my father could map my mind. Letting your parents know you still remember and cherish such memories is a wonderful gift. This not only helps those who are moving on, it helps those who are left behind.

The day before my father passed away, he became disoriented and occasionally delusional. But even in that state, he was still the same sweet man he'd always been. He wasn't very lucid, and he didn't know who I was when he said, "We need to send home those people in the other room." He was referring to the caregivers tending to his physical needs who, like so many others, had fallen in love with my parents.

"Why do we need to send them home?" I asked.

"There was a problem, but it's fixed now, so they can go home and be with their families."

I said to Dad, "That's awfully nice of you to be thinking of them like that. You're a really nice guy!"

My father beamed. "I am a nice guy!"

At that moment, my father and I were just two people sitting in a room, talking and enjoying each other's company. We could have been two strangers who had just met. Instead of detracting from the poignancy of the moment, it actually added to it. I was struck, once again, by how truly kind and generous my father was. It was a wonderful moment of meeting, one of our last, and I'll never forget it.

LET'S DO THIS!

The more we learn about our brains, the more it becomes increasingly clear that we are all interconnected, right down to the neurons and neurochemicals in our heads. When I was growing up, the pithy educational slogan of the day was "A mind is a terrible thing to waste." As traumatic mind mapping receives more attention, perhaps this will change to "A mind can be a terrible thing, period." Buddhists have said this very thing for centuries. Maybe we'll be more willing to listen as this is increasingly supported by brain science.

Simultaneously, is there anything on earth as sublime as the human mind? All good deeds, all acts of valor, and all seeds of compassion are born there. Offering someone a quiet place in your mind and letting others know they live inside us are perhaps two of the greatest humanitarian acts we can give others and ourselves at the same time.

We humans are remarkably resilient creatures. No matter how much traumatic mind mapping you've endured, your cause is not hopeless. The tools in this book can help you resolve its negative impacts. In doing this, you not only benefit yourself, you improve the lives of everyone around you. Hopefully reading this has expanded your mind; deepened your humanity, prosocial empathy and compassion; and your level of functioning has started to rise. Now it's time to put this into action.

In the 1970s, the Beatles sang, "There will be an answer, let it be. Let it be."[8] A half century later, it's clear you have to do more than hope things happen. You must *be* the answer. Understanding and utilizing mind mapping is something you can and must do on your own. I'm hoping my words guide and support your efforts. Let's start on our respective journeys together. Let's make parenting a richer experience for children and parents alike. Let's interact with our kids so their brains stay healthy and they seek our counsel. Let's strengthen our romantic relationships and deepen our capacity to love each other. Let's make sex more peaceful, more enjoyable, and, yes, sexier, too. Let's treat older people with more dignity, respect, and compassion. Let's make the world a better place.

It's time to get down to business. You need to be part of making this happen. Get moving.

Let's *do* this!

Appendices

Appendix A

MIND MAPPING

Brain *Talk* is a cross-over book designed for both the general public and mental health practitioners and academics. To keep the chapters readable and engaging, much of the underlying scientific research has been placed in four Appendices. They are organized around four main topics: Mind Mapping, Traumatic Mind Mapping, Antisocial Empathy, and Creating Neuroplasticity. Here you'll find further discussion of important ideas, results of brain-scan studies, and the scientific underpinnings of Crucible Neurobiological Therapy. The Appendices provide a step-by-step development of the logic and scientific support for this approach.

There's a progression in style across the four Appendices, that's not of my choosing. The appendices on traumatic mind mapping, antisocial empathy, and creating neuroplasticity are highly readable and engaging. The last one also gets into neurophysiology in a significant way, but it's still fairly readable and easier to digest because it's integral to an ongoing discussion.

This first Appendix, however, presents you with an amazing amount of information about neurophysiology and neurochemistry where the thread of conversation is "how does the brain actually create mind mapping." It is the shortest of the Appendices, because the whys and wherefores of mind mapping appear in the body of this book.

I fear that this may lead you to conclude that the remaining Appendices are as overwhelming and impenetrable as you may find the very first thing you read here. Don't try to memorize all the brain parts. Read to get an overview, and enjoy how much the best minds in the field have figured out. Subsequent sections are easier to digest. But if you're interested in brain science, prepare to be amazed.

PARTS OF THE BRAIN INVOLVED IN MIND MAPPING

Many people are surprised to learn that scientists have figured out the neuroanatomy of mind mapping. The results of hundreds of studies are summarized in Table 1.

Brain Regions Involved in Mind Mapping	
Brain Region	**Brodmann's area**
Posterior regions	
Temporoparietal junction (including Inferior parietal lobe) (TPJ and IPL/pSTS)	39/40
Posterior cingulate/precuneus (PCC/PCun)	31/7
Superior temporal sulcus (STS)	21/22
Limbic-paralimbic regions	
Orbitofrontal (OFC)	
Ventral medial prefrontal cortex (vMPFC)	10/32
Anterior cingulate/paracingulate cortex (ACC/PrCC)	24/32
Temporal pole (TP)	38
Amygdala	Subcortical
Striatum	Subcortical
Frontal regions	
Dorsal medial prefrontal cortex (dMPFC)	8/9
Dorsal lateral prefrontal cortex (DLPFC)	9/46
Inferior lateral frontal cortex (ILFC)	44/45/47

Table 1. Brain regions involved in mind mapping. From Abu-Akel and Shamay–Tsoory (2011).

There was a time when just identifying the brain anatomy involved would have sounded like science fiction. But there's a lot more to know about mind mapping: For example, how does the brain discern whether the mind you are mapping is someone else's or your own? How does it map out emotions versus detecting thoughts and knowledge? What parts of the brain are involved with these specific activities? How does neurochemistry fit into this picture?

HOW THE BRAIN TRACKS WHOSE MIND YOU'RE MAPPING

In 2011, Abu-Akel and Shamay-Tsoory published a neurobiological integration of the neuroanatomy and neurochemistry of mind mapping.[1] It's pretty amazing to see how far the science of mind mapping has progressed.

Their model synthesizes voluminous research into functionally organized networks that underlie your ability to map other people's minds as well as your own. Because so much is actually known about the neuroanatomy of mind mapping, this model seems complex. But once you grasp its

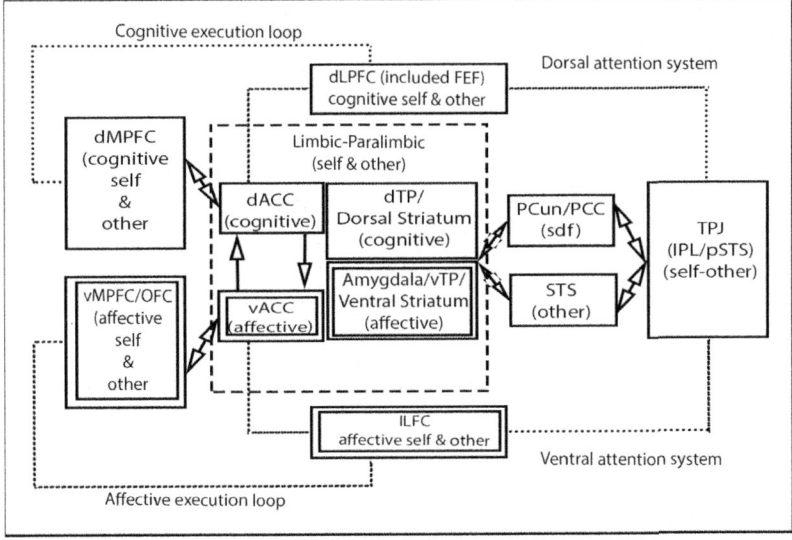

Figure 5. Brain systems differentiating mental maps of self and other people. From Abu-Akel and Shamay-Tsoory (2011).

overall organization, it's pretty simple.

Abu-Akel and Shamay-Tsoory organized these brain parts into three functional neural networks (see Figure 5). In this diagram, the forebrain is on the left, the mammalian (emotional) brain is in the middle, and the back of the (reptilian) brain is on the right. All arrows represent bi-directional connections.

Whether you are mapping your own or someone else's mind, the process actually starts in the most primitive part. Representations of your own and other people's mental states (the mental map itself) are first processed in the same area: the temporoparietal junction (TPJ) in the back of your brain.

VENTRAL AND DORSAL ATTENTIONAL SYSTEMS

These two mental maps are constructed by two structurally and functionally different brain attentional systems (the ventral and dorsal systems), which determine whether the incoming information pertains to you or someone else. The "top-down"-biased dorsal network, located on both sides of the brain, focuses on other people. The ventral system is lateralized on the right side of the brain and tracks information about you. The ventral network also controls reorienting responses and filtering information, Interestingly, research indicates your ventral network shows increased activity when you're mapping very familiar people. [2]

DIFFERENTIAL ROUTING THROUGH THE BRAIN

By selectively turning themselves on and off, this flexible interaction between the ventral and dorsal attentional systems sends different streams of information to the TPJ, depending on whose mind you're mapping.

- If the information regards another person, it is sent back to the TPJ through the posterior portion of the STS (pSTS) to the STS.
- If the information is about you, it is sent back through the inferior parietal lobule (IPL) to the PCun/PCC.

Moreover, automatic engagement and disengagement of

the ventral and dorsal attentional systems differentiates self-relevant or other-relevant information within the anterior cingulate cortex (ACC).

MAPPING OUT THOUGHTS VERSUS FEELINGS

It's pretty amazing that scientists have been able to figure this out, but get this: While this is going on (and also contributing to this process), your brain has highly specialized ways of mapping particular mental content.

The brain actually makes two maps for you and two maps for other people: one for emotions and feelings (affective map) and another for thoughts and knowledge (cognitive). Represented mental states formed in the TPJ are relayed through the STS or PCun/PCC to mammalian brain regions to be assigned cognitive or affective values.

- This limbic-paralimbic system (involving the amygdala, striatum, temporal pole [TP] and the anterior cingulate cortex [ACC]) determines whether representations are cognitive or affective.
- The dorsal stream of information processes cognitive mental states, and the ventral stream processes affective mental states.

MAPPING EMOTIONS AND FEELINGS

In Figure 5, the boxes indicate whether they are involved in mapping emotions (double line boxes) or thoughts (single line boxes).

- Affective mind mapping (double boxes) is mediated by a network of the ventral striatum, amygdala, ventral temporal pole (vTP), ventral anterior cingulate cortex (vACC), orbitofrontal cortex (OFC), ventral medial prefrontal cortex (vMPFC), and inferolateral frontal cortex (ILFC).
- Affective maps are relayed from the ventral stream through the ventral medial prefrontal cortex and orbitofrontal (vmPFC/OFC) complex to the inferior lateral frontal cortex (ILFC) for execution/application decisions.

MAPPING THOUGHTS AND KNOWLEDGE

Figure 5 indicates other parts of the brain are involved in mapping out other people's thoughts, knowledge, beliefs, and values.

- Cognitive mind mapping (single line boxes) is mediated by a network of the dorsal striatum, dorsal temporal pole (dTP), dorsal anterior cingulate cortex (dACC), dorsal medial prefrontal cortex (dmPFC), and dorsal lateral prefrontal cortex (dlPFC).

- Self or other cognitive maps are sent through the dorsal medial prefrontal cortex (dmPFC) to the dorsal lateral prefrontal cortex (dlPFC) for "executive" decisions (what you're going to do with your maps).

NEUROCHEMISTRY OF MIND MAPPING

Thus far we've focused on the neuroanatomy underlying the brain's mind-mapping ability. But there's a lot of neurochemistry involved too. It turns out there's a neurochemical system, similar to the ventral-dorsal attentional system, that allows your brain to keep track of whether you're mapping someone else's mind or your own. Mental maps of yourself and other people depend on a neurochemical system involving two neurotransmitters, dopamine and serotonin.

Figure 6 depicts a somewhat-overwhelming integrated neuroanatomical-neurochemical model of mind mapping. Figure 6 contains two panels:

- Panel A (left side) depicts cognitive maps of yourself and other people.

- Panel B (right side) represents affective maps of your own emotions and the emotions of other people.

These two sub-networks share the posterior regions (TPJ, STS, and PCC/PCun) for representing and distinguishing self from other mental states through the reorienting functions of the ventral and dorsal attentional systems.

In Figure 6, dopamine and serotonin are shown by two

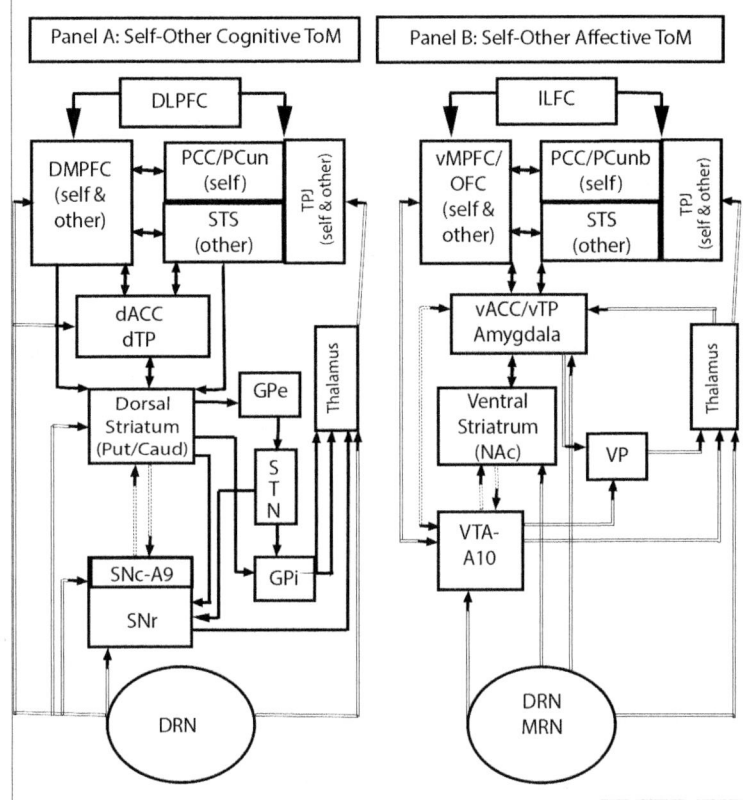

Figure 6. Serotonin–dopamine (DS) system in cognitive and affective mind mapping networks. From Abu-Akel and Shamay-Tsoory (2011).

ACC = anterior cingulate cortex	PCC/PCun = posterior cingulate/precuneus
DMPFC = dorsal medial prefrontal cortex	STS = superior temporal cortex
DRN = dorsal raphe nucleus	SN = substantia nigra pars compacta
DLPFC = dorsolateral prefrontal cortex	SNc = substantia nigra pars reticulata
GPi = globus pallidus interna	STN = subthalamic nucleus
GPe = globus pallidus externa	TPJ = temporoparietal junctions
ILFC = inferolateral frontal cortex	vMPFC = ventral medial prefrontal cortex
MRN = media raphe nucleus	VP = ventral pallidum
NAc = nucleus accumbens;	VTA = ventral tegmental area
OFC = orbitofrontal cortex	

different types of dotted lines (dashes [serotonin] and dots [dopamine]). Even if you don't take the time to trace how serotonin and dopamine selectively interact to create mind mapping, it's comforting to know that smarter people than you and I have figured this all out.

Dopamine and serotonin operate in an integrated fashion within the mind mapping network to create a highly sophisticated dopamine-serotonin (DS) system. The DS system has two functions:

- Regulating participation of the prefrontal brain circuits in constructing cognitive and affective maps of your own and other people's mental states.
- Maintaining a fine balance between cognitive stability and flexibility in maintaining and updating these mental maps.

Disorders Impacting Mind Mapping Ability	
Psychiatric and personality disorders	*Genetic disorders*
Autism	22q11.2 deletion syndrome
Asperger's Syndrome	Down's syndrome
Anorexia Nervosa	Martin-Bell Syndrome
Bipolar Disorder	Phenylketonuria (PKU)
Psychopathy and Antisocial Personality	Attention Deficit Hyperactivity Disorder
Schizophrenia	Sotos Syndrome (Cerebral Gigantism)
Social Anxiety	Spinocerebellar Ataxia
Neurological disorders	Ullrich-Turner Syndrome
Frontotemporal Dementia	William's syndrome
Alzheimer's Disease	Basal ganglia disorders
Multiple Sclerosis (MS)	Prader-Willi Syndrome
Traumatic Brain Injury (TBI)	Parkinson's disease
	Huntington's disease

Table 2. Disorders impacting mind-mapping ability.
From Abu-Akel and Shamay-Tsoory (2011).

Disorders Impacting Mind-Mapping Ability

Now that you know how the brain pulls off mind mapping, other questions become more pragmatic. As noted repeatedly in *Brain Talk*, your mind wants to believe that important people in your life don't have mind-mapping ability. To confront this, it helps to know what the research says about disorders that could possibly impair someone's mind-mapping ability. These include a range of psychiatric, neurological, and genetic problems such as autism, Asperger's Syndrome and schizophrenia. (See Table 2).

The mind-mapping network can be disrupted at the neuroanatomical or neurochemical or genetic levels, giving rise to three different areas of problems:

1. Representational problems (ability to form cognitive and affective mental maps).
2. Attributional errors (ability to attribute mental states to self or other).
3. Execution errors (how mental maps are applied).

Mind mapping is not an all-or-nothing capacity. Disruption of the neurochemical or anatomical networks–by a DS system malfunction, for instance–can lead to varying degrees of impairment.

The DS system can explain abnormalities associated with attribution and application of mental states, ranging from lack of spontaneously applying mental states in Asperger's Syndrome, to uncontrolled application of mental states (overattribution) in paranoid schizophrenia. Research has shown some patients with Asperger's Syndrome have significantly reduced serotonin receptor binding. This has also been found in the progression of some schizophrenia. Damage to your TPJ could wipe out your whole ability to create maps of minds. Impaired ventral and dorsal attentional systems could diminish your ability to distinguish between self and other mental states. Impaired lateral PFC structures could impair your attributions of other people's mental states.

A word of caution is appropriate in understanding possible

DS disruption in Asperger's Syndrome. This applies to diagnosed patients with clear-cut Asperger's symptoms. It doesn't apply to people who appear socially inept, "clueless," or who do cruel things to selected people, but who otherwise function well in other aspects of their lives. This also does not apply to the common tendency to identify oneself "somewhere on the autism spectrum," or to presume your antagonist doesn't know what he/she is doing or recognize the impact he/she is having.

Psychopaths' Mind-Mapping Ability

Because this is such an important issue, let's pursue this one step further: Do mind-mapping impairments explain people's antisocial or psychopathic behavior? One study tested people on a standardized measure of psychopathy[3] and compared this with their performance on a standardized measure of mind-mapping ability called the 'Reading the Mind in the Eyes' Test.[4] This test requires identifying of people's mental states from photographs of the eye region alone. Results indicated that people who scored high on psychopathy didn't demonstrate impairments in general mind-mapping ability.[5]

Subsequently, Shamay-Tsoory did a more detailed study to examine whether psychopaths have a specific impairment in reading other people's emotions (affective mind mapping) rather than general mind mapping per se. She and colleagues used a task that differentiated affective versus cognitive mind mapping processing separately. They examined hard-core criminals diagnosed with antisocial personality disorder and high psychopathy, and patients with frontal lobe damage. Both groups showed very similar patterns of mind mapping impairments: impaired emotional mind mapping, but not cognitive mind mapping. This supports an ongoing theory of amygdala-OFC dysfunction in psychopathy.[6]

What are the implications of this? Let's assume your antagonist, who does inconsiderate, cruel, or hurtful things, has the same problem as convicted and incarcerated criminal psychopaths who habitually do harmful antisocial things. In other words, let's speculate that your antagonist has a milder form of the same OFC damage. At what point does this lead to a demonstrated behavioral malfunction? Is this an "early onset" problem where a little neurological or neurochemical difficulty has a big behavioral impact, or do behavioral impacts show only in more advanced damage? Let's assume a little damage creates a big impact because that's the argument your mind is predisposed to make.

It boils down to this question: Are people less responsible for their behavior if they truly have this problem? Shamay-Tsoory's study of convicted criminals says society's tendency is to hold people accountable. Mbembe Jabbi and Shamay-Tsoory's research on disgust reactions, and my clinical work on traumatic mind mapping, suggest there's good evidence that people who repeated trigger disgust reactions are, indeed, harming other people through their behavior and thoughts.

But when all is said and done, it boils down to the immutable reality that our brain, itself, holds people accountable when they trigger moral disgust in the people around them. When you're dancing around whether or not to "hold your antagonist responsible," you're fighting with all the interoceptive cues he or she is triggering in your body. As long as you do that, the traumatic mind mapping they inflict continues to create problems, because it forces you to dissociate from all the SAM memories this person triggers in you. (See Appendix C regarding VAM and SAM memory in trauma).

Appendix B

Traumatic Mind Mapping

Perhaps the easiest and quickest way to grasp traumatic mind mapping is to speak of it as a form of post-traumatic stress disorder (PTSD). Virtually everyone knows PTSD exists, and many soldiers have it. My father had PTSD from being a prisoner of war during World War II, and he never received treatment until late in life. I have no doubt Crucible Neurobiological Therapy will help many soldiers coming back from serving in the Middle East who experienced traumatic mind mapping during their tours of duty.

To most laymen, PTSD is synonymous with having been negatively impacted by something traumatic. You may not know that PTSD means something very different to mental health professionals who have been refining PTSD diagnosis for decades. When it was first recognized in the 1950s, PTSD focused on war-related trauma in men but ignored "domestic trauma" experienced by women and children such as rape and sexual abuse. Today, PTSD diagnosis has evolved to include a broader and less dramatic range of symptoms as society is increasingly willing to recognize these problems.

Traumatic mind mapping fits the current framework of PTSD as a disorder that develops in some people who experience a shocking, scary, or dangerous event. This needn't involve a dangerous event like wartime combat because experiences like the sudden unexpected death of a loved one can also cause PTSD.[1] Traumatic mind mapping also fits trauma expert Michael Finkelhor's definition of interpersonal victimization as "harm caused by someone behaving in ways that violate social norms, wherein malevolence, betrayal, injustice, and immorality are more central than in accidents, diseases, and natural disasters."[2] Traumatic mind mapping often triggers involuntary disgust brain reactions when mapping other people's social norm violations.

People who suffer from traumatic mind mapping display some but not necessarily all of the current behavioral criteria for a PTSD diagnosis. They also have other subtle symptoms not currently included. This is fully in keeping with the history of PTSD diagnosis and its continually evolving criteria. As society becomes more willing to recognize antisocial empathy and disgusting parenting, PTSD diagnoses will no doubt expand to take this currently invisible epidemic into account.

We need to assist combat veterans with PTSD, in part, by destigmatizing it. But whereas men are most frequently traumatized by war, accidents, assaults, and natural disasters, women are far more frequently traumatized by childhood abuse.[3] If you survey the general population, 17 to 33 percent of women report histories of sexual or physical abuse. If you ask people going for mental health services, the rates increase from one-third to half.[4] Ten percent of women report being raped as adults, but more than twice as many report childhood sexual abuse.[5] In the U.S.A. 60 percent of all rapes occur before victims are 18 years old. And 30 percent of forcible rapes occur before age 11, usually by family members.[6] These reportable offenses also involve traumatic mind mapping.

Women are more likely than men to be traumatized in an intimate relationship. In the United States, over half of four million annual assaults on men are by strangers. But over half of five million reported attacks on women are by people they know.[7]

The more you know the mind of your attacker, the greater the traumatic mind mapping this produces. It's also more difficult to recover from an assault by someone who continues to menace you than from a stranger you won't see again.

Stop and make yourself convert these awful statistics into actual events by visualizing them. You'll suddenly realize lots of people have traumatic mind mapping experiences. These statistics document the intergenerational transmission of subtle brain damage through cruelty, abuse, neglect, emotional reactivity, punitiveness, abuse of authority, and abdication of responsibility. It's not surprising that studies of traumatized children report problems with attention and dissociation, anger and impulse control, and difficulty maintaining relationships with caregivers, peers, and subsequent marital partners.

Unfortunately, we're only seeing the tip of the iceberg in three important ways. First of all, these statistics greatly underestimate their actual frequency because the vast majority of incidents go unreported.

Secondly, traumatic mind mapping is not limited to reportable and legally actionable behavioral events. All it takes is someone's deplorable-but-not-reportable thoughts and actions. Most adults have automatic disgust reactions to imagining a child's being raped. The same holds true for visualizing a man beating his wife. So imagine looking into that person's mind on a daily basis–especially if this is your parent, your spouse, or another trusted person.

PTSD involves stress or fright even when people are not in danger. But traumatic mind mapping highlights that danger is not confined to other people's behavior. It can be what's in their minds. Even though physical, sexual, or emotional abuse may not be daily events, people who engage in these sorts of things are often thinking disgusting thoughts daily or hourly. Anticipatory traumatic mind mapping is often constant for children and adults living in homes where these sorts of things happen.

Thirdly, although diagnostic criteria have become more inclusive of domestic trauma, this is not reflected in corresponding increases in people being diagnosed with PTSD. An original 1990s study estimated PTSD prevalence in the general population at eight percent, with five percent among men compared to ten percent of women (using DSM-3 criteria). A 2003 resampling of over five thousand adult Americans yielded an estimated lifetime PTSD prevalence of seven percent, with four percent among men and ten percent among women (using DSM-4 criteria).[8] These results coincide with other estimates that only nine percent of people experiencing or witnessing traumatic things develop PTSD symptoms.[9] No population-based epidemiological studies have actually examined prevalence of PTSD among children.

Clearly, not everyone who experiences traumatic events develops PTSD. The 1988 National Vietnam Veterans Readjustment Study interviewed 3,000 American Veterans, and estimated their lifetime PTSD prevalence at 31 percent for men and 27 percent for women.[10] Their actual PTSD incidence at the time was 15 percent for men and 8 percent for women. A subsequent 2003 study of 11,000 Gulf War Veterans estimated their current incidence was 12 percent.[11] A similar 2008 study of almost 2,000 service members deployed to Afghanistan and Iraq found their actual prevalence was 14 percent.[12]

How Traumatic Mind Mapping Fits With PTSD

Many traumatized people who clearly have post-traumatic stress disorders don't qualify for PTSD diagnoses. Studies of women exposed to prolonged interpersonal violence, and physically and sexually abused children, consistently demonstrate serious psychological problems without meeting the criteria for PTSD.[13] Rape victims,[14] battered women,[15] and concentration camp survivors[16] show numerous long-term problems (e.g., controlling their feelings, focusing their attention, and maintaining relationships) without the dramatic flashbacks and intrusive memories of combat veterans.[17]

Post-traumatic stress disorder isn't the most common psychiatric diagnosis received by people with histories of abuse and neglect. A study of over 300 abused children found their most common diagnosis was separation anxiety disorder (an attachment disorder). Other diagnoses (in order of decreasing frequency) were oppositional defiant disorder, phobic disorders, PTSD, and attention-deficit hyperactivity disorder (ADHD).[18] In other words, PTSD is among the least frequent diagnoses traumatized children receive.[19]

DESNOS SYMPTOMS

The Diagnostic and Statistical Manual, published by the American Psychiatric Association, is used by clinicians and researchers to diagnose and classify mental disorders. From its inception, PTSD diagnostic criteria haven't focused on the post-traumatic problems many people experience. The original PTSD diagnostic criteria in DSM-3 (published in 1952) was woefully inadequate. The revised DSM-4 (published 1994) was better but still problematic.

Trauma pioneer Bessel van der Kolk demonstrated how people exposed to chronic interpersonal trauma consistently demonstrate psychological problems that didn't fit the revised PTSD diagnostic criteria. From 1990 to 1992, field trials of the DSM-4 PTSD diagnosis studied 400 traumatized people seeking treatment and 128 traumatized people simply residing in the community.[20] Victims of chronic interpersonal trauma like child abuse and domestic battering frequently displayed symptoms not included in the revised PTSD criteria. These symptoms were called "disorders of extreme stress not otherwise specified" (DESNOS). The DESNOS symptoms included difficulty with:

1. *Regulation of affect and impulses.* Affect dysregulation, unmodulated anger, self-destructive behavior, excessive risk-taking, suicidal preoccupation, difficulty modulating sexual involvement.

2. *Attention or consciousness.* Amnesia, transient dissociative episodes and depersonalization.

3. *Somatization.* Digestive system, chronic pain, cardiopulmonary symptoms, conversion symptoms, sexual symptoms.

4. *Self-perception.* Feeling permanently damaged, guilty, responsible, or ashamed. Nobody can understand. Minimizing problems.
5. *Perception of perpetrator.* Adopting distorted beliefs, idealization of perpetrator, preoccupation with revenge.
6. *Relations with others.* Inability to trust. Revictimizations or victimizing others.
7. *Systems of meaning.* Despair and hopelessness, loss of previously sustaining beliefs.

The field trials highlighted the important of DESNOS symptoms. Many people with childhood trauma meet the criteria for DESNOS but not for PTSD. Moreover, the community sample and the treatment-seeking sample had about the same prevalence of PTSD symptoms, *but almost half of the treatment-seeking sample also had DESNOS.* In other words, DESNOS symptoms rather than PTSD may be what drive people to seek treatment.[21]

The problem with the DSM-4 PTSD diagnosis was that it conceptualized memory imprints of particular experiences (flashbacks) as the central psychological consequence of trauma. This focused treatment on the impact of specific past events and processing specific traumatic memories. In contrast, patients with histories of early abuse and DESNOS symptoms needed treatment focused more on ongoing interpersonal, systemic, and situational traumas.

This study also found chronic interpersonal trauma from an early age gave rise to greater DESNOS symptoms than interpersonal traumas later in life or traumas from accidents and natural disasters. The younger the onset of interpersonal traumas, the more likely people have DESNOS symptoms. The longer they are exposed to traumatic events, the more likely they develop both PTSD and DESNOS. Interpersonal trauma can significantly affect psychological functioning beyond causing PTSD, including affect dysregulation, aggression against self and others, psychosomatic symptoms, dissociation and character disorders.

Not everyone who experiences interpersonal trauma, even early in life, develops PTSD. But research indicates victims of interpersonal trauma frequently experience DESNOS symptoms that destroy their quality of life without incapacitating them. If you do develop PTSD, treatment is less likely to succeed if you have DESNOS symptoms too.[22] All things considered, DESNOS symptoms, rather than PTSD per se, may be more significant to victims of interpersonal trauma.

TRAUMATIC MIND MAPPING AND PTSD DIAGNOSTIC CRITERIA

The current diagnostic criteria, DSM-5 (published 2013), broadened the "acceptable" triggers of PTSD to include exposure to actual or threatened death, serious injury or sexual violation. This includes directly experiencing the trauma, witnessing it in person, or learning of a traumatic event from close family members or friends, including repeated exposure to aversive details. If this involves death, it must be an actual or threatened violent or accidental death and not "merely" death of a loved one from illness.

To be diagnosed with PTSD, an adult must have the following for at least one month: At least one *re-experiencing* symptom and one *avoidance* symptom, plus at least two *arousal and reactivity* symptoms, and two *cognition and mood* symptoms.

Re-experiencing involves flashbacks, spontaneous memories, and recurrent dreams of the traumatic event, including sensations, images, thoughts, or perceptions, and accompanied by intense or prolonged psychological distress. *Avoidance* means avoiding distressing memories, thoughts, feelings or external reminders of the event, including "emotional numbing" and dissociation. Negative *cognitions and mood* refers to things like acute depressions, distorted self-blaming, estrangement from loved ones, and inability to remember key aspects of the event. *Arousal and reactivity* refers to "fight or flight" reactions, aggressive, reckless or self-destructive behavior, sleep disturbances, and hypervigilance.[23]

DSM-5 took a positive step forward by including two

criteria for childhood PTSD, one for children younger than six years (PTSD Preschool Subtype), and PTSD with prominent dissociative symptoms (feeling detached from one's own mind, body or experiences).[24] DSM-5 recognizes that children and teenagers can have extreme reactions to trauma that differ from adult symptoms.

In children under the age of six, for instance, this can include bed-wetting after having learned to use the toilet, forgetting how to or being unable to talk, acting out the event during playtime, and being unusually clingy with a parent or other adult.

DSM-5 now incorporates many DESNOS features reported by people who experience physical and sexual abuse, and domestic violence. But there are still many impacts of trauma that don't qualify for a PTSD diagnosis.

Beyond DSM and DESNOS

In writing about traumatic mind mapping, I'm not describing people who qualify for a traditional diagnosis of PTSD. My clients rarely suffer from intrusive flashback recollections accompanied by panic attacks or dissociation. Perhaps the closest some come to this is having intrusive visions of their mates' extramarital escapades.

DSM-5 specifies the "recognized circumstances" for a PTSD diagnosis: actual or threatened death, serious injury or sexual violation. A handful of my clients had traumatic mind mapping involving violent or accidental deaths, a small number had partners seriously threaten suicide. One had a parent die through the parent's refusal to be taken to a doctor. Other than this, the vast majority of my clients wouldn't qualify for a PTSD diagnosis.

Brain Talk describes many clients who might qualify under "sexual violation," but not really. This refers to things like rape—meaning behavioral sexual violations, the kinds of things that can be proven in a court of law. If you end up in an emergency room for rape these days, most ER staff understand that you've been traumatized and you may develop PTSD. But DSM-5 PTSD criteria doesn't cover being shown pornography by your parent, or your sibling coming to your room frequently to display his or her genitals in your face. The brain impacts are not as acute or immediately disabling as with rape, but they occur nonetheless.

My clients have the same symptom constellations observed in PTSD, only less extreme. When I specifically inquire, all my clients report "hot images" of past traumatic mind-mapping events. Whether or not they qualify for formal diagnosis of PTSD, they certainly have traumatic memories. This isn't like a flashback of a kid wearing a suicide vest who looks at you before blowing himself up. But the day you realize your father has an inappropriate sexual interest in you, or wants to drink himself to death, you don't forget these interactions. Many clients with histories of traumatic mind mapping have intrusive thoughts of worthlessness (despite notable success), agonizing ruminations, constant nervousness and occasional anxiety attacks.

DESNOS was an important attempt to recognize, legitimize, and treat "peace time" traumas like rape and domestic violence. Because folks like Bessel van der Kolk pushed for developmentally appropriate trauma diagnoses, DSM-5 diagnoses now address symptoms observed in abused children.[25] But currently there aren't trauma diagnoses for people who experience more nuanced sexual and emotional abuse. It's terribly important to understand how "lower intensity" traumas like traumatic mind mapping actually impact the human brain, if you want to be able to correct this.

My clients who experienced repeated, severe, traumatic mind mapping have what appear to be brain-based problems, less dramatic than typical PTSD symptoms, but significant nonetheless. The problems I've identified don't fit within DSM-5 PTSD criterion, or the behavior-based DSM diagnostic system. They fit better with the brain-based diagnostic system being developed by the National Institutes of Health (NIH) under the current name of "*R.DOC*" with the ultimate goal of promoting brain-based psychotherapies and research.

Varieties of Interpersonal Neurobiological Problems

What once were "disorders of extreme stress not otherwise specified," now are part of how therapists understand and diagnose trauma per se. With this in mind, I have identified a new series of DESNOS symptoms: Interpersonal neurobiological problems resulting from traumatic mind mapping. Perhaps these will also become mainstream ways of approaching trauma as traumatic mind mapping becomes more widely recognized and understood.

I've grouped a large number of impairments in my clients into five categories of problems. No doubt time will show some should be merged into a single impairment, just as others will certainly be added to this list.

MIND MAPPING IMPAIRMENTS

1. *Gaps in mind-mapping radar.* Narrow-band mind-blindness to significant people, often coupled with general hyper-vigilance.
2. *Holes in autobiographical memory.* Retrieval errors for prior personal events.

NON-PSYCHOTIC THINKING DISORDERS

3. *"Spaghetti Brain."* Sudden short-term decrease in cognitive functioning, foggy thinking, loss of precision, overly concrete thinking, or overlooking the obvious.
4. *Thought blocking.* Narrow-band restrictions in thinking (strictures), critical self-censorship in thoughts or communications, usually regarding a particular person, organization, or topic.
5. *Psychotic thinking patterns in non-psychotic people.* Thought patterns of a psychotic individual, wired into another person's mind/brain through repeated traumatic mind mapping.
6. *Uncontrollable tangential thinking.* Inability to remain focused on difficult topics, producing rambling conversations.

7. *Repeated reflexive thought patterns.* Distorted or manipulative thought patterns endemic to family interactions, e.g., reflexive defensive "batting away' response to all questions.
8. *Acute loss of self-awareness.* Saying and doing things under stress with complete lack of self-awareness. Often accompanied by confabulated inner experience.
9. *Non-psychotic mental "voice."* A single malevolent mental entity that preys on insecurities, usually presenting itself as an ally or the individual's own mind. The "voice" is often derived from personality and thought patterns of disgusting parents.

NON-PSYCHOTIC EMOTIONAL REGULATION DISORDERS

10. *Acute emotional regression.* Precipitous loss of emotional equilibrium, feeling overwhelmed by one's own feelings and emotions, acute feelings of anxiety or panic, accompanied by profound worthlessness and loss-of-time awareness. Can occur in seconds and resolve spontaneously in minutes, hours, or days.
11. *Steady-state (long-term) regression.* Reduced cognitive and/or emotional functioning, subsequent to acute regressions, lasting weeks, months, or years. Often mistaken as limited ability, ADHD, depression, or personality trait.
12. *Reactive regression.* An acute regression triggered by improvement in steady-state regression, wherein the person's functioning improves enough to permit new upsetting recognitions. This sign of progress is often misinterpreted as "just one more regression."
13. *Emotional "dead spot."* Lack of emotional responsiveness or indifference to negative impact on others, in someone who possesses empathy and

mind-mapping ability. Differs from true sociopathy in being treatable.

14. *Reflexive angry crying.* Uncontrollable crying when angry, coupled with complete inability to express anger directly.

DISGUST REACTION IMPAIRMENTS

15. *Reduced disgust reaction.* Blunting of physical or moral disgust reaction, e.g., physically abusing or emotionally exploiting other people.

16. *Eroticized disgust reaction.* Getting turned on by doing disgusting things, e.g., sexually abusing children, rape, coprophilia.

17. *Absent disgust reaction.* Total and complete loss of disgust reaction, e.g., necrophilia, serial killing.

ADDITIONAL IMPAIRMENTS

18. *Masked face.* Fixed facial expression, devoid of emotion, even during self-experienced feelings. The expression looks like a joyless mask or fixed smile glued on the face.

19. *Verbal, facial, and behavioral tics.* Sudden repetitive motor movements (eye blinking, head jerking, or shoulder shrugging), simple phonic tics (throat clearing, sniffing, or grunting) or complex tics like echolalia (repeating words just spoken by someone else) and palilalia (repeating one's own previously spoken words).

20. *Noxious touch.* Experience of touch as physically and emotionally irritating. Uncontrollable ticklishness.

21. *Troublesome orgasm trigger fantasies.* Dependence on upsetting or humiliating sexual fantasies in order to reach orgasm.

22. *Amusia.* Deficit in fine-pitch recognition and processing. Tone deafness manifesting as irritating unmelodic voice.

23. *Anosognosia.* Severe deficit in self-awareness, particularly regarding one's own serious personal disability or impairment.

Many of these disorders appear in combinations, wherein someone may have a variety of thinking problems. Another person may have multiple difficulties regulating their emotions. Some people have both types of problems and an impaired disgust reaction as well. I've also noticed relatively rare disorders, usually associated with damage to the anterior insula, showing up in combinations within severely dysfunctional families.

For instance, in Chapter Eight, I showed you the in-session physiological recordings of Michael and Irene. I documented that Michael had anosognosia in the course of their recorded family sessions. Anosognosia is a *severe* deficit of self-awareness, in which someone is unaware of the degree of their impairment, like denying being blind or paralyzed and unable to see or walk. Anosognosia shows up in clients saying and doing things with complete lack of awareness or memory.

In our family session, Michael and Irene and their boys were discussing an explosive topic that usually led to nasty arguments and hard feelings. Usually Michael was defensive and handled himself poorly. This time, however, they seemed to be getting somewhere. As the discussion approached an amicable conclusion, Michael encouraged his family to put off making an important final decision. He seemed to be functioning so much better, so everyone agreed.

Afterwards, Michael's sons asked him when the family was going to decide about what they discussed. Michael insisted that the decision had already been made in-session, although he couldn't remember what the decision was. The boys insisted that no decision had been made, largely because that's the way Michael wanted it. Michael was sure they were mistaken.

When Irene agreed with the boys, Michael was so sure they were "gaslighting" him (trying to make him feel crazy), he contacted me to complain about this. When I confirmed their description, Michael didn't believe me either. He only later accepted this when I confronted him with videotape proof.

Anosognosia typically results from physical damage to the anterior insula, but apparently you can also develop it from severe traumatic mind mapping and repeated disgust reactions (located in the anterior insula). Clients, who come out of homes where disgusting behavior is a daily event, often have no idea how impaired they are. They usually mistakenly think they are the healthier half of their marriage because they think (mistakenly) they came from a better family.

When Michael and Irene first came to see me, Michael maintained a second family with another woman whom he'd brought into the business he'd built with Irene. When he was growing up, both of Michael's parents frequently had affairs. He regularly slept with his mother, and on at least one occasion he had sex with her. In addition to anosognosia, Michael had an impaired disgust reaction, as well as problems thinking clearly, controlling his emotions and containing his abusive behavior. Michael also had a permanent wax-like smile ("masked face").

For her part, Irene was generally hyper-vigilant and suspicious, but she had a narrow-band gap in her mind-mapping radar, which rendered her mind-blind to her husband, Dieter. She also had holes in her autobiographical memory and severe emotional blunting. Irene's mother was self-indulgent and uninvolved, and she pushed Irene to raise herself and her siblings. Irene's father was volatile and emotionally abusive, and he tried to have sex with her on one occasion. Despite her difficulties, Irene worked hard to be a responsible parent to her two teenage sons.

Michael and Irene's younger son demonstrated a shocking amusia. Amusia is a deficit in fine-grained pitch discrimination, associated with loss of ability to produce melodic speech. (In some cases it interferes with enjoying the emotional impact of music.) In the polarized family dynamics, this younger son was most closely aligned with Dieter. The older son, who was more aligned with Irene, showed no similar impairments.

People who have amusia have incredibly irritating voices, like nails on a chalkboard. Just listening to them is enough to set one's teeth on edge. The younger son's almost-inaudible gravelly voice made it unpleasant to converse with him, although he was extremely intelligent and talented. This exacerbated his frequent panic attacks and deep depression, leaving him seriously withdrawn.

Despite this frightening array of impairments, Michael and Irene demonstrated marked improvements as a result of treatment. They now function much better as individuals, as a married couple, and as a family. Michael's disgusting behavior has stopped, his cognitive and emotional functioning has improved, his self-awareness has markedly increased, and his attempts to be a good husband and father have become more frequent and effective.

Irene is more emotionally responsive and expressive, and she seems more alive with greater emotional depth. She is also no longer mind-blind to Dieter, and better able to handle occasions when he becomes manipulative or abusive.

Michael and Irene are respect-worthy examples of how even severely troubled people can repair themselves, when the best in them finally stands up. They have become more involved and effective parents to their sons, who are now showing the benefits of Michael and Irene's protracted efforts.

IMPACTS OF TRAUMATIC MIND MAPPING ON THE BODY

As far as I can determine, these various interpersonal neurobiological problems arise from the impacts of traumatic mind mapping on the central and autonomic nervous systems. All the terrible-but-common life experiences described earlier–being sexually exploited as a child, adolescent, or adult, or being physically assaulted by someone you know–all involve traumatic mind mapping. You'd have traumatic mind mapping had these experiences been inflicted by a complete stranger. But when your tormentor is someone you know–and love–the power and depth of traumatic mind mapping is exponentially greater.

This is born out in research studies of vivid persistent "flashbulb memories" in nonclinical populations. It's noteworthy that a minority of the negative memories in these studies meet the DSM-4 criteria for a traumatic event.[26] Flashbulb memories are a particular type of emotional memory (rather than a special class of memory) possessing a highly perceptual "live" quality.

Flashbulb memories come from events having three main characteristics: (1) surprise, (2) high levels of emotion, and (3) significant personal consequences.[27] Memories of circumstances in which you learn about important emotion-arousing events tend to be persistent and accompanied by vivid visual images.[28] Vivid memories are more likely to concern personally important events associated with strong emotions.[29]

Brain-scan studies suggest increased retention of emotionally arousing material involves an interaction of two amygdala-controlled processes: (1) arousal during encoding at the time of the event, and (2) the subsequent effects of stress hormones (e.g., epinephrine and cortisol).[30] These processes enhance storage for emotional material and increase or decrease recall of separate aspects of the event.[31] The overall amount of detail in a memory may be less important than the presence of particularly salient, individual details.

Traumatic mind mapping occurs when your brain/mind experiences "catastrophic interference," meaning you are unable to rapidly digest new information contradicting your preexisting mental maps. Your brain responds by ignoring relevant details.[32] Unfortunately, this leaves you with alternative competing representations in your memory, some containing and some omitting important information about the event. (See the dual-representation model of memory in Appendix D.)

TRAUMA AND THE CENTRAL NERVOUS SYSTEM

Anthony Damasio, who studies the neurobiology of the mind, emphasizes how our mind removes awareness of our inner body states, particularly when we're stressed. He proposes that the mental content of our minds often masks the inner states of our bodies, including visceral components of unprocessed trauma, allowing us carry on.[33] This is born out in neuropsychology and neuroimaging research demonstrating that traumatized people have problems with sustained attention and working memory. This makeses it hard for them to maintain focused concentration and be fully engaged in the moment.

According to Bessel van der Kolk, traumatized people "lose their way in the world." They react to sensory information with subcortically-initiated responses that decrease central nervous system (CNS) activity in brain regions involved in (a) integrating sensory input with motor output, (b) modulating physiological arousal, and (c) communicating subjective experiences in words.[34] Neuroimaging studies of traumatized people repeatedly demonstrate that, under stress, areas of their brains involved in "executive functioning" become less active.

Most of the time the most evolved parts of your brain run the show, giving you maximum adaptability to your environment. This is called "executive" or "top down" functioning. In high-anxiety, stressful situations, however, the more primitive limbic parts of your brain take over, usurping control from your more evolved prefrontal cortex. This is called "bottom up" functioning.[35]

Traumatic mind mapping flips this "circuit breaker." When you map out someone's mind—particularly someone important to you—if what you see is really "dark," it "blows your mind" and impacts your brain. Usually you don't freak out on the spot. You might feel dazed or confused, or find it difficult to say what's on your mind, or refute things you know aren't true. But when you've repeatedly experienced traumatic mind mapping, you're more prone to have your executive functioning shut down and your more primitive "fight or flight" reactions take over. Traumatized people show decreased activity in their medial prefrontal cortex (mPFC), which includes the anterior cingulate cortex (ACC) and orbitofrontal cortex (OFC).

Research consistently implicates the ACC in PTSD. The ACC is involved in experiencing emotions and integrating them with cognitions. It is functionally connected to the hypothalamus, amygdala, and brain stem, and helps orchestrate the autonomic, neuroendocrine, and behavioral components of emotions. The larger mPFC regulates stress hormone cortisol production, and mPFC dysfunction likely contributes to arousal dysregulation in PTSD. The mPFC also suppresses the brain's HPA-axis-mediated stress response, which we'll discuss momentarily.[36]

PTSD patients often show verbal memory deficits, presumed to result from hippocampus damage due to excessive cortisol.[37]

During moderate cortisol increases, the hippocampus acts as a negative feedback mechanism to reduce additional release. However, this mechanism breaks down when the hippocampus is flooded with excessive cortisol during trauma. Hippocampus damage impairs explicit memory and interferes with nerve regeneration to revitalize its negative feedback functions. For example, coal mine accident survivors with severe PTSD showed short term memory problems and reduced response in their ACC, IFO and mPFC.[38]

MRI studies of combat soldier with PTSD, and children with histories of physical and sexual abuse, show dramatically reduced hippocampus volume and associated memory deficits.[39] Reductions in hippocampus size are generally proportional to the severity of trauma experienced.[40] These anatomical changes are associated with reduced synaptic plasticity, inhibited neurogenesis (growing new cells), and neuron death attributed to high cortisol levels. Reduced hippocampus volumes appear over time rather than right after the trauma.[41]

People with repeated traumatic mind mapping often experience stress and anxiety between actual traumatic incidents. *Anticipatory traumatic mind mapping* causes you to reexperience danger when it is not actually present or long after it passes.

Anticipatory traumatic mind mapping probably occurs through a process called "neural priming." The brain uses hormones (e.g., arginine, vasopressin), neurotransmitters (e.g. catecholamine), and corticotrophin-releasing hormone(CRH) to "prime" the neuro-circuits that respond to stress and danger. These brain circuits are then hyper-sensitized and over-respond to stress, creating hypervigilance.[42] This amygdala-driven response results in greater release of ACTH and cortisol, which further exacerbate the impacts of chronic stress and trauma.[43]

Fortunately these trauma-induced impairments are potentially reversible.[44] Our brains are capable of neurogenesis throughout our lives, including in the hippocampus and PFC.

TRAUMA AND THE AUTONOMIC NERVOUS SYSTEM[45]

When treating traumatic mind mapping, I conduct in-session

real-time physiological monitoring of my clients and myself. This not only allows ongoing monitoring of everyone's stress levels in the moment, it also checks for the physiological effects of trauma known as *allostatic load*. Allostatic load is "wear and tear on the body," which accumulates over time when you're repeatedly exposed to heightened neural or neuroendocrine responses that result from chronic stress. Allostatic load is generally measured by indicators of cumulative strain on several organs, especially the cardiovascular system.

The amygdala, the critical brain structure in conditioned fear memories, distributes its output to brain stem areas that control the autonomic nervous system (ANS). The amygdala's connections with the hypothalamus controls the response of the hypothalamic-pituitary-adrenal (HPA) axis, a major neuroendocrine system composed of the hypothalamus, pituitary and adrenal glands. The HPA axis controls reactions to stress and regulates digestion, the immune system, mood and emotions, sexuality, as well as heart rate and heart-rate variability.[46]

Heart-rate variability (HRV) provides the best available measure of brain-stem regulatory integrity.[47] Heart-rate variability is the number of milliseconds between beats of the heart. Low HRV is correlated with coronary vascular disease and increased mortality, while high HRV is associated with resistance to stress.[48]

Your parasympathetic nervous system also influences your heart rate through the vagus nerve (the tenth cranial nerve) coming out of your brain stem, creating almost instantaneous shifts in your emotional and behavioral responses to stress.[49] The parasympathetic system consists of two branches, the ventral vagal complex (VVC) and dorsal vagal complex (DVC) systems. The DVC controls digestion, taste, and hypoxic responses in mammals, plus the lowest viscera of your abdomen and contributes to ulcers via excess gastric secretion and colitis. The VVC controls upper body organs including your larynx, pharynx, bronchi, esophagus, and heart.

Vagal tone refers to activity of the vagus nerve, the key part of the parasympathetic branch of the autonomic nervous system (ANS). Vagal tone is a good measure of the health of the

ANS. Healthy vagal tone is indicated by a slightly increased heart rate when you inhale, and decreased heart rate when you exhale. Deep diaphragmatic breathing—with a long, slow exhale—is key to stimulating the vagus nerve and slowing heart rate and blood pressure in times of anxiety.

In psychophysiological research, vagal tone influence on heart rate is taken as an index of the functional state of the entire parasympathetic nervous system.[50] Cardiac vagal tone is an accepted psychophysiological marker of emotional regulation and psychological adjustment. Higher vagal tone isgenerally linked to physical and psychological well-being.

For instance, higher vagal tone and vagal reactivity in infants and children correlates with emotional expressiveness, greater heart-rate acceleration and behavioral reactivity to painful stimuli (like having blood drawn), greater interest and attention toward novel stimuli, less distractibility, and more attachment security, empathic responding, social competence, greater ability to self-soothe under duress and fewer symptoms of psychopathology.[51]

Likewise, studies of adolescents and adults link lower baseline vagal activity to hostility, aggression, bulimia, anorexia, depression, anxiety, and panic attacks.[52] In adults, higher vagal tone correlates with decreased negative emotional arousal in the face of moderate-to-high stress, and better self-regulation.[53] High vagal tone seems to buffer marital hostility.[54]

Here's how this all comes together in treating traumatic mind mapping: PTSD involves fundamental dysregulation of arousal modulation at the brain stem level. People with PTSD show more arousal and less vagal control over their heart rate when facing emotional challenges.[55] Other studies suggest poor vagal tone may play a significant role in PTSD.[56] Extreme threat, particularly early in life, can significantly impair your ability to modulate your sympathetic and parasympathetic nervous systems in response to subsequent stress.[57] PTSD patients suffer from higher baseline autonomic hyperarousal and lower resting HRV, which suggests decreased parasympathetic tone.[58]

When cumulative allostatic load reaches a tipping point from repeated traumatic exposure, the ANS "fight or flight" response to stress wears out. Your heart stops accelerating to prepare you for action. Some PTSD patients do not have elevated heart rates.[59]

Moreover, lower vagal tone is associated with less facial expression of emotion and fewer expressions of interest and joy.[60] I have observed many clients with histories of severe traumatic mind mapping who have markedly reduced facial expression of emotion ("masked face."). They often appear joyless and disinterested in most things, but some also develop a mask of a pleased expression. A range of studies suggest people with lower vagal tone are less open to new experiences.

Low vagal tone in adults is correlated with inflammation, depression, negative moods, loneliness, heart attacks, and stroke. Low vagal tone is also associated with impaired immune system response. Adults with PTSD show increased levels of immune-system-damaging IL-6 cytokines in their cerebral spinal fluid.[61] Children who experience trauma have elevated IL-6 and evening cortisol, which predisposes them to developing PTSD.[62] This is particularly relevant because elevated IL-6 cytokine disrupts HPA axis function.[63]

The good news is there is growing evidence you can combat inflammation by engaging the vagus nerve and improving vagal tone. Recently, clinical trials demonstrated that stimulating the vagus nerve with a small implanted device significantly reduced inflammation by inhibiting cytokine production for patients with rheumatoid arthritis.[64] You can also achieve some of these benefits through daily yoga and meditation. Given all this research, there is every reason for brain-based psychotherapies to monitor heart rate variability during sessions.

Appendix C

ANTISOCIAL EMPATHY

Understanding antisocial empathy involves incredible cutting-edge brain research you've probably never heard about, and dealing with aspects of human nature you probably don't want to think about. This Appendix lays out the scientific basis for a difficult reality that is completely absent from modern psychotherapy and commonplace pop psychology. It covers how the brain creates empathy, the role of empathy in enjoying other people's suffering, what kinds of people enjoy this, and what they get out of it.

Another batch of brain research provides a conceptual reorganization that will challenge your picture of empathy itself. It offers a sophisticated picture of how emotional and cognitive empathy differ and how they fit with people's differentiation (personal development).

The final section of research on the brain's hard-wired disgust reaction will blow your mind, period. You'll be amazed how much you can learn about interpersonal neurobiology by studying disgust. By the time you finish reading this, you'll find it easier to comprehend and recognize how antisocial empathy plays out in many people's homes.

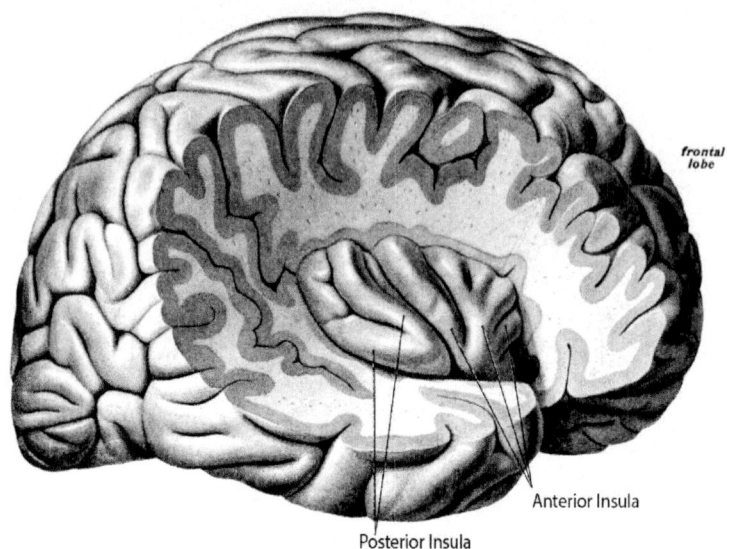

Figure 7. Right Anterior and Posterior Insula. From Sobotta's Textbook and Atlas of Human Anatomy (1909). Public Domain.

How The Brain Creates Empathy

To explain how your brain supports empathy, I'm going to focus on one specific part: your right anterior insula. This is not the only brain part involved in this universal human experience. But it's a very important part of empathy, and I don't want to confuse you by covering all the brain anatomy involved. By the end of this Appendix, where we'll discuss the human disgust reaction, you'll understand why I'm focusing on the anterior insula now.

Here's a picture of the human insula. (Figure 7.) Your insula is multifaceted, participating in awareness of gut reactions and body sensations, autonomic regulation of your gastrointestinal tract and heart, and control of your behavior.[1] Emerging evidence suggests it is part of a "salience detecting network" in your brain. Your insula is located at a key intersection between your frontal cortex and other "salience sensing" areas deep in your brain.[2]

You actually have two insula, one on either side of your brain. Each one has anterior and posterior (front and back) portions. Posterior portions of your insula detect actual stimulus intensities, whereas your anterior insula (AI) tracks your feelings and perceptions of your bodily states, and creates your subjective evaluation of your body's condition.[3] For example, the posterior insula are activated when research subjects receive painful stimulation, but their AI activate when they witness a loved one receiving pain.[4]

The right anterior insula (right AI) is more developed in humans than in any other species. It works with your prefrontal cortex to interpret and regulate your emotional experiences. Your right AI is particularly involved in emotional intelligence and self-awareness, integrating your body sensations with your physical reactions to produce conscious experience of your emotions, and awareness of yourself and other people and your environment.[5]

It may interest you to know that your *left* AI activates more than your right during positive emotions, when viewing happy faces, and affiliative moments of maternal or romantic love. These "attachment behaviors" simultaneously deactivate regions in your right AI associated with negative emotions, social judgment and assessing other people's intentions and emotions. Human attachment seems to employ a push-pull mechanism that overcomes social distance by deactivating networks used for critical social assessment and negative emotions.[6] In other words, maternal love and infatuated love are mind-blind.

Now let's get to empathy: Have you ever had gut feelings about another person's emotional state? Your anterior insula largely gives you that capacity, which is empathy. You probably never heard about the AI when people talk about empathy. But there's some chance you've heard about mirror neurons, because they've generated lots of buzz among therapists. Mirror neurons were initially discovered in monkeys, when researchers discovered that specific neurons fired in one monkey's head replicating what another monkey was doing.

Mirror neurons were probably a critical survival mechanism in human evolution, allowing your forebears to predict other people's goal-directed behaviors. This is an adaptive advantage if someone means to harm you. On the positive side, mirror neurons enhance your capacity for prosocial empathy and making other people feel good. Mirror neurons exist in various parts of your brain that make up your "social brain network," which includes your insula.

However, you're less likely to have heard about another "socially sensitive" neuron found in your insula (especially the right side): "Spindle cells" (von Economo neurons [VENs] named after the neuroanatomist who discovered them in 1929) are largely absent in other species.[7] Spindle cells probably emerged relatively recently in our specie's evolution.[8] In humans, these cells appear in small numbers around the thirty-fifth week of gestation, and adult numbers typically emerge by four years of age (when explicit mind mapping emerges).

It's no coincidence that spindle neurons are 30 percent more numerous in your right anterior insula than in your left.[9] Functional neuroimaging studies show the right AI is involved in subjective awareness of feelings, including disgust, anger, sexual arousal, and judgments of trustworthiness.[10]

Spindle cells respond quickly in emotionally-charged situations, allowing you to have sudden insights and make fast intuitive assessments ("snap judgments") in complex social situations, especially those involving uncertainty. They give you the capacity to adapt to changing conditions, or recognize when you've committed an error, and adapt to reduce further errors.[11] The unusually long VANs rapidly relay information processed within the AI to other parts of the brain.

Spindle cells and mirror neurons allow us to feel what other people feel emotionally (i.e., experience empathy). AI activation is positively correlated with people's scores on empathy scales.[12] The AI probably contributes to empathy by mapping other people's bodily feelings onto the internal bodily states of the observer.[13] Abnormal spindle cell development may contribute to reduced empathic and emotional responsiveness in autism spectrum disorders,[14] which is one of its core symptoms.[15]

Chronically traumatized people typically have reduced right insula activity compared to non-traumatized people. They have difficulty feelings their bodies and differentiating their emotions. They report feeling "dead" or "empty." Conversely, people with dramatic hyper-arousal typically have greater right insula activation. This is probably why repeatedly traumatized people either feel too much or not enough emotion.[16]

NEUROANATOMICAL BASIS OF ANTISOCIAL EMPATHY

Brain science suggests a neuroanatomical basis to antisocial empathy. A study by Mbembe Jabbi, who does neuroimaging studies on disgust, also explored how empathy for positive and negative emotions emanates from the "gustatory cortex." Jabbi and his colleagues showed that participants' empathy scores predicted their AI activation while viewing pictures of people with pleased or disgusted facial expression.[17] Previously, the IFO (largely consisting of the AI) had been reported to process other people's negative emotions, and support empathy for their negative experiences like pain. Jabbi's research proved that the IFO also involves empathy for intense positive feelings too.

In other words, Jabbi's research supports the view that the AI contributes more broadly to empathy by mapping other people's bodily feelings onto the internal bodily states of the observer. This is how the brain creates empathy. Empathy is an affective knowing of another person's emotions or sensory states.

With this understanding of empathy in place, now we can talk about *antisocial* empathy. Tanya Singer, a well-known empathy expert, points out that empathy differs from sympathy in ways that highlight antisocial empathy.

She writes: "Empathy is not necessarily linked to a prosocial motivation, whereas there is such a link from sympathy or compassion to prosociality. Empathy can have a dark side, for example when it is used to find the weakest spot of a person to make her or him suffer, which is far from compassion. Empathy has to be transformed into sympathy or empathic concern to elicit prosocial motivation."[18]

This is certainly a head-turning viewpoint of empathy for a lot of people: Empathy, like mind mapping, is neither inherently good nor bad. Empathy and mind mapping are both survival mechanisms. Whether they are prosocial or antisocial depends on how they are deployed. If you understand empathy from a brain science perspective like Tanya Singer does, antisocial empathy doesn't sound so blasphemous or off the wall.

When you discover that a majority of brain studies of empathy address empathy for pain perceived in another person, you realize that scientists have long been linking empathy with perceiving other people's negative states. Results suggest there's a brain 'pain matrix'–comprised of your anterior insula (AI), anterior cingulate cortex (ACC), brain stem, and cerebellum–that activates when you experience pain yourself or perceive someone else's pain.[19] Documenting exactly how the brain perceives another person's pain offers an unexpected vantage point for approaching a very difficult topic: enjoying other people's pain.

This is what we'll consider next. But when your brain/mind wants to argue *How could anyone enjoy seeing their child suffer?!* remember this moment, because you have just been handed the scientific answer: They couldn't enjoy their child's suffering if they couldn't detect it, and the human brain is wired to detect it.

Schadenfreude!

Antisocial empathy encourages people to do nasty, hurtful, destructive things to one another. One of the most difficult aspects of antisocial empathy is recognizing someone's enjoyment of another person's discomfort or anguish. This has a number of levels, each one having its own deplorable contribution: There is the experiential component of experiencing someone else's negative feelings and emotions, particularly their discomfort, unhappiness, disappointment and/or suffering. There's also the behavioral component of doing (or omitting) things to inflict these negative mental states. There's somatic awareness of pleasant sensations in your own body, and cognitive awareness of enjoying this person's negative experience. This all comes together in consciousness with an enjoyable self-awareness that you're not supposed to do this or like this.

Brain science offers shocking insights into antisocial empathy. Some of the biggest ones come from the scientific study of *schadenfreude*, the enjoyment of other people's unhappiness or misfortune. Schadenfreude is often studied together with envy, because they are complex "social" emotions that arise in social situations. They are also considered "competitive" emotions, because they involve comparisons with other people. They are sometimes called "fortune-of-others" emotions, because they concern what happens to other people.

Schadenfreude involves comparing your situation with another person's negative situation, and perceiving his or her negative emotions as well as your own resulting positive emotion. Envy is a negative personal experience in relation to another's good fortune. Schadenfreude is a positive experience regarding another's misfortune. It shouldn't surprise you that social emotions are mediated by the mind-mapping network.

Schadenfreude and envy are often studied together for other reasons. Superficially schadenfreude and envy look like opposites, because schadenfreude feels good and you seem one-up, whereas envy feels bad and you seem one-down. But one theory holds that envy is necessary for schadenfreude: Envy involves perceiving another person's superiority in some coveted way. Envy, which is painful, creates pleasure when the envied person suffers or experiences misfortune.[20] Schadenfreude is frequently experienced if misfortune befalls someone you envy.[21]

Unfortunately, many parents compete with their children and envy their greater opportunities in life. Another factor in antisocial empathy is dispositional envy, the proneness to feel envy. Dispositional envy predisposes someone to make envious responses to another person's perceived superiority.[22] If your parent(s) are predisposed to be envious of you, they're more likely to engage in antisocial empathy. Research indicates schadenfreude is often directed at high achievers.[23]

According to my clients' descriptions, their parents were often extremely bright people who were competitive with their own children's achievements and current and future opportunities, whether they well-educated or not. Many of them punished their children if the kids didn't achieve or succeed—and then competed with them if they did. Some parents bragged about their kids' successes to enhance their own reflected sense of self, while never praising their children or encourage them to feel good about themselves.

A commonly reported pattern was a mother who was very smart but who didn't go to college or pursue a career. Whether by design or unplanned pregnancy, raising children filled her time. Opportunities to pursue her own talents and interests were limited, especially compared to her children. Her independent earning potential was often meager, and she had few opportunities to leave her dysfunctional marriage and start a life on her own. Many of these women said to their children, "Don't think you're smarter (or better) than me!" Obviously parents are supposed to want their kids to have things better than they did—but some are competitive and envious. When kids figure this out it creates traumatic mind mapping.

But antisocial empathy isn't limited to competitive feelings and dispositional envy. As case examples in virtually every chapter of *Brain Talk* highlight, it often surfaces in two parts: One is enjoyment of inappropriately sexualizing a relationship, particularly from positions of trust. And the other is enjoying the other person's resulting discomfort and dismay. Both aspects contribute to the resulting traumatic mind mapping for the recipient. There's a competitive level here too, as in "winning a battle." There's also the comparison aspect of who feels good and who feels bad.

But the most important aspect of this—particularly from a traumatic mind mapping standpoint—involves empathy, *the ability to know and appreciate another person's feelings and experience.* Knowing your abuser *feels* your experience makes it worse. One way or another, knowing he or she is having an embodied experience of you increases the emotional and visceral impact on you. This further delights your abuser. Your abuser starts to gloat. This enjoyment of another person's suffering, which lies at the heart of antisocial empathy, offends our moral sensibilities and triggers hard-wired involuntary "primary" emotions of disgust and anger in our brains.

Mind-mapping researcher Simone Shamay-Tsoory (whose work I describe in Chapter Three) conducted the first neuroanatomical study of schadenfreude.[24] Being Israeli, she knew a similar term in Hebrew (*Simha La-aid*). However, she chose to language her research in terms of *gloating* (pleasure at another's misfortune). Shamay-Tsoory knew competitive emotions, such as gloating and envy, involve perspective taking capacities in addition to mind mapping. Whereas empathy involves taking another person's perspective, these 'fortune of others' emotions additionally involve comparing their emotional or mental states with your own. She also considered other research showing that the medial prefrontal cortex (mPFC) is activated more in competitive than in cooperative conditions.[25]

Shamay-Tsoory and her colleagues conducted neuroimaging studies to explore how understanding gloating and envy is impaired in patients with damage to their ventromedial prefrontal cortex (vmPFC). They discovered that patients with vmPFC lesions had intact mind-mapping ability, but they didn't recognize gloating or envy. Impaired gloating recognition involved additional lesions in the inferior parietal lobule (IPL). Patients with lesions in the left hemisphere were more impaired in recognizing gloating (a positive emotional experience). Patients with right hemisphere lesions were more impaired in recognizing envy (a negative emotional experience). Patients' ability to identify gloating and envy was related to their perspective-taking and mind-mapping abilities.

Here's where things connect back to our earlier discussion of how the brain creates empathy, which emphasized the importance of the anterior insula (AI): The vmPFC has strong connections to the AI and other parts of the limbic system (e.g., temporal pole, IPL and amygdala). This positions the vmPFC to regulate and evaluate limbic input, and integrate this with information from the prefrontal cortex, allowing comprehension of complex social situations through affective identification of basic emotions and 'fortune of others' emotions. Shamay-Tsoory and colleagues concluded that the mind-mapping network, which includes the vmPFC and the IPL, plays a fundamental role in competitive emotions like schadenfreude and envy.[26] This is part of the neurobiological basis of antisocial empathy.

COGNITIVE AND EMOTIONAL EMPATHY: A BRAIN-BASED PERSPECTIVE

Now that your perspective encompasses antisocial empathy, there are other aspects of empathy you're probably still not aware of. For example, it turns out there are two types of empathy, emotional and cognitive empathy. Emotional empathy is about experiencing what other people are feeling. Cognitive empathy is about mapping out other people's minds. If you are like most people, you probably consider empathy to focus more on people's feelings than their thoughts. Moreover, you probably consider detecting people's feelings to be more important, harder to do, a higher calling, and a sign of greater personal development than mapping out what they're thinking. This is why we need to consider one final study conducted by Simone Shamay-Tsoory.

In 2009 she and other colleagues conducted another neuroimaging study to explore recent evidence suggesting there might actually be two systems for empathy: a basic "emotional contagion" system and a more advanced cognitive perspective-taking system.[27] However, the neuroanatomical bases of these systems were largely unknown. It also wasn't clear whether these two systems, if they existed, were part of a single two-part empathy system or were two independently-operating systems. She suspected there were two distinct systems, with emotional empathy involving the mirror neuron system, and cognitive empathy involving the mind-mapping system.

To explore this, Shamay-Tsoory studied people with lesions in their vmPFC or their inferior frontal gyrus (IFG), together with two control groups (with lesions in their posterior cortex or no lesions at all). Participants were also given self-administered tests measuring cognitive and emotional empathy. Results revealed a "behavioral and anatomic double dissociation" between cognitive empathy (involving the vmPFC) and emotional empathy (involving the IFG). Precise anatomical mapping revealed that a small section of the brain known as "Brodmann area 44" is critical for emotional empathy, whereas Brodmann areas 11 and 10 are necessary for cognitive empathy. This is the first direct evidence that emotional and cognitive empathy are two distinct systems differing in anatomical location, developmental synaptic hierarchy and phylogenetic age.[28]

In other words, there are two behavioral systems for understanding others: The first is an early-developing emotion-matching system involving the mirror neuron system (MNS) in your motor cortex. Other research has demonstrated that activation of the MNS, especially involving the IFG, is particularly important for emotional empathy.[29] It was no coincidence that people's scores on emotional empathy tests correlated with activity in their IFG.[30] The IFG is the core structure of emotional empathy, the source of emotional contagion, emotion recognition, empathic concern and compassionate distress. Overall, the MNS is essential for emotional empathy.[31]

On the other hand, there's also a developmentally and phylogenetically more advanced system for cognitive understanding of mental states that is unique to primates and humans.[32] The vmPFC plays a core role in mind mapping and cognitive empathy (and also participates in emotional empathy).[33] Cognitive empathy involves perspective taking, fantasy imagining future outcomes, envisioning situations, and mind mapping. Other fMRI studies report increased vmPFC activation during mind mapping activities.[34]

Simone Shamay-Tsoory's intimate knowledge of the brain allowed her to tease out powerful insights that mental health professionals can appreciate when they are pointed out, but would never arrive at on their own: She recognized that the emotional and cognitive empathy systems are not just distinct; they involve different levels of sophistication.

This is why she referred to the emotional empathy system as "emotional contagion." Emotional contagion is the lowest common denominator of all empathic responses, which we share in common with rodents and birds. In contrast, perspective-taking abilities are only found in more phylogenetically advanced mammals such as great apes.[35]

Many mental health professionals lack the sophistication or differentiation to make this discernment. In today's attachment therapy heyday, therapists are biased towards emotional

Emotional Empathy	Cognitive Empathy
Simulation system Emotional contagion Personal distress Empathic concern Emotion recognition	Mind mapping system Perspective taking Imagination (of emotional future outcomes) Theory of Mind
Core Structure	
IFG BA 44 Unimodal Dysgranular cortex	VM BA 10, 11 Heteromodal Granular cortex
Development	
Infants	Children/adolescents
Phylogentics	
Rodents, birds	Chimpanzees

Table 3. Differences between emotional and cognitive empathy. From Shamay-Tsoory, Aharon-Peretz and Perry (2009).

empathy as "the real stuff," and cognitive empathy as the bastion of emotionally-stunted engineers and computer jocks.

But Shamay-Tsoory's work highlights that human empathy is more than emotional contagion, more than non-volitional embodied knowing of another person's experience. It involves cognitive perspective taking, which requires deliberate effort, focused attention, decoupling from your own perspective, mapping someone else's mind, and constructing an alternative reality.

There's a lot more to empathy than getting into someone's feelings because your mirror neurons are firing. (See Table 3.) She writes: "...the second empathy system involves more complex cognitive functions, including empathic perspective-taking and [mind mapping]. Understanding another person's perspective, (cognitive empathy), (e.g., 'I understand what you feel') involves higher cognitive functions like cognitive flexibility."[36]

Shamay-Tsoory's conceptualizations provide a differentiation-based picture of empathy. Among its many implications, it turns out Murry Bowen's view of relationships was right: The major problem is *not* people's "failure to connect."

Poorly differentiated people naturally go into (unfortunate) connection because of emotional contagion.

The real issues involve buffering emotional contagion through personal differentiation, which boil down to (1) keeping straight who you are (with a partner who's too ready to tell you), (2) keeping your emotions under control during heated conflict, (3) not over-reacting to your partner's over-reactions, and (4) resolutely working at problems despite failure and disappoint.

I've just enumerated the Crucible Four Points of Balance, my way of operationalizing differentiation. Notice that they all involve perspective-taking *and* emotional self-regulation. It's no coincidence that the best forms of self-soothing (empathy and compassion for yourself) involve giving your dilemma meaning (cognitive empathy).

Before leaving Shamay-Tsoory's study, her findings also concurred with other research indicating that the *insula*, amygdala, right somatosensory cortex, and right temporal pole are also involved in emotional empathy.[37] I've highlighted the reference to the insula, because this region will surface again later in brain research on disgust reactions.[38] Knowing the close integration of mind mapping and cognitive and emotional empathy helps you discern these abilities in people who, at first glance, appear to lack them.

Machiavellianism, Narcissism, Sociopathy, and Sadism

Now that we've considered the neurobiological underpinnings of antisocial empathy, let's consider this from the character and personality-side of things. In addition to facilitating cooperation, empathy and mind-mapping ability can be used to manipulate other people for exploitive personal advantage. As previously stated, it all depends on how you use it. There's some noteworthy research indicating toddlers' malicious teasing increases with age-related increases in awareness of other children's internal states. Children's growing mind mapping abilities are sometimes associated with negative outcomes like ringleader bullying and increased sensitivity to criticism.[39]

This sets the stage for daily traumatic mind mapping by many children, adolescents, and adults who realize their seemingly totally-self-preoccupied parent (or other antagonist), who is engaging in inappropriate behaviors and thoughts, is actually mapping them. The second "hit" is that the parent is enjoying their plight. Recognizing the *malevolence* in what your parents are doing creates further trauma. This is what I'd characterize as someone as having a "dark" mind.

Several times in *Brain Talk* I've referred to some people's minds being "dark," implying they have malevolent qualities. There happens to be a large body of research on this topic.[40] Psychologists Delroy Paulhus and Kevin Williams coined the term "Dark Triad," referring to three aversive or offensive personalities considered to be within the normal range of functioning (i.e., non-pathological): Machiavellianism, narcissism, and psychopathy.[41] All three personalities entail a socially-malevolent character with behavior tendencies toward self-promotion, emotional coldness, duplicity, aggressiveness, interpersonal hostility, callous-manipulative interpersonal style and lack of prosocial empathy.[42]

The hallmark of Machiavellianism is using other people as pawns to achieve your personal goals.[43] Machiavellian people behave in cold, manipulative ways in laboratory and real-world studies.[44] They enjoy manipulating and exploiting others for personal gain, and have tendencies towards narcissism and psychopathy.[45] They also have a cynical disregard for morality, a taste for deception, and believe interpersonal manipulation is the key to success in life.[46]

Narcissism involves pursuit of gratification from vanity or egotistic admiration. Narcissism is characterized by grandiosity, dominance, pride, egotism, and lack of prosocial empathy.[47] Four common aspects of narcissism involve arrogant superiority, self-absorption and self-admiration, an exploitive sense of entitlement, and needing positions of leadership and authority. Narcissism is more than egocentrism and is correlated with psychopathy.[48]

Psychopathy, the most malevolent of the Dark Triad, involves persistent antisocial behavior, bold impulsivity and thrill-seeking, selfishness, callousness, remorselessness, and low anxiety. Psychopaths demonstrate pervasive disregard for violating other people's rights.[49] However, they are not always short-term-oriented or impulsive. Some prevail over their households (and workplaces) for decades, by methodically traumatizing everyone through intimidating and threatening gestures without committing a reportable offense.

Here's the important thing to remember about the Dark Triad: We're talk about "normal range" personalities (i.e., normal people) with mind mapping ability. What do you think these people do with that ability? Especially when they have control of "reality" within their households?

Machiavellianism, narcissism, and psychopathy are usually thought of as patterns of behaviors. But traumatic mind mapping doesn't require behavior. Machiavellianism involves a mental world.[50] Machiavellian people attribute negative intentions to others. They don't expect cooperation, and don't offer collaborative alliances. The offer collusive alliances dressed up as collaborative. They start from the assumption others will exploit them if they don't do it first.[51] Machiavellian people excuse their actions by claiming others are unreliable and would similarly resort to cheating, deception, and exploitation. In other words, Machiavellian people think they're just protecting themselves.

Machiavellian people are mind mappers (and mind maskers). At the very least, Machiavellian people certainly think they're mapping other people's minds. Moreover, they believe their maps are accurate. They believe what's in their own minds. But if you're a young kid, and I just described how your mother talks about other people when the two of you are alone, you're going to have traumatic mind mapping. The content of your mother's mind is upsetting enough, but the fact she believes whatever is in her mind is even more frightening.

Machiavellian people go out of their way to exploit other people regardless of the consequences they inflict–including when they want this outcome. The lengths to which they will go creates traumatic mind mapping. One study found that Machiavellian people take serious revenge for other people's real or imagined offenses, even if the offender already confessed.[52] Imagine having a Machiavellian mother or father who thinks you've been disrespectful. The fact you've already apologized–even if you've done nothing wrong–isn't going to protect you from the punishment that follows.

The apparent lack of empathy in Machiavellian people has made researchers question their mind-mapping abilities. Two studies found Machiavellian children, who were skillful social manipulators, had neither impaired nor advanced mind mapping development. The difference was that, although Machiavellian children had good mind mapping ability, by late childhood *their mental attributions of others had a decidedly negative social bias.* During preschool years, Machiavellian tendencies correlated with "being on the outs" with one's peers. In other words, how people view social situations may more strongly determine their antisocial behavior than their mind mapping abilities.[53]

Other research further indicates Machiavellians are excellent mind mappers.[54] One study found bullying children who regularly mock, harass, and intimidate their peers, achieve very high scores on mind mapping tests while ignoring their victims' daily sufferings.[55] Rather than mind-mapping ability being below par, Machiavellians are good at controlling their emotions and masking their minds when it facilitates their goals.

You can't successfully manipulate other people without good-mind mapping ability, in part because this often involves recognizing weak points they themselves might not be aware of. Obviously, Machiavellian people, narcissists, and psychopaths are prone to antisocial empathy. They can readily place themselves into other people's perspectives and understand others' intentions, views, and knowledge. This lets them efficiently achieve their goals and predisposes them to manipulate and exploit others simply because they can. Being good at manipulating and exploiting lowers the opportunity costs of cheating and misleading.[56]

Recently, Delroy Paulhus and colleagues expanded the Dark

Triad into the Dark Tetrad by including "everyday sadism."[57] (This is akin to my concept of Normal Marital Sadism applied more broadly.[58]) They felt this was necessary to fully cover the human urge and willingness to make another person unhappy. This addition is important for realizing how people, especially children, are exposed to antisocial empathy involving aggression.

Paulhus suggests aggression from Dark Triad people tends to be context dependent.[59] Psychopaths have no qualms about hurting others, but their goals are largely instrumental.[60] Their impulsiveness limits their aggression to low-investment, short-term responses.[61] Narcissists are unlikely to bother with aggression unless their ego is threatened.[62] Machiavellians are too calculating to risk retaliation or punishment for gratuitous aggression without sufficient benefits.[63]

However, sadists are a different story. Sadism refers to a predisposition to be cruel. It's common for people to attack others when provoked or for revenge. But sadists hurt innocent people without provocation.[64] Most people get upset if they hurt innocent people, but sadists don't. They enjoy it.

For sadists, cruelty is its own directly-reinforcing motive. They inflict cruelty without additional motivation.[65] This goes beyond getting angry to accomplish some situational goal,[66] psychopathic antisocial aggression,[67] or narcissistic callous entitlement.[68] This self-sustaining, pleasure-driven aggression makes sadism especially morally disturbing. This is why it's particularly traumatic when you finally recognize this in your parent or partner or child. Many parents and caretakers have sadistic tendencies.

Only sadists engage in unprovoked aggression that requires time and effort. They crave cruelty enough to invest in harming an innocent child or vulnerable person, even when there are no other discernible benefits.[69] Cruelty is pleasurable, exciting, and even sexually arousing for these folks. People with sadistic tendencies seek out opportunities to be brutal and cruel.[70] Instead of seeking to alleviate suffering, they use their empathy and mind mapping radar to zero in on their targets. One study found a paper-and-pencil sadism test predicted people's unprovoked aggression in the laboratory, independent of psychopathy.[71]

To test whether "everyday sadism" warranted being added to create a new Dark Tetrad, Paulhus and colleague ran psychology students through two laboratory procedures designed to capture acts of cruelty. The first study actually involved killing bugs. Students who scored higher on sadism killed more bugs, and preferred bug killing over other tasks. These students reported the visceral experience of personally killing the bugs was particularly appealing to them. These findings held up even when analysis controlled for disgust sensitivity. Although murder horrifies the average person, some people commit murder for the sheer pleasure of it. Some sadists find doing this through direct physical contact particularly gratifying.[72]

The second study took things a step closer to the real world: It examined participants' willingness to harm innocent people, based on the unexpected discovery in an unrelated study that found some participants blasted their opponents with gratuitous, loud noise without provocation.[73] During what looked like game research, participants could administer very loud, sustained noises to a non-aggressive bogus opponent in the next room.

Students who scored high on sadism or Dark Triad measures were more likely to be bug killers. However, only sadist students tried to hurt another person. In fact, they went out of their way–actually working overtime–to victimize and hurting an unknown, innocent person. Participants with low sadism scores preferred having ice water poured on them rather than hurt someone else. Paulhus and colleagues concluded they needed a new Dark Tetrad that included sadism to acknowledge and address its everyday nature and surprising commonness.

The Dark Tetrad provides a framework for understanding intergenerational transmission of traumatic mind mapping via cruelty and disgusting behavior. Let yourself visualize what it's like to marry someone with these characteristics. (Or, imagine living in a house where this is your father or mother.)

People who exhibit the Dark Tetrad often have "accelerated mating strategies." This involves multiple sex partners, casual sex,[74] low standards for partners,[75] and a pragmatic and game-playing love style.[76] They steal or poach other people's partners,[77] abuse drugs and alcohol,[78] engage in impulsive behavior, and exercise limited self-control.[79] These traits characterize a "fast-life strategy," enacted through an exploitative and opportunistic approach to life.[80] Unfortunately, lots of people won't have to imagine very much—all they need to do is remember.

It's traumatic to recognize a Machiavellian, narcissistic, sociopathic or sadistic mind in someone you care about! Experiencing the behaviors they do is a double whammy! No one wants to believe someone is doing cruel things because they're mapping you and enjoying the impact they are having. It's easier to keep your mind from turning to spaghetti if you tell yourself they're oblivious to what they're doing. In this way, scientists are no different than you or I. The research literature on the Dark Triad reveals they also want to picture these folks as "not feeling other people's emotions," "tuned out," "not in touch with their own feelings," or "lacking good perspective-taking skills."

"Gaslighting"

Gaslighting is a particular form of emotional and verbal abuse, which involves manipulating someone so effectively and completely that the person comes to doubt his or her ability to accurately perceive reality. The term comes from a popular novel and subsequent movie about a malevolent husband who manipulated his wife so completely that she began to doubt her sanity.

Gaslighting is more than just lying and getting away with it. You have to be so good at implanting false beliefs, and do this so systematically over time that other people doubt their own perceptions of reality, particularly their map of you. You can't gaslight someone if you don't have excellent mind-mapping ability and antisocial empathy.

Gaslighting usually refers to adult relationships. Gaslighting an adult requires some effort and a consistent game plan. For instance, gaslighting your wife involves using what you know about her fears and insecurities to manipulate her. When she objects, you accuse her of being too sensitive, neurotic, or crazy. You tell her she doesn't know what she's talking about. You exploit her love for you and her emotional dependencies to keep her from leaving, so you can do this repeatedly. Machiavellian, narcissistic, psychopathic, and sadistic people get a physical and emotional charge out of gaslighting other people. Their antisocial empathy gives them schadenfreude.

It's much easier to gaslight a child because the relationship is so lopsided. Parents control the little world a child lives in—like the rules at the dinner table—and whether it will be stable, chaotic or scary. They dictate how experiences and events in that world and outside the house are to be interpreted. Children are hardwired to look to parents to understand how the world works. Gaslighting a kid is like shooting fish in a barrel.

Here's a description of what it's like. One person wrote: "... one thing terrified me most of all: Being crazy. Just the thought of it made my stomach lurch and my throat tighten, because I knew with the kind of certainty that only a small child can muster that if I were crazy, then no one would ever like or love me. I was probably around six or seven when the crazy thing really worried me, because I knew that either my mother was right and I was the crazy one, or I was right and she was the crazy one—and the first thought was much easier to consider as a possibility than the second.... Worrying about being crazy kept me up at night. I was an only child at the time and my obsession with being crazy was in part because I was also sure I was the only little girl in the world whose mother didn't love her."[81]

This woman described four examples of gaslighting from her childhood:

1. I'm carrying a platter of food into the dining room and accidentally drop it all over the floor. I explain the plate was slippery. My mother says: "You did that on purpose. Why do you always do stuff to make me angry?"

2. I'm being bullied by my older brother and ask my mother to intervene. Her response is, "When you stop bothering him, he'll stop hitting you."

3. I'm walking down the street with my mother, feeling happy. My mother says, "Stop skipping. Can't you ever be normal? Your skipping is making my heels

catch in the cobblestones and you will ruin my shoes. Do you have to ruin everything?"

4. My mother tells me that if I play quietly and let her work, she will take me out for ice cream. I spend the afternoon playing and then ask my mother when we're going for ice cream. She responds: "I never promised you ice cream." When I protest, my mother says, "Stop making things up. No one likes a liar." [82]

She also described a typical scene in families where parents are gaslighting children: "My father always insisted that my mother was the final authority. And my two brothers—one older and one younger—always called me the 'cuckoo bird' because what I said or did was supposedly so crazy. When I confronted my mother, she'd simply deny what she'd said or make up a reason for why she acted as she did. I was a bad person, an ungrateful person, and I believed it up until I left home. It was only then that I realized, *No, I wasn't the crazy person after all.* That said, now I'm 30 and, from time to time, I still wonder if my view of things is skewed. It's hard getting my family out of my head."[83]

Why do parents gaslight their children? For one thing, it feels good. The parent enjoys feeling the child's displeasure, disappointment, shock, and confusion. That's what antisocial empathy is all about. It's how they take their child prisoner: This is how they instill a combination of disgust and hatred that creates emotional super-glue that welds the child to them for life.

Parents with Dark Tetrad personalities are predisposed to gaslight their own children, in part, because lying and manipulating is how they handle other adults. But there's also something about children's naiveté, innocence, and trust that calls forth an urge in sadistic folks to screw with their children's reality. They do it for the self-sustaining pleasurable aggression in it. Machiavellian people do it because the child will eagerly do their bidding. Narcissistic people do it because they love the child's adoration despite being maltreated in a one-sided relationship. Psychopathic people do it because it gratifies their remorseless, callous, and selfish enjoyment of antisocial behavior. Maybe the most common reason why parents gaslight their children is that they have so much opportunity and it's so easy. It's not hard to make a love-deprived, insecure child doubt his or her reality.

And what do you think happens when you realize that the people charged with helping you master skills, manage emotions, and become sure of your own worthiness and solidity, are actively undermining you by saying you're crazy? The resulting traumatic mind mapping devastates you emotionally and negatively impacts your brain.

Incredible Things We Can Learn From Disgust

Given how difficult and distasteful it is to think about people doing morally disgusting things, it's amazing how much the brain science of disgust has to offer. Far beyond discovering which tiny three dimensional brain coordinates (voxels) disgust actually emanates from, the scientific study of disgust is unlocking core secrets about how the brain actually works. Understanding disgust gives you a functional map of the brain that organizes your thinking, and lets you make sense of things rather than trying to remember lots of anatomical names.

To start with, the study of disgust goes back to Charles Darwin, who wrote in 1872, "Disgust refers to something revolting, primarily in relation to the sense of taste, as actually perceived or vividly imagined; and secondarily to anything which causes a similar feeling, through the sense of smell, touch and even eyesight."[84]

But if you're going to really understand disgust, you need to know more about the anterior insula of your brain. There are two men who know all about that: One is Mbembe Jabbi, mentioned earlier for his neuroimaging studies of empathy. The other is Bud Craig, a functional neuroanatomist, whose deep understanding of the brain helps other scientists (and mere mortals like you and I) understand what Jabbi's sophisticated research really means.

ANTERIOR INSULA AND DISGUST

The insula is an integral part of the representational system that allows you to be aware of internal bodily and feeling states. As mentioned earlier, the insula has two main functional subdivisions: an anterior 'visceral' sector and a multi-modal posterior section. Your anterior insula (AI) receives information from your olfactory and gustatory centers, as well as your anterior ventral STS, which respond to seeing faces. Your AI also generates a representational map of your own body.

The AI has been implicated in various conditions, from low heart rate variability to bowel distension, orgasm problems and cigarette cravings, and common things like decision making, sudden insight, and maternal love. Its role in representing the physical condition of the body (interoception) suggests it is involved in all subjective feelings.

Research suggests the AI (particularly the VEN "spindle cells" it contains) has a fundamental role in basic conscious awareness.[85] The AI plays a key role in conscious experience of emotions and awareness of feeling states arising from your body. Your amygdala responds if you unconsciously perceive a threat, but your AI activates when threats are consciously perceived.[86] Research subjects' AI responses are positively associated with awareness of their own heartbeat.[87] Anterior insula impairments are associated with impaired, and in some cases catastrophic loss of, self-awareness.

Now that you begin to understand the AI from this perspective, let me tell you about some incredible research on disgust.

Besides his studies of empathy, Mbembe Jabbi, also conducted landmark neuroimaging studies on disgust. Jabbi investigated the AI and adjacent frontal operculum as an integrated unit called the inferior fronto-occipital cortex (IFO). Neuroimaging studies have shown that people's IFO activates when they view or become aware of other people's delight,[88] pain[89] or disgust.[90] The IFO reacts to other people's emotions, and these reactions are correlated with people's empathic tendencies.

It turns out that IFO impairments disrupt patients' experience and recognition of disgust, underscoring the IFO's role in emotion simulation and understanding.[91] Research also indicates the IFO plays an important role in interoceptive awareness (sensing the inner state of your body),[92] emotionally-laden autobiographical recall,[93] and imagining the taste of pictured foods.[94] These results point to this region's functional role in self-awareness, per se.[95]

Jabbi and colleagues examined the IFO's involvement in processing other people's disgust reactions. They conducted functional magnetic resonance imaging (fMRI) studies while participants viewed facial expressions of people having disgusted, pleased and neutral reactions to food. They compared participants' self-reported empathy with their IFO activation. Participants' empathy scores predicted their IFO activation while witnessing both *pleased and disgusted* facial expressions. Prior investigators thought the IFO processed empathy for other people's negative emotions and negative experiences like pain. But Jabbi's findings demonstrated that the IFO responds to other people's positive feelings too.

Once he discovered the IFO encodes personal experience and social observation of disgust, Jabbi wondered if the IFO also responds to vividly imagined disgusting experiences. Research showing IFO involvement in personal experience and observation of other people's pain suggested that IFO response may not be limited to disgusting tastes and smells. Other evidence suggested parts of the IFO were also involved in imagining basic emotions and sensations such as taste.[96] This raised the possibility that the IFO could be more generally linked to simulation of bodily feeling states during social cognition.[97]

Jabbi and colleagues subsequently scanned people's brains while they (a) experienced disgust, (b) viewed someone else experiencing disgust, and (c) imagined experiencing disgust (through script-driven imagery). They scanned people while they tasted disgusting liquids, watched actors drink from a cup and look disgusted, and read and imagined disgusting scenarios. In all three modalities of disgust, the anterior insula and adjacent frontal operculum (the IFO) were involved. Shared IFO circuitry activates when you personally experience disgust, imagine your own disgust, or observe other people's disgust. Whenever you have any of these experiences your IFO (anterior insula) lights up.[98]

Jabbi suspected that imagination and social perception of emotions shared the same neuroanatomical underpinnings, similar to findings in mirror neuron research showing common neural representations for perceived, executed and imagined motor actions. But Jabbi's research provided the first direct evidence that two distinct forms of mental simulation (social perception and imagination) actually share a common neural substrate in the IFO.

Jabbi then analyzed his results differently. He wanted to see if he could distinguish the functionally connectivity inside the IFO that differentiates imagining, observing and experiencing disgust. Over the last two decades, neuroimaging has advanced enough to go from studying functional segregation in the brain, to becoming a *systems* neuroscience that looks at distributed processing and connectivity to identify the brain's functional architecture and operating principles. Functional connectivity research studies the dependencies among remote neurophysiological events to discern the integration within a distributed system.[99] Effective connectivity is dynamic (activity-dependent), wherein one neural system exerts influence over another.

Jabbi hypothesized that the IFO is connected to three distinct functional circuits, which is why observing, imagining and experiencing emotions feels so different. By analyzing his results this way, Jabbi discovered that the IFO changed its effective connectivity with other parts of the brain during participants' experiences.[100] The IFO changed its influence on the (1) somatosensory (posterior insula), (2) gustatory/motivational (basal ganglia, orbitofrontal cortex), and (3) motor output brain regions (ACC and primary motor cortex). Increased connectivity with the hippocampus probably reflected autobiographic memories being triggered.[101] Jabbi's hypothesis about how the IFO produces three distinctly different sensory experiences (observing, imaging and experiencing disgust) turned out to be true.

Moreover, when observing disgust, the IFO showed stronger connectivity with the ipsilateral right BA45 region (known to be involved in general observation and imitation of facial expressions).[102] Jabbi realized that if emotional facial expressions, particularly disgusting ones, increased the effective connectivity between this region and the IFO, this would link (1) a simulation of the bodily feeling state of disgust with (2) the simulation of the disgusted facial expression. Other research supported this possibility: Lesions in the IFO create deficits in general perceiving of facial expressions, with particular deficits in disgust recognition.[103]

Jabbi recognized he had discovered something fundamental about how the brain works. He concluded that the IFO contributes to empathy by mapping other people's bodily feelings onto the internal bodily states of the observer, using the interoceptive functions of the IFO. IFO involvement in observing and experiencing disgust *and* pleasure provides the simulation mechanism that supports your awareness of other people's emotional states.[104]

Isn't this amazing?! Who'd guess that the human disgust reaction provides a fundamental basis for understanding interpersonal neurobiology? If you're interested in learning how the brain functions–particularly if you're a therapist interested in brain-based therapy–stop and realize what this means. The implications for treatment are enormous:

Understanding the brain's mechanisms for overlaying personal experience with observed and imagined emotional experiences with other people, by anchoring them on a visceral (interoceptive) map of your own body, facilitates therapies particularly focused on generating positive neuroplasticity.

Moreover, this can promote methods integrating interoceptive cues, personal experience, and observation of others to best advantage. As a case in point, this understanding underlies the visualization techniques, written dialogs, and other perspective-taking methods of Crucible Neurobiological Therapy. (Appendix D).

SPINDLE NEURONS AND ANTISOCIAL EMPATHY

By now you're probably accepting the bad news: There are people who enjoy doing disgusting things, and who know what they're doing is disgusting, if for no other reason than they can detect other people's disgust reactions. This visceral knowledge of other people's unpleasant and unhappy mental and physical states is what triggers involuntary moral disgust in the brains of those around them. Especially in their children. Even if they're not enjoying other people's disgust reactions, at the very least they know about it and do nothing to stop it. They trigger moral disgust reactions just the same.

Given how anticipatory traumatic mind mapping develops in homes were parents do disgusting things, Jabbi's research on the interpersonal neurobiology of disgust demonstrates how disgust reactions are daily traumatic events in some families. It makes no difference whether your father and mother are doing disgusting things to you, or you observe them doing them to your siblings (or each other), or imagine them doing disgusting things to each other, or you, or your brothers and sisters. In your brain, mind, and body it's all the same. Repeatedly activating your anterior insula with disgust reactions is not good for your brain.

The good news is that knowing how disgust reactions work allows us to mitigate the harmful impact of disgusting parenting. The negative impact of repeated disgust reactions upon the AI can be cured. I'll actually describe some ways to accomplish this in Appendix. D.

But our efforts must also focus on getting people who do disgusting things to stop doing them. The first step, as Delroy Paulhus said about sadism, involves acknowledging its everyday nature and surprising commonness, particularly in homes that show no outward signs of trouble. Research on the Dark Tetrad indicates there are lots of people who are prone to antisocial empathy and doing disgusting things. They display no major signs of psychopathology. They fit right in to mainstream society. They are normal. They are your neighbors. (You could easily be one of them.)

Thinking about disgusting behavior, traumatic mind mapping, and the prevalence of antisocial empathy opens your mind to identifying and confronting people who enjoy doing disgusting things. This can be hard to do, so it helps to know that research says they probably know what they're doing, because of how the AI maps out other people's body states on top of the AI-generated interoceptive map of your own body.

Focusing on the anterior insula's role in processing highly emotional information, rather than the amygdala, is probably new to most readers. You're also probably more use to hearing about mirror neurons than VEN spindle cells. But your anterior insula plays a pivotal role in experiencing other people's emotions, whether we're considering prosocial empathy for their emotional states, or tormenting them by doing (and thinking) disgusting things, and enjoying their disgust reactions. Your anterior insula ties this whole prosocial/antisocial empathy system together. Just ask Bud Craig.

If you really want to know about the AI, you should read Bud Craig's frequently cited 2009 paper about the AI and how it works together with the ACC. Scientists didn't understand much about the AI until Craig provided a succinct integration of an amazing range of convergent functional, anatomical, and clinical evidence.[105] Craig knew that the AI, the ACC, and mPFC co-activate in many studies of behavior and virtually all research on emotions. He proposed that the insula is the limbic sensory unit that engenders the feelings in every emotion, and the ACC is the motor cortex that provides the motivation (agency) every emotion contains. He believes the AI is the site of human awareness (where feelings from the body are represented), and the ACC is where behaviors are initiated.[106]

Craig arrived at this conclusion through his understanding of evolutionary neuroanatomy. To follow his reasoning, let's pick up our prior discussion of those large spindle-shaped von Economo neurons (VENs). The human AI and ACC both have unique concentrations of them. VENs are plentiful in adults, progressively rarer in infants, gorillas, bonobos and chimpanzees, and completely absent in macaque monkeys. VENs have been reported in elephants and whales. (Rats and lizards don't have corresponding brain anatomy.[107]) Patients with VEN degeneration lose awareness of their emotions and behaviors.[108]

Apparently, VENs provide fast interconnections between the physically-separate limbic sensory (AI) and motor (ACC) cortices. Craig proposes that at an earlier time in human brain evolution they were a single unit. The ACC evolved first, as a motor-control region in mammals (guided by sensory input from the amygdala and hippocampus), to propel olfactory-driven group behavior. The insula evolved later for cortical processing of homeostatic sensory activity in individual animals.

Craig proposes that as mammals evolved, the insula grew as limbic activity became more aligned with the autonomic nervous system rather than with olfactory activity.[109] VENs enable the brain to make highly integrated representations of emotional experiences and behaviors, and probably underlie the joint AI and ACC activity reported in most studies.

What does this mean for traumatic mind mapping of antisocial empathy and brain reactions to disgusting behavior? We're describing how the AI plays a core role in how this whole biobehavioral brain-based system fits together. The AI supports the interpersonal neurobiological transmission of subtle emotional trauma. It also motivates people to do disgusting things because they have humankind's most unique abilities: Emotional empathy and mind mapping—with an antisocial bent.

The AI also has an important role to play in resolving traumatic mind mapping and the impact of repeated disgust reactions. This is discussed in detail in Appendix D.

As for dealing with people who enjoy other people's suffering, this discussion should facilitate identifying those who do disgusting things. You'll discover them hiding right out in the open, because no one wants to acknowledge their "surprising commonness." Hopefully, your reading this will change that.

Appendix D

CREATING NEUROPLASTICITY

Prior Appendices laid out the scientific basis for subtle interpersonal trauma via traumatic mind mapping, antisocial empathy and disgusting behavior. This final Appendix focuses on resolving the brain problems these all-too-common experiences create. Here you'll find the science underlying the final section of *Brain Talk*, and of Crucible Neurobiological Therapy as a whole.

Traumatic mind mapping is a complex disorder involving a variety of biological and psychological disturbances in mind-mapping radar, autobiographical memory, focused attention, emotional self-regulation, disgust reactions, identity, world view, and behavior. Fortunately, trauma treatment increasingly benefits from collaborations between neuroscientists, neuropsychologists, physicians, and psychotherapists trying to develop brain-based therapies to ameliorate these impacts. There's a lot of anatomy and chemistry to cover: The brain contains roughly 100 billion neurons, each maintaining 10,000 connections to other neurons and communicating through 100 different types of neurochemicals.

Neuroplasticity and neurogenesis are two good reasons to be hopeful about people's ability to recover from trauma.

Neurogenesis is your brain's ability to grow new neurons throughout life. You can increase neurogenesis directly through aerobic exercise, consuming fewer calories overall, fasting, and eating omega-3 fatty acids. But there's a lot more you can do than exercise and watch your diet.

Neuroplasticity means the brain is not hard wired. It turns out the brain remains malleable ("plastic") all through life. It rewires itself based on your personal experience.[1] Traumatic mind mapping is an example of negative neuroplasticity. Crucible Neurobiological Therapy focuses on creating positive neuroplasticity to resolve interpersonal trauma. Positive neuroplasticity results from growing new neurons, synaptic connections, changes in synaptic shape and efficacy, and increased glial cells, neurotransmitters, and receptor density and activity.

Therapies attempting to create positive neuroplasticity generally expose clients to traumatic memories as a framework for change-oriented interventions, rather than focusing on insight or catharsis.[2] The goal is to establish new neural circuits, remodulate hyper- or hypo-activated circuitry, and integrate interoceptive bodily cues with memories of traumatic events, to produce new thoughts, emotions, and behavioral patterns.

TYPICAL PTSD TREATMENT

Crucible Neurobiological Therapy (CNT) developed from working with people who experienced repeated traumatic mind mapping early in life. The resulting methods happen to follow PTSD treatment recommendations from many international consensus panels, including the United States Institute of Medicine, the United Kingdom National Institute of Clinical Excellence, and the International Society for Traumatic Stress Studies (ISTSS).

In 2008 the ISTSS issued new PTSD practice guidelines, the first update in eight years. These confirm the recommendations of other practice-related organizations such as the U.S. Department of Veterans Affairs and the Department of Defense, the American Psychiatric Association, and Great Britain's and Australia's national health care guidelines.[3] According to the latest research, two treatments make a clear difference: Prolonged Exposure Therapy (PET) and Cognitive Processing Therapy (CPT).

In PET treatment, therapists guide people through recalling traumatic memories in a controlled fashion to eventually regain mastery of their thoughts and feelings around the incident. The primary focus of exposure therapy is the intense fear, helplessness, or horror experienced at the time of the trauma.

In Crucible Neurobiological Therapy (CNT) treatment focuses on traumatic mind-mapping content and all physical reactions occurring at the time, including disgust and sexual arousal. Whereas exposure therapies mostly focus on calming an overactive amygdala, CNT focuses on the anterior insula (AI).

Like most exposure treatments, CNT utilizes elements that were part of the traumatic experience (through revisualization and written dialogs, for example). The treatment activities are deliberate and specific, and many of the resulting changes are automatic (rather than requiring insight). Changes are accompanied by beneficial insights and emotional expression, but these are more the result than the cause of reduced anxiety and improved mood, energy level, alertness, cognitive acuity, memory recall, and emotional stability.

CNT methods do more than just provide exposure to traumatic mind-mapping memories. They force the client's brain/mind to engage in mind mapping and perspective taking of the antagonist. Besides reducing anxiety associated with these memories, this adds mind-mapping information that's missing from the verbal narrative; retrieves situationally accessible, unprocessed kinesthetic sensations and promotes reorganization of a more accurate and coherent autobiographical narrative.

Cognitive Processing Therapy (CPT), a form of cognitive behavioral therapy was specifically developed to treat PTSD in sexual assault victims. It is now applied to PTSD treatment in general.[4] It has two integrated components: writing and reading about the traumatic event (exposure therapy) and addressing erroneous thinking and ongoing reactions to the event (cognitive therapy). Cognitive methods excel at altering higher-order cognitive appraisals.

The cognitive behavior therapy portion of CNT uses mind-mapping information to reconstruct the world of trauma present before, during, and after the traumatic mind-mapping event. This facet of treatment is guided by each client's unique history and memory access. Insight and emotional integration of new information are important, but the process is driven by right-brain visualizations rather than left-brain deductive logic and introspection. This includes integrating body sensations emerging from revisualizations.

Prior research has compared the effectiveness of PET and CPT. One study compared treating PTSD and depression with cognitive processing (CPT) or prolonged exposure (PET) therapies versus no treatment at all. Results indicated both treatments were highly effective and superior to no treatment. Both therapies produced similar results, except CPT was better at reducing guilt.[5] A different study compared treating female veterans with PTSD using PET versus present-centered supportive therapy. PET was better at reducing PTSD symptoms.[6]

Considerations for More Effective Therapy

Developing new treatments for subtle interpersonal neurobiological problems caused by traumatic mind mapping is a new horizon for therapists. There's much to be gained by using PET and CPT as reference points. Likewise, brain research on disgust and self-awareness suggests treatment should focus on particular parts of the brain. Understanding how trauma disrupts synchrony between the implicit and explicit memory systems can further guide these efforts.

But converting these insights into effective action plans is uncharted waters. How therapists actually operationalize these good ideas determines the effectiveness of different approaches. There's a big difference between doing brain-changing therapy and explaining the brain to clients, or referencing brain anatomy when you're trying to stop their in-session arguments.

In a moment I'll describe the methods I've developed and refined over a decade to resolve my clients' problems. Here are some things I've found important to keep in mind.

INSIGHT IS NOT ENOUGH

Treatment must move beyond focusing on insight and talk therapy. Bessel van der Koch offers a cogent critique of the limitations of talk therapy in treating trauma. He proposes insight and understanding are usually insufficient to help traumatized people because the rational executive brain that prevails in talk therapy has limited capacity to squelch sensations, control emotional arousal, or change fixed-action patterns.

Van der Kolk thinks neither cognitive-behavior therapy nor psychodynamic psychotherapy pay sufficient attention to experience and interpretation of clients' disturbed physical sensations and preprogrammed physical reactions. He emphasizes alternative methods involving the body and subjective inner experience. Van der Koch says effective treatment must involve: (1) learning to tolerate feelings and sensations by increasing interoception (awareness of internal body states), (2) learning to modulate arousal, and (3) taking effective action.[7]

According to van der Kolk, resolving trauma requires reprogramming resultant automatic physical responses and learning to regulate physiological arousal including heart-rate variability (HRV). He encourages techniques like mindfulness training, and his colleagues conducted pilot studies into the effects of yoga on HRV. "Normal" participants significantly changed their HRV over eight sessions of hatha yoga. PTSD patients significantly improved their symptomatology. Other colleagues conducted an fMRI imaging study of people engaged in mindfulness meditation. Results indicated experienced meditators had thicker brain regions associated with interoception, sensory processing, and attention, including the right anterior insula and PFC cortex. The largest difference for meditators was the thickness of their right AI.[8]

For the last three decades, Crucible Therapy has used conjoint body-oriented activities (like Hugging 'till Relaxed, Heads on Pillows, and Eyes-Open Sex) to resolve couples' sexual problems and produce personal growth (differentiation). Crucible Neurobiological Therapy utilizes these same action-driven conjoint somatosensory activities to produce positive neuroplasticity to reduce interpersonal trauma and increase interoceptive awareness of body cues, while simultaneously helping people learn to modulate their emotional arousal.

FOCUS ON INNER EXPERIENCE

Consensus is emerging among trauma experts regarding the need to deal with internal residues of old traumas to keep them from intruding into current experience. Traumatized people often have difficulty attending to their inner sensations and perceptions. They feel overwhelmed by them when they do and often avoid doing so. They also report feeling self-disgusted, helpless, or panicked when re-experiencing trauma-related images and physical sensations. They lose the typical "fight or flight" defense and respond to perceived threats by becoming immobilized.

Mind-mapping-based Crucible Neurobiological Therapy helps people develop interest in their internal experience. This is important for people who don't want to know what's going on inside them or what happened to them. CNT focuses on physical aspects of emotional experiences and increases body self-awareness, rather than just working off preexisting narratives and interpretations of the past. This helps people identify their physical sensations, integrate them with their feelings and perceptions, and translate this into communicable language that lets them understand themselves better.

CNT attempts to activate the AI, anterior cingulate (ACC) and mPFC by focusing people's attention on their internal experience. This is what happens when you revisualize traumatic mind-mapping events and interweave their cognitive, emotional, and sensorimotor elements to make it as vivid as possible. Becoming aware of body sensations can enhance control over feelings and emotional arousal in general.[9]

SYNCHRONIZE IMPLICIT AND EXPLICIT MEMORY

People who experience trauma often report autobiographical memory problems (explicit memory deficits) attributable to changes in their hippocampus.[10] Traumatic events are remembered with gaps in details, and memories are often vague, unclear and disorganized.[11] One study found traumatized adolescents had autobiographical memory deficits, and their memory loss correlated with the number and severity of their traumatic events.[12] Working memory deficits are associated with decreased left dorsolateral PFC activation.[13]

Dysregulated memory systems are major contributors to flashbacks. Traumatic memories are vulnerable to being triggered and re-experienced because they are fragmented and disorganized. Flashbacks result from the implicit (emotional and sensory) memory systems becoming unsynchronized with explicit memory (conscious understanding) due to actions of the hippocampus and amygdala during and after trauma. Explicit memory (driven by the cortex and hippocampus) becomes overwhelmed by dysregulated emotions and anxiety (controlled by the amygdala).

To explain traumatic mind mapping, Appendix B discusses research on "flashbulb memories." Scientists suspected flashbulb memories resulted from a special memory system being brought into play when a person encountered a surprising, consequential, and emotion-arousing event. Unlike quickly forgotten ordinary memories, this "flashbulb" memory remained fixed in mind for a long time. They speculated that the memory itself was not of a verbal or narrative form but might consist of an image.[14]

Other researchers investigated the existence of two memory systems. *Implicit memory*, in which memories are expressed through images, behaviors, or emotions, is present at birth and is operational throughout life. *Explicit memory*, which emerges during preschool years, encodes in narrative form (i.e., words) those events which are actively thought about or mentally processed.[15]

This laid the foundation for Chris Brewin's studies of involuntary imagery and memory at the Institute of Cognitive Neuroscience in England. Brewin developed what is known as the "dual-representation theory" to capture the dynamic relationship between these two memory systems.[16] According to Brewin, someone's narrative (explicit) memory of a trauma consists of the content in their verbally accessible memory (VAM). VAM memories can be retrieved at will and described in words. They make up someone's personal life history. VAM memories contain information about what people attended to before, during, and after the traumatic event. These memories receive sufficient conscious processing to transfer them into long-term memory for later retrieval.

When people think about the trauma event or its consequences and implications afterwards, or discuss them with other people, they are accessing VAM memories.[17] VAM memories are subject to modification and changes in meaning, but they are often vague, disorganized, and full of gaps. The amount of information in VAM memories is limited because they only record what the person consciously attended to at the time. This is restricted due to the effects of high arousal and anxiety, collapse of mind-mapping functions, and diversion of attention.

Brewin proposes that PTSD flashbacks come from activities of the situationally accessible (implicit) memory system (SAM). The SAM system contains extensive information from low-level perceptual processing of the traumatic event. This includes sights, smells, sounds, gut reactions, sexual "vibes," mind-mapping information, and empathic "contagion" too briefly apprehended to be integrated into conscious VAM (explicit) memory or processed into long-term memory. The SAM system stores information about bodily response to the trauma, such as changes in heart rate, flushing, temperature, and pain. Experiments indicate that unattended objects or items not consciously seen are nevertheless encoded and analyzed in considerable detail by the brain and can unconsciously effect responses on tests of indirect memory.[18]

Far more details, particularly sights, sounds, smells, movement and feelings, are encoded in the SAM system than in verbally accessible memory (VAM), which records conscious experience of the trauma. The longer and more drawn out the traumatic mind-mapping event, and the more frequently this occurs, the more likely flashbacks become. When the SAM and VAM systems become desynchronized (contain different information), people re-experience things encoded only in SAM memory.

The image-based SAM system encodes unattended sensory information but doesn't contain verbal language, making it difficult to process them through talk therapy or to communicate to others. SAM memories don't get updated by other autobiographical knowledge. Dual-representation theory maintains that the original traumatic memories are not altered in any way but remain intact in SAM, ready to be vividly re-experienced as a flashback if the person unexpectedly encounters reminders of the trauma.

Clients are able to retrieve mind-mapping information about their antagonist because implicit memories of the event, initially missing from their verbal narrative (VAM memories), remain intact. Mind mapping and empathic contagion information about the traumatic event resides in SAM. In Crucible Neurobiological Therapy, revisualizations are used to trigger SAM memories. Written dialogs assist processing this information into verbal form (VAM).

CNT's focus on disgust, a hard-wired primary emotion, fits well with Brewin's view of how SAM and VAM memories involve different emotions. According to dual-representation theory, the emotions accompanying SAM memories, experienced during the trauma or in subsequent moments of intense arousal, are primary emotions. They are "primary" in the sense of occurring at the time, as well as being automatic emotional and visceral responses requiring no cognitive processing.[19] This describes disgust reactions during traumatic mind mapping. Childhood disgust reactions are readily stored in SAM, remaining there outside conscious awareness or recall, yet influencing the individual's personal life.

VAM memories, in contrast, are accompanied by "secondary emotions" because they were not experienced at the time of the trauma. They are also "social emotions" like embarrassment, shame, and guilt, generated by retrospective evaluation of the event.

Trauma-related desynchronization of implicit and explicit (SAM and VAM) memory systems surfaces another way: People with PTSD also suffer from a fragmented sense of self. Their emotions become fragmented from their cognitions, and their emotional experiences become overgeneralized beyond their actual context. Therapeutic memory reconsolidation provides the basis for people regaining their sense of self and capacity to regulate emotions. When they deliberately attend to the content of visual images, information present only in the SAM system becomes re-encoded into the VAM system where memories are assigned spatial and temporal contexts.[20] This must happen repeatedly because lots of SAM information must be transferred into VAM.

Reprocessing SAM memories involves creating new trauma memories in the VAM system that become more permanent and easily accessible by reviewing them repeatedly. This produces competing memories of the trauma, an original one associated with fear, helplessness, horror or disgust, and more recent ones recalled in a better place as something belonging to the past. These new memories are accessed when people encounter reminders of the trauma, which inhibits inappropriate amygdala activity. Reintegrating implicit and explicit memories makes them more coherent and reduces unwanted intrusions. Research indicates that reactivating traumatic memories through exposure therapy, and reincorporating them into more accurate pictures, makes them less distressing and more easily managed.[21]

EXPLOIT THE IFO (ANTERIOR INSULA)

Bud Craig's seminal integration of anterior insula research (Appendix C) proposes that we could develop better psychotherapies by broadening how we understand feelings. Research demonstrates people's expectations have a dampening effect on their IFO response while tasting aversive things, suggesting this region regulates reflective/cognitive processes relevant to homeostatic maintenance.[22]

But Craig says we need to move beyond looking at feelings as homeostatic mechanisms reflecting the inner state of the individual. He proposes we can exploit what feelings from the body and feelings about objects or people or cognitions have in common. In other words, Craig suggests new forms of therapy might emerge from exploiting how the anterior insula operates.[23]

Here's where Mbembe Jabbi's research on disgust reactions comes in. His findings support a functional relationship between the IFO (largely consisting of the AI) and interconnected regions during social cognition, imaginary and actual emotional experience. Jabbi's work suggests the

IFO starts with visceral interoceptive (body) experience, and superimposes mPFC functioning involving visualized memories, imagined experiences, observing other people, and witnessing their experiences. This and other research suggests trauma treatment can exploit the IFO's unique participation in self-awareness and embodiment of other people's observed and imagined feeling states.[24]

Jabbi's disgust studies provide insights into the neural basis of captivating experiences like reading a book or watching movies of other people's experiences.[25] The combination of movies and written material in Jabbi's experiments demonstrate that mental imagery (imagining what you're reading), as well as watching or imagining other people's experiences, recruits basic brain regions involved in experiencing emotions. The IFO appears to provide key simulation functions that make feeling an emotion, seeing that emotion on someone else's face, and imagining that emotion, share similar feeling components. Jabbi wrote:

> Humans can achieve vivid emotional feeling states in the absence of actual emotional encounters in a myriad of ways, including the recall of past experiences, the imagination of hypothetical experiences, reading a good book, watching a good movie or witnessing a friend's experience.... The IFO is a convergence zone where bodily feeling states relevant for the emotion of disgust are coded according to a common code regardless of stimulus modality.
>
> Our findings of IFO involvement in all three modalities supports the idea that simulation through both pre-reflective (viewing someone else's disgust) as well as reflective (deliberate mental imagery and language) routes may therefore be complementary rather than independent of each other."[26]

Jabbi's discovery of a common neural basis to personal experience, imagery, and observation of social emotions lays the scientific basis for new clinical methodologies to treat traumatic mind mapping. This includes utilizing visualization imagery and mental dialogs (particularly involving shifts in first-person and third-person perspectives) to integrate implicit SAM memory with explicit VAM knowledge. Moreover, this allows therapists to develop interventions that activate targeted brain areas and functions. For instance:

1. Focusing on your emotions, feelings, and reactions while visualizing powerful pictures stimulates your AI. This particularly happens when you view other people's faces.

2. Reporting your inner bodily states activates your mPFC. Doing this while revisualizing traumatic mind-mapping experiences integrates SAM interoceptive cues with frontal cortical processes.

3. You can stimulate your AI and ACC through vicarious processing of your own and other people's feelings through revisualizations with an interoceptive focus. In revisualizations and dialogs, reflecting on your antagonist's inner experience uses pathways normally used for your own bodily representations to simulate his or her bodily states. This creates simulated reflective representations that help you map your antagonist.

Watching traumatic mind-mapping experiences–mapping out what your antagonist is thinking, feeling, desiring, and knowing–triggers empathic transformation of the antagonist's bodily states into pre-reflective representations in your own brain of similar self-states. This provides intuitive understanding of what's going on in your antagonist's body and mind.

This offers several advantages: One is processing SAM into VAM. Another is overcoming trauma victims' reluctance to look into their antagonist's mind, and tolerate the short-term dysregulating effects of fully appreciating the trauma. Studying what you see, when you look at your antagonist in the revisualization, often provides nonverbal kinesthetic cues that trigger important-but-difficult new information and non-deductive realizations, together with a visceral sense of certainty and emotional release.

METHODS OF
CRUCIBLE® NEUROBIOLOGICAL THERAPY

Exposure methods, writing therapy, and recall of traumatic memories have all been done before. What makes the methods of Crucible Neurobiological Therapy unique is their focus on mind mapping, traumatic mind mapping, disgust reactions, emphasis on right-brain visualizations, and dialog construction. These methods are consistent with trauma treatment trends towards (1) reactivation of trauma-related circuits via exposure therapy, (2) creating neuroplasticity through moderate anxiety and meaningful stress, (3) helping trauma victims tolerate their feelings, and (4) attending to their somatic experiences.

VISUALIZATION AND REVISUALIZATION

Tanya Singer, who directs the Department of Social Neuroscience at the Max Planck Institute in Leipzig, Germany, is well known for her brain studies of empathy. She even proposed an interoceptive model of empathy, suggesting that cortical representations of bodily states have a dual function. Like Jabbi, Singer proposes that the AI and ACC structures, which play crucial roles in representing our own feelings, also vicariously process other people's feelings too.

First, cortical re-representations allow us to develop subjective representations of feelings, which not only help us understand our feelings about emotional stimuli but also predict the bodily effects of anticipated emotional stimuli to our bodies. Second, cortical re-representations trigger visceral correlates of prospective empathic simulations of how things may feel to others. This helps us understand the emotional significance of a stimulus and its likely consequences.[27]

CNT revisualization methods allow people to develop subjective representations of feeling states connected with the traumatic interaction. While visualizing traumatic mind-mapping experiences from first-person and third-person perspectives, they are directed to watch their antagonist's face and map his or her mind. This allows them to experience the bodily effects of current or anticipated things happening in the visualization. It also provides the visceral correlates of the antagonist's experience. Understanding their antagonist's mental content and physical reactions helps clients understand the emotional significant of what happened to them and its impact on them.

Clients are asked to relive the trauma as vividly as possible through revisualization. This evolves from a detailed narrative account involving everything they recall seeing, hearing, touching, smelling, feeling, and thinking throughout the whole event. Many clients close their eyes. Negative memories are more likely than positive memories to have very salient, specific features, although they may be few in number.[28]

The reliving process provides additional image-based, detailed information to be elaborated and incorporated into the narrative. Through repetition, the most disturbing elements are desensitized. Visualization techniques are particularly good—and often better than conventional talk therapy—for resolving traumatic mind mapping because the amygdala has far more connections with the visual areas (versus the auditory areas) of the brain.[29] The amygdala is strongly connected to almost all brain regions involved in visual processing.[30]

CNT's use of revisualization and dialog construction is similar to virtual reality exposure techniques that also emphasize visual stimulation. Researchers are adding virtual reality to prolonged exposure therapy, whereby participants experience 3-D imagery, sounds, and smells corresponding to a traumatic event. One study with Iraq War veterans gives participants a virtual reality that simulates combat conditions.

Likewise, researchers at the Program for Anxiety and Traumatic Stress Studies at Weill Cornell Medical College are using virtual reality to treat victims of the 2001 World Trade Center attacks. They are enhancing cognitive behavioral therapy with virtual reality. First responders, firefighters, and civilians in nearby buildings see increasing vivid versions of the Twin Towers scenario, starting with simple images of buildings on a sunny day, and progressing to include horrific sights and sounds.[31]

Revisualization is somewhat like "imagining exposure therapy," which exposes clients to memories in a guided imagery format. Therapies involving imaginative reconstruction of traumatic events have shown good results in treating PTSD. Trauma-related images are manipulated in the imagination to produce a different or more reassuring outcome. In some approaches, childhood sexual abuse survivors replay traumatic moments and imagine their adult-self intervening to comfort the child and prevent the trauma from occurring. These techniques reportedly achieve rapid reduction in anxiety without repetitive reworking of the trauma narrative.[32]

However, CNT methods don't involve visualizing ideal outcomes. Instead, written dialogs allow incremental working through of mind-mapping interactions with the antagonist. Ultimately these dialogs resolve through clients progressively making gold-standard responses, which in turn require removing gaps in their mind-mapping radar. Often this takes repeated attempts so that, even in mental dialogs, clients are able to keep up with their antagonist and keep him/her from getting around them.

Typically, people start off using a first-person view. Some clients initially report being unable to visualize the traumatic mind-mapping scene or see their antagonist. If their first-person view of the traumatic memory is blocked, clients are often able to see the scene and map their antagonist by changing to a third-person view. Similarly, those whose third-person view is blocked (which happens less frequently) are able to visualize the scene by adopting a first-person view. Research indicates the left temporal lobe is involved in third-person perspectives, especially in emotional contexts, whereas the somatosensory cortex handles emotional processing in first-person perspectives.[33] Deliberate shift in perspective is not inherent in revisualizations, nor is this a commonly tried solution. In other words, clients have to be prompted to do this. Table 4 outlines additional aspects of visualization methods in this approach.

WRITTEN DIALOGS

In contrast to revisualization, writing dialogs between oneself and one's antagonist inherently require repeated shifts in perspective-taking. Revisualization involves a wide purview of the traumatic mind-mapping event, which some clients attempt to misuse to avoid mapping their antagonist's mind. However, mental dialogs cannot move forward without this taking place. Mentally shifting back and forth between first-person and third-person views, and forcing the brain to accomplish perspective-taking of the antagonist appear to have their own beneficial impacts.

Aspects of Visualization Activities in Crucible® Neurobiological Therapy
• Involves "taking action" and circumventing avoidance.
• Recall of traumatic mind-mapping events.
• Triggering and focusing inward to interoceptive cues.
• Visualization (rather than "insight" or "introspection") forces right-brain involvement.
• Mind mapping the antagonist.
• Integrating visuospatial memory with mind–mapping information.
• First-person and third-person views.
• Forcing perspective taking.
• Facilitating articulation of unexpressed affect.
• Seeing the whole instead of focusing on parts.
• Moderate levels of anxiety.
• Pushing "mental blocks."
• Keeping brain functioning under stressful re-experiencing of traumatic mind mapping.
• Putting mind mapping information back into traumatic memories, dramatically.
• Deepening or radically changing formerly inaccurate or absent attributions.
• Integrating information, sensations, and memories under meaningful stress.

Table 4. Aspects of visualization activities in Crucible® Neurobiological Therapy

Shifting between first-person and third-person views, or perspective-taking of an antagonist, are not inherent parts of writing therapies. For instance, an important part of cognitive processing therapy (CPT) involves rape victims writing detailed accounts of the event, including what the rape means to them. They are instructed to relive their emotions in full while they are writing. They then reread their writings daily as a form of prolonged exposure therapy.

In comparison, CNT asks trauma victims to have an action-oriented mental dialog with their antagonist, in which they say, ask, or do whatever they feel is necessary to achieve closure. They are told to mind-map their antagonist throughout this interaction and to describe his or her ongoing demeanor, thoughts, feelings, and responses. The client is supposed to handle whatever the antagonist says or does as best as possible.

Written dialogs allow the therapist to see an accurate picture of the client's mental map of the antagonist, which can be referenced in subsequent discussions. These usually contain information not available in client's VAM (explicit) memories. Clients' implicit memories and non-conscious understandings are often embedded in the dialog; and once recognized by the therapist, they can be pointed out to the client for further study. Moreover, by visualizing the client's description of interacting with the antagonist, rather than listening to the client's report of the person's characteristics, dialogs give the therapist new modes of accessing a client's mental maps of the antagonist.

The therapist's written responses to clients' written dialog interactions frequently identify gaps in their mind-mapping radar and provide corresponding alternative pictures. Often this challenges the accuracy of the client's map of the antagonist. Other comments focus on how the client handled the interaction, highlighting underlying beliefs and dynamics embedded in it. Sometimes the feedback contains a completely different—and often more difficult—interpretation of the interaction and how it turned out.

Rather than being presented as "homework" or a writing assignment, written dialogs are an ongoing vehicle for working through interpersonal trauma. Written dialogs are a process requiring multiple iterations during which people modify how they organize their responses, express themselves, and handle their antagonist. Over the course of these iterations, the antagonist's attitudes and responses in the dialogs usually change based on the client's making better responses.

Eventually, people evoke in the dialog the responses they fear most from their antagonist and handle themselves in the face of this. They work to a point—much like having repeated interactions with someone in real life—where they feel they can handle their antagonist, regardless of what he or she does. Unlike "fantasy visualizations" with idealized outcomes that occur by magic, the evolution of written dialogs mirrors real life and takes repetition, making mistakes, and learning from experience. Additional aspects of written dialogs in Crucible Neurobiological Therapy are outlined in Table 5.

CONJOINT NEUROPLASTIC ACTIVITIES

Hugging 'till Relaxed (HTR) and Heads on Pillows (HOP) tap into the brain's "social network" to enhance close and trusting relationships. Roughly stated, HTR is a long-duration (ten-minute) hug, and HOP involves long-duration eye-gazing. Developed over 30 years of doing marital and sex therapy, HTR and HOP reduce negative input from dysfunctional relationships and repair the relationship-straining effects of PTSD. This is important because people impacted by interpersonal trauma often have sexual difficulties. HTR and HOP offer similar benefits:

- They allow physical contact in a nonsexual setting.
- They signal partners reaching a turning point in their declining relationship.
- They establish expectations of future additional change.

However, to really utilize these activities to full advantage, you have to understand the underlying automatic and volitional brain processes they engage. Deliberate and purposeful eye gaze like HOP triggers a special reaction in the brain associated with bonding. A long-duration focused hug like HTR operationalizes everything scientists understand about how the anterior insula accomplishes physical and emotional

APPENDIX D

Aspects of Written Dialogs in Crucible® Neurobiological Therapy

- Involves "taking action" and circumventing avoidance.
- Requires self-soothing and giving your dilemma meaning.
- Mind mapping the antagonist.
- Perspective taking.
- Moderate level of arousal.
- Draws on explicit memory.
- Triggers implicit memory.
- Draws on existing mind map of antagonist.
- Triggers interoceptive experience.
- Constructing statements, questions, strategies, and responding to questions.
- Requires flexible responding to heated, threatening interpersonal interactions.
- Reexperiencing traumatic mind mapping but keeping brain together under stress.
- Practice making gold standard responses.
- Repetition until success is achieved.
- Conversion of failure experiences to success.
- Examination of errors and bad strategies, and overlooked aspects of antagonist's responses and strategies.
  - Increased accuracy in mind mapping of antagonist.
  - Identification of gaps in mind mapping radar.
  - Uncovering hesitancies to accurately map or respond to antagonist.
  - Forcing brain to think in previously restricted ways.
- Ready transfer of learning to real-time actual situations with antagonist.
- Changing anticipation of future events.

Table 5. Aspects of written dialogs in Crucible® Neurobiological Therapy.

self-awareness and awareness of other people's mental states.

Remember back to Mbembe Jabbi's studies demonstrating that identical neural responses are elicited in the AI when subjects view pictures of disgusted faces or smell disgusting odors themselves. The same happens when subjects view videos of people sampling pleasant or unpleasant tastes and sample the different tastes themselves. Tanya Singer notes that identical activation of secondary somatosensory (touch processing) cortices occurs when subjects watch videos of people being touched or when they themselves are touched.[34]

This happens during HTR as partners focus on their own interoceptive cues (e.g., body tension, breathing) while also being acutely aware of the state of each other's mind and body, experienced through their own body. This coordinated brain activation through touch exploits the right and left AI to create profound self-awareness and awareness of the partner at the same time. This also occurs during Heads on Pillows, particularly when partners touch each other's faces. It specifically transpires during Eyes-Open Sex and Eyes-Open Orgasms.

One of Tanya Singer's empathy studies discovered that activity in the posterior insular only occurred when participants actually experienced pain themselves. In contrast, AI activity occurred when people experienced pain themselves and observed (vicariously simulated) it in others.[35] Other fMRI studies demonstrate that anticipation of pain activates more AI regions, whereas actual pain activates more posterior insular regions. The AI plays an important role in mental imagery of tactile and proprioceptive sensations and probably gets activated in imagery of actions and sensations in general.[36] HTR, HOP, and Eyes-Open Sex build on indications that the posterior insular processes sensory-specific primary representations, and the AI creates secondary representations of anticipated experience.[37]

Trauma victims frequently complain of feeling cut off from their bodies ("disembodied"). This is why HTR, HOP, and Eyes-Open Sex can be particularly helpful in treating trauma. These activities put them back in touch with themselves. One way to think of HTR, for example, is as putting your arms around your partner, focusing on yourself, and calming your body and mind. But a subsequent part of HTR involves mapping your partner while you're doing this. Doing this involves your brain's using the AI to make a metamap of your partner's physical body and then overlaying this on a similar metamap of your own body.

Some researchers focus on pre-reflective versus reflective meta-representations in the brain, analogous to the way Chris Brewin approached SAM and VAM dual-representation systems. They propose that the right AI handles pre-reflective (SAM-level) simulation and the vmPFC handles reflective (VAM-level) mental simulations. When processing one's own self-awareness, the right AI represents pre-reflective bodily states.[38]

To do HTR, clients are initially told to focus their attention on their own physical and mental self-states to calm their mind and body down. As they become better able to do this, they gradually become aware of their partner's body and mind without losing this self-awareness. This creates self-regulation during close physical and emotional contact with one's partner, which is the definition of differentiation. Doing this repeatedly probably stimulates and integrates AI and mPFC activity and recruits additional brain circuitry through "body learning."

Why is this important? Dysfunction in the mPFC contributes to arousal dysregulation in PTSD.[39] The fact that the mPFC can directly influence emotional arousal has enormous clinical significance, suggesting that increased interoceptive awareness can enhance control over emotions and compensate for mPFC dysfunction. Both HTR and HOP involve top-down regulation requiring mPFC participation. The research literature on breath regulation contains concerns about clients misusing breadth control training to avoid self-awareness of their own anxiety. Because of the intimate presence of the partner, HTR and HOP don't facilitate avoidance. These activities require repetition for full benefit, preferably several times a week.

Pay attention to the fact that the majority of studies on empathic brain responses focus on pain. Previously I described research indicating that the AI, ACC, brain stem, and cerebellum–the so-called pain matrix–activate when people experience pain themselves or perceive a loved one experiencing pain. These areas process the affective component of pain–the unpleasant, subjective, felt experience. Personal experience and knowing a partner's experience activates the same affective circuits. This pain network activates when subjects see an unknown-but-likable person suffering pain, observe painful facial expressions, view body parts in painful situations, or watch hands being pricked by needles.[40] There are no empathy studies of joy.

But treatment needn't be limited to revisiting or anticipating negative experiences. Treatment should stimulate the AI, ACC, brain stem and cerebellum through happy affects too, including feeling relaxed in your body and feeling your partner relax too. The positive meta-meanings of these kinesthetic cues have their own positive impact on the right AI, which is centrally involved in this process. Jabbi's studies indicate that the AI processes one's own and other people's *positive* physical and mental states too. Treatment needs to utilize empathy for positive–if not joyous–experiences as well. Here's where Hugging 'till Relaxed, Heads on Pillows, Eyes-Open Sex and Eyes-Open Orgasms really shine.

Aspects of Hugging 'Till Relaxed, Heads on Pillows and Eyes–Open Sex
• Involves "taking action" and circumventing avoidance.
• Focus on interoceptive cues.
• Body awareness combined with awareness (mapping) of partner.
• Body learning.
• Learning to control anxiety and strong feelings.
• Relaxing in the presence of the partner.
• Going from moderate anxiety to calm.
• Self-soothing: giving dilemma meaning.
• Breath slowing.

Table 6. Aspects of Hugging 'till Relaxed, Heads on Pillows, and Eyes-Open Sex.

MODIFY CURRENT RELATIONSHIP WITH ANTAGONIST

Along with exposure to traumatic memories through imagery or observation, patients in PTSD-exposure therapy are encouraged to participate in real-life situations linked to the trauma. The aim is to provoke fear reactions and extinguish them by staying in place, focusing on the trauma reminders, and awaiting realization of safety to overcome expectations of danger. Incorporating more adaptive cognitions into these real-life interactions enhances cognitive restructuring.[41]

In CNT this occurs by clients changing their interactions with the antagonists responsible for their traumatic mind-mapping experiences. (When the antagonist is deceased, this is accomplished through written dialogs.) Remember, one clear sign predicting treatment failure with my clients was their maintaining a relationship with a dysfunctional parent; in those cases the current equilibrium could only be maintained by my client's remaining impaired. I now tell such clients they will probably fail in treatment if they leave the existing relationship intact. It's not necessary to cut off the relationship with the antagonist, but it is necessary to change it.

What needs to be done is fairly predictable: For instance, you have to:

1. *Take action.*

2. *Show your antagonist that you can see him or her.* You probably don't want to do this because you know this will predictably trigger their bad behavior and more traumatic mind mapping for you.

3. *Keep your antagonist from getting around you, sliding underneath you, or walking over you.* You don't have to "pin" your antagonist or extract a confession, admission, or apology.

4. *Make "gold-standard" responses,* as described in the text of *Brain Talk*.

5. *Shift the locus of control from your antagonist to yourself.* Do this by controlling what you're going to

do, not by foolishly issuing ultimatums or threats, or telling your antagonist what he or she must do.

6. Ultimately, *your level of engagement is determined by–and must vary according to–your antagonist's level of functioning.*

How clients go about this is unique, as is just how far they have to go because both are dictated by the antagonist's responses. This is evident in various cases described in *Brain Talk*.

INCREASE CLIENTS' DIFFERENTIATION

The process of modifying the relationship with one's antagonist usually requires and results in increased differentiation, which is subsequently experienced as post-traumatic growth. Handling yourself in real time requires mind mapping, perspective-taking, empathy, and cognitive flexibility in handing manipulative moves while you're fighting off "spaghetti brain."

This is not an impossible task. It's how everyone gets the job done. Doing this helps you develop (1) a clearer sense of who you are, (2) more ability to soothe your emotions, (3) better control of your emotional reactivity, and (4) forbearance in the face of adversity, but also knowing when to quit. These are the Crucible Four Points of Balance, which you can enhance through revisualizations, written dialogs, and conjoint neuroplastic activities in addition to real-life interactions with your antagonist.

ROLE OF THE THERAPIST IN CNT

Many cases involving interpersonal neurobiological problems don't start out that way. Typically, people seek treatment for any number of individual, marital, or family problems. In the process of therapy for these presenting problems, neurobiological problems often surface in the midst of heated therapy. Such problems also come to light when people's prior history of trauma and abuse become known. In order to treat people for their traumatic mind mapping and resulting neurobiological problems, a therapist must be able to competently treat clients' presenting problems too. In this way, Crucible Neurobiological Therapy builds nicely on top of Crucible Therapy, which integrates marital and sexual therapy within a differentiation framework.

So, strictly speaking, what role does the therapist play in Crucible® Neurobiological Therapy? The therapist's initial task is to find out as much as possible about details of the traumatic mind mapping and possible reactions, distortions, or misinterpretations that may have developed. In some cases this may be explicit rules or assumptions the client can articulate. In others, the distortion may be associative in the form of negative thoughts or images that come spontaneously to mind and influence their experience. In both cases, the therapist tries to determine the chains of reasoning that support these misinterpretations. This is accomplished through a variety of modalities:

1. Guiding the client through revisualizations of traumatic mind-mapping experiences, while carefully avoiding implanting false beliefs or assumptions.

2. Visualizing the scene together with the client. I have to understand the scene even when the client doesn't. And even when they do, I often realize what's happening before they do. Besides developing traumatic mind mapping in session from scenes my clients describe, sometimes it happens if they ascribed meanings that are tragically distorted or if they display neurobiological problems without apparent awareness of this.

3. Reading clients' written dialogs, providing relevant written commentary, and returning these to the client for further study. The therapist also suggests subsequent dialogs for the client.

4. Suggesting and explaining conjoint neuroplastic activities to clients, and debriefing their experiences with HTR, HOP, and Eyes-Open Sex. This includes interpreting clients' physical experiences as "elicitation windows" in which their mental maps, attributions, emotional blocks, and differentiation issues are evident.

These and other activities provide helpful stimuli to the client and a host of challenges to the therapist. In guiding clients through revisualizations, the real-time stimulus provided by the therapist's actions impinge directly on the client's senses as she attends to the traumatic image. This contributes to her encoding a distinctive set of features with the new VAM memory of the trauma. This is often the first time the client has tried to process traumatic mind-mapping experiences with a helpful person. The relationship between client and therapist in CNT probably activates the "attachment circuits" in the client's (and the therapist's) brain.

This carries over to the process of identifying interpersonal neurobiological problems, which are often only evident when the client's attention is directed to overlooked, avoided, or misinterpreted aspects of the visualized scene. When the client is not given the freedom to continue dodging, this is when interpersonal neurobiological problems–like gaps in mind-mapping radar, mental tics, thought strictures, and emotional dysregulation–become manifest. This can look like the therapist is giving the client "a hard time" or "making things worse."

Crucible® Neurobiological Therapy is a "brain-to-brain" therapy requiring the therapist to track the client's brain/mind at an unusually granular level which allows detection of topic-related inconsistent thinking or repeated attempts to change topics, gaps in observations, and other non-volitional and deliberate cognitive and emotional distortions. This ongoing "brain-to-brain and mind-to-mind" connection produces a constant stream of profound "moments of meeting," which is vitally important to facilitating positive brain change. Although written mental dialogs and revisualizations are important clinical tools, the most powerful ingredient in CNT is probably the therapist's sustained ability to map the client's brain/mind when dealing with traumatic mind-mapping events.

Until the purpose of the therapist's actions subsequently become clear, the therapist's dedication and non-reactivity gets tested if clients become hostile and reactive. The therapist also has to buffer clients' initial shocked realizations of having subtle brain impairments, even though they are subsequently thankful to no longer have them.

Reviewing clients' written dialogs presents equally difficult but slightly different demands. The therapist's written comments often involve a move-by-move analysis of what the antagonist and client are doing. This requires the therapist to map the antagonist according to the client's depictions of the antagonist's verbal and physical behavior.

But sometimes the most revealing observations come from stepping back from the dialog itself, and recognizing shocking details hidden right out in the open. For example, one highly traumatized client habitually manifested an inappropriate sardonic grin ("masked face"). I was startled to recognize that his written mental dialogs with his highly abusive mother contained constant mutual facial posturing, like smiling displays of contempt by both of them.

Even the most implausible interactions in revisualizations, and irrational thoughts in written mental dialogs, have a logic that makes sense for that particular person's history. Even though written dialogs appear to be text-based, the therapist must visualize the interaction and follow the developments in order to decode what the client is describing. The skill of the therapist lies in uncovering visually encoded, nonverbal logic. This helps clients discover alternative ways of understanding events that make equally good (if not better) sense of their own descriptions. But unlike in conventional talk therapy, this emerges out of their written work product, rather than from something they just said in session.

The therapist's personal differentiation comes into play in multiple other ways. One involves his or her resilience to traumatic mind mapping. Therapists get "spaghetti brain" from visualizing clients' materials, and you have to be willing to be traumatized to some degree in the process of doing treatment. This comes under the Crucible Fourth Point of Balance: Tolerating pain for growth—only in this case, it's the therapist's pain and the clients' growth.

Another variable is the therapist's ability to recover from traumatic mind mapping. How is the therapist's personal life going these days? Is he getting enough rest, making time for exercise, eating well, and living a balanced lifestyle? Is his clinical practice out of control and stretching him too thin? In CNT, the therapist's differentiation plays out directly and indirectly in myriad ways.

I've learned that I can handle a lot of traumatic visual imagery without becoming overwhelmed. Occasionally I will be briefly staggered—or more rarely, impacted for days—by things I visualize in my clients' histories. I now take this as an inherent part of my job.

What allows me to remain enthusiastic about working with deplorable, disgusting, and horrifying material? I've discovered I don't get seriously traumatized if my clients get better. As long as there's utility to the process, I'm OK. Something good has to come out of this. In a synergistic way, this motivates me to always do my best. This is not treatment a therapist wants to do poorly, for his own welfare.

IN-SESSION PSYCHOPHYSIOLOGICAL MONITORING

Crucible Neurobiological Therapy employs in-session psychophysiological monitoring of clients and the therapist in a variety of ways. First off, this allows quick, accurate, and meaningful initial assessment of clients, allowing therapists to figure out what kind of shape their new clients are in. Psychophysiological monitoring demonstrates impairments and impacts from past and current traumas, and negative readings often contrast sharply with reported positive childhood histories.

In-session physiological monitoring brings an undeniable *gravitas* to therapy. Some clients give more truthful and complete histories, thinking that the therapist has placed a wireless lie detector on them. Real-time readouts allow therapists to pace sessions more precisely, allowing them to turn the intensity up or down to maintain an optimal level of stress and arousal. It also helps clients modulate their arousal if and when they get upset in sessions.

As progress assessment and positive reinforcement goes, it's hard to find something better than psychophysiological monitoring. At selected times, therapists can turn the readout towards the clients, allowing them to note of their own progress (or lack thereof). It allows clients and therapists alike to chart progress over time, using empirical data exempt from subjectivity and bias.

New observations and applications of real-time monitoring are emerging all the time. This increases therapists' motivation and investment in treatment, and changes clinical practice from routine to an experimental "happening." It increases the intensity of therapeutic moments of meeting and brings a sobering reality to what's being discussed.

HEART-RATE VARIABILITY BIOFEEDBACK

Since psychologist Paul Lehrer first experimented with heart-rate variability (HRV) biofeedback in the 1990s,[42] support has grown for using HRV biofeedback to treat a variety of disorders.[43] A 2013 review found favorable outcome research treating asthma, cardiac rehabilitation, COPD, hypertension, irritable bowel syndrome, chronic muscle pain, cyclic vomiting, induced hypertension, recurrent abdominal pain, depression, fibromyalgia, anxiety, pregnancy, insomnia, and PTSD.[44]

A large amount of evidence suggests people are more resilient–physically and emotionally–when their HRV is higher and more complex. People with low HRV tend to be physically sick,[45] emotionally ill,[46] and less aerobically fit.[47] They are at greater risk of dying when their physical health is compromised.[48] Athletic performance improves after HRV biofeedback training.[49] HRV biofeedback also restores suppressed, autonomic, nervous system function when people are exposed to inflammatory cytokines in experiments.[50]

During HRV biofeedback, heart-rate oscillation amplitude grows while the pattern becomes simple and sinusoidal. This pattern, often achieved within a minute on a person's first attempt, largely involves the baroreflex.[51] The baroreflex is the fastest reflex in the human body, kicking into gear in less than a single cardiac cycle. It is the body's homeostatic mechanism for maintaining blood pressure upon standing (postural hypotension). Six cycles of diaphragmatic deep breathing per minute create an optimal effect.

Only recently has the effect of vagal tone (strength) on HRV been considered.[52] The vagus nerve connects the insula, amygdala hippocampus, locus coeruleus, and orbitofrontal cortex with the lower viscera of the body.[53] It also constantly sends sensory information about the state of the body back up to the brain, giving rise to gut instincts, visceral feelings, and fear responses.[54] Research indicates high vagal tone is part of a feedback loop between positive emotions, physical health, and positive social connections. Simply reflecting on positive relationships and working to improve close bonds improves vagal tone.[55]

The vagus nerve closely interacts with the inflammatory system. Increased vagal tone produced via electrical stimulation reduces inflammatory cytokines.[56] Surgically implanted vagus nerve stimulation (VNS) devices can reduce cytokine levels and inflammation in extreme cases of rheumatoid arthritis and other disorders.[57] A non-invasive transcutaneous vagus nerve stimulator (tVNS), that sits in the ear like a pair of earphones, significantly reduced symptoms in patients with major depressive disorders.[58]

Voluntary breath control can affect your autonomic nervous system via your vagus nerve through your brain stem to your heart. Branches of the vagus nerve passing through the thalamus can quiet your prefrontal cortex and reduce anxious worrying.[59] A preliminary study trained 12 participants in HRV biofeedback for 4 sessions and compared them to 13 who received relaxation training. The HRV biofeedback group substantially improved their HRV, whereas the relaxation training group did not.[60]

Combining HRV biofeedback with prolonged exposure therapy (PET), cognitive processing therapy (CPT), or cognitive behavior therapy (CBT) improves their efficacy treating PTSD.[61] An anecdotal report cited significant clinical improvement in 24 of 27 consecutive PTSD patients using HRV biofeedback (88%).[62] Several studies suggest HRV biofeedback can reduce depression and anxiety.[63] Directly producing the interoceptive cues of relaxation via HRV biofeedback may induce subjective feelings of being relaxed, thereby directly countering the effects of stress. Affordable and practical wireless biomonitoring devices such as the Zephyr system I use open the door to new treatments for interpersonal trauma utilizing HRV biofeedback.

HIGH AROUSAL AND "SAFE EMERGENCIES"

Treating interpersonal trauma always raises concerns about clients "feeling safe" and avoiding inadvertent retraumatization. This evokes an aura of mental fragility, suggesting such clients must be handled delicately. Certainly, first and foremost, therapists must always do no harm. But they also have responsibility to be effective agents of change. This requires dialing in how stressful an optimal treatment should be and how insecure or anxious clients should feel.

Here's where the therapist's picture of how much stress clients should feel in the process is important. How hard should clients push themselves? Should they take baby steps or big bites? What if people start to get upset or anxious? Is this counter-productive, or a sign of ineffective pacing? Is this exactly what they need?

Issues of safety and security are terribly important in resolving interpersonal trauma, since this often involves realizing you're not safe with people you're supposed to trust. In domestic violence situations, clients' physical safety is always first and foremost. There's no point attempting to resolve traumatic mind mapping if domestic violence is ongoing. It's also dangerous and foolhardy to use methods described here to maintain clients' ability to tolerate physical and emotional abuse.

More commonly, questions of "safety and security" translate into making yourself think about or do things you know or suspect will upset you. How hard should you push yourself? If you start to get upset, should you keep going? If you get anxious, how much anxiety should you expect yourself to tolerate? What's too much and how much is just enough?

Psychologist Fritz Pearls, founder of Gestalt Therapy, talked about "safe emergencies" to describe feeling safe in psychotherapy while being challenged to go beyond what feels safe.[64] Therapists traditionally emphasize the "safe" part of therapy, but PTSD treatments increasingly utilize the "emergency" side. For clients to rewire their brains, they have to get out of their comfort zones. This is very much like bodybuilding: They repeatedly need to deal with more than they can easily manage.

For decades I've conducted "outside the comfort zone" Crucible Therapy, which harnesses the inevitable emotional gridlock and natural two-choice dilemmas permeating adult love relationships. Crucible Therapy occurs at a higher level of anxiety and pressure than conventional talk therapy. This provided a natural platform for brain-changing therapy as I developed Crucible Neurobiological Therapy.[65]

I know from experience how clients want to wait until they feel totally confident before facing their anxieties. However, moderate anxiety and discomfort actually work better to rewire their brains and raise their differentiation. Clients must initially challenge themselves with tasks of sufficient difficulty, and increase this as they progress. They have to engage in tasks of sufficient intensity and frequency, preferably every day but at least several times a week for several weeks.

Clients rewire their brains by creating neuroplasticity. To make their brains do this, they have to accept a moderate degree of discomfort. They not only have to do things that might make them feel awkward, they have to be deeply engaged and maintain heightened attention and alertness (which increases neurotransmitter levels).

Clients can't wait until they feel no anxiety about what they might revisualize or realize. If they insist on taking baby steps–or no steps–until they feel secure, they'll be waiting forever. They need to make themselves do things they don't feel comfortable doing, exposing themselves to safe but anxiety-provoking revisualizations and written mental dialogs, in order to function better subsequently in traumatic mind-mapping situations.

WINDOW OF TOLERANCE

There's always a treatment question of how much is too much anxiety. A therapist can help clients decide. If there's a serious question in the client's mind, a consultation is always appropriate. (If you're wondering whether it's worth your time, money, and effort, do yourself a favor and err in the direction of getting a professional opinion.)

My clients don't "freak out" from doing revisualizations, written dialogs, conjoint neuroplastic activities, or dealing with their antagonists. But here are some things to keep in mind that can help clients pace themselves: There's no point in "blowing yourself up." Research indicates extreme stress produces dendritic atrophy and debranching in the hippocampus, and simultaneous, enhanced-dendritic branching in the amygdala.[66] However, these high-stress, adverse effects don't occur with moderate stress because different hippocampus receptors are involved. High stress fills the glucocorticoid cortisol receptors with neurotransmitters. Low-to-moderate levels of stress fills the mineralocorticoid receptors instead.[67]

Once clients reach a moderate arousal level which can help them the most, they can have intense and upsetting visual realizations. This can seem like things are going wrong, but trauma experts hope this happens. It's difficult to confront deeply upsetting feelings, thoughts, and sensations associated with traumatic mind mapping. But when traumatic memories are reactivated through exposure, and reconsolidated by reincorporating interoceptive cues and mind-mapping information, they become less distressing, more easily managed, and less likely to produce unwanted intrusions.[68] These newly-consolidated explicit memories and insights can dramatically diminish the effects of traumatic mind mapping.

Psychologists Robert Yerkes and John Dodson developed the Yerkes-Dodson law that says performance increases with physiological and mental arousal—but only up to a point. When arousal is too high or too low, performance decreases. In this case, if arousal levels are too low, traumatic images and SAM memory (interoceptive cues, mind-mapping information, and emotional contagion remnants) are not accessed. If arousal levels are too high, clients can lose contact with their immediate situation and surroundings (briefly dissociate).[69]

The "window of tolerance" refers to how much input clients can absorb and remain receptive to therapy. PTSD treatment typically focuses on helping clients stay within the top boundary of their window of tolerance, consistent with the Yerkes-Dodson law. Effective treatment often exceeds the level of anxiety and pressure clients think is prudent. Clients often need to be told that a moderate degree of anxiety is most productive, and waiting to feel comfortable is counterproductive.[70]

(In case you are not a therapist: Clients often experience powerful shifts in understanding during revisualizations, written dialogs, and conjoint neuroplastic activities. It's possible you could experience "flashback" imagery, kinesthetic sensations in your body, shifts in your taste and hearing, odd feelings in your head, and even "sparks" in your visual field. Once you know this can happen, these symptoms don't have to frighten you. You're not going insane. Trauma treatment experts think this occurs from the long-overdue processing of SAM memory into VAM.

On the other hand, what if you become so upset you can't go about your daily routine? Or you haven't slept for days? Or you're seriously considering harming yourself or someone else? In these unlikely events, you should definitely seek professional help in your community. When in doubt, visit a local mental health or medical expert. Services are available in most communities through your place of religious worship, local public health department, or community mental health center. I greatly doubt things will come to this from doing revisualizations, having mental dialogs, and hugging your partner. But if they do, stop doing these activities, and talk to a professional.)

POST-TRAUMATIC GROWTH

Treatment of traumatic mind mapping often develops clients' sense of self-efficacy and encourages an optimistic outlook on the future. Positive growth after traumatic mind mapping is a valuable potential people shouldn't pass up. While "post-traumatic growth" may sound odd, it's actually not unusual. Half of trauma victims report some post-trauma positive outcome.[71]

A systematic review of 39 studies found positive changes commonly reported by 30 to 70 percent of survivors of traumatic events, including transportation accidents, natural disasters, medical problems, and interpersonal experiences like combat, rape, sexual assault, child abuse, relationship breakdown, parental divorce, bereavement, and immigration.[72] Heart attack patients who experience personal benefits immediately after their first attack have reduced reoccurrence and morbidity eight years later.[73]

Post-traumatic growth results from struggling with highly challenging life crises and manifesting the courage to engage in making sense of what happened to you. Post-traumatic growth is the clearer sense of self this leaves you with, the positive revamping or reinvesting in your philosophy of life and the constructive changes in relationships that come with both.

Post-traumatic growth is characterized by leaps in personal development to higher levels of psychological sophistication and emotional functioning, otherwise known as differentiation. This emerges from lots of cognitive and emotional processing of the mental maps and world views threatened or nullified by traumatic mind-mapping events.

Coming to grips with traumatic mind mapping involves seeing a darker perspective of what happened with your antagonist and accepting what really occurred. Previously we discussed how "giving your dilemma meaning" is humankind's most effective self-soothing method. Moving beyond traumatic mind mapping to psychological resilience involves developing new meanings for the traumatic events and understanding how they've shaped your life.[74]

This produces an increased sense of personal agency, purpose, and resilience, beneficial changes in priorities, a wider perspective of one's place in the world, a richer existential or spiritual life, and increased appreciation for life in general. This shapes one's life narrative and produces life wisdom. The resulting increased cognitive complexity and more sophisticated self-organization can bolster people's resilience to stress.[75]

Post-traumatic growth is actually an ongoing process rather than a static outcome. Richard Tedeschi and Larry Calhoun of the University of North Carolina offer five principles about post-traumatic growth that can help people get the most from their experiences:

1. Growth occurs when your old psychological schemas are destroyed by traumatic events and replaced by new schemas. Some assumptions are resistant to disconfirmation. They buffer distress but reduce the possibilities of schema change and growth.

2. Positive self-valuations must follow for growth to occur. This can be as basic as "I am a survivor." But deep appreciation of what you're gone through, and what you've accomplished despite this, works much better.

3. Different types of events cause different types of growth. Was the person giving you traumatic mind mapping very important to you? How emotionally powerful were the particular topics or circumstances? How far-reaching were the ramifications? The greater the challenge involved in coming to grips with this, the greater the potential for post-traumatic growth.

4. Personality characteristics like fortitude, forbearance, optimism, and self-efficacy allow people to persevere through unfamiliar and destabilizing situations and turn them into opportunities for growth and success.

5. Growth occurs when the trauma becomes a pivot point for radical shifts in perspective, ushering in a new era of living, based in realistically seeing and accepting what's possible and what is not.[76]

Having been traumatized countless times by visualizing my clients' terrible experiences, I am a walking demonstration of post-traumatic growth. Conducting Crucible Neurobiological Therapy has brought out the best in me rather than draining my energy, demoralizing me, and filling me with despair. It's motivated me to write a book with over 400 scientific references. What more evidence could you ask for?

I hope reading the text and Appendices of *Brain Talk* leaves you with bushels of post-traumatic growth. I hope this inspires the best in you and motivates the rest of you to heed what brain science says about how we influence the world around us: The world we co-create has as much to do with what's in our minds, as it does with how we behave.

End Notes

Chapter 2

1 The triune brain model was originally formulated by the American physician and neuroscientist Paul MacLean in the 1960s. He later discussed it at length in his book The Triune Brain in Evolution (MacLean, 1990).

2 Gervais & Robinson (producers) (2009).

3 The inferior parietal lobule (IPL).

4 The superior temporal sulcu1s (STS), responds selectively to sounds and movements of hands and face and detects other people's attention.

5 The temporal and parietal posterior brain regions involved in mind mapping include the IPL and the STS.

6 The limbic-paralimbic system includes the amygdala, the orbitofrontal cortex (OFC), the ventral medial prefrontal cortex (MPFC), and the anterior cingulate gyrus (ACG).

7 The prefrontal region includes the dorsal MPFC and inferolateral frontal cortex (ILFC).

8 The temporoparietal junction (TPJ), anterior cingulate cortex (ACC), and prefrontal cortex.

9 Posterior cortical regions, which include the TPJ, the STS, and the posterior cingulate cortex/precuneus (PCC/PCun) are reciprocally connected with limbic and frontal paralimbic regions including the amygdala, the temporal polar cortex (vTP), the vACC, the OFC, the vMPFC and the ILFC.

10 Domes et al. (2007).

11 Bednya et al. (2009).

12 Whenever I refer to your nemesis or "antagonist," you can substitute "parent," "spouse," "child," "sibling," "co-worker" or "boss" to put a specific face on this generic term.

Chapter 3

1 Wellman, Cross & Watson (2001).
2 Doherty (2009).
3 Wimmer & Perner (1983).
4 Baron-Cohen, Leslie, & Frith (1985).

Chapter 4

1 See http://www.nbcnews.com/science/weird-science/study-shows-dogs-dont-people-who-are-mean-their-owners-n374786
2 See http://www.nbcnews.com/today/pets/canine-cognition-center-yale-studies-what-your-dog-thinking-1D80135075.
3 See http://www.canidae.com/blog/2009/12/what-is-left-gaze-bias-how-does-it-relate-to-dogs./
4 See http://www.today.com/health/gazing-our-dogs-eyes-releases-cuddle-chemical-t15746
5 See http://www.sciencemag.org/news/2016/08/video-your-dog-understands-more-you-think
6 Dunn et al. (1991A).
7 Meins et al. (1998).
8 Meins (1998).
9 Meins et al. (1998).
10 Ibid.
11 Meins et al. (2001).
12 Ibid.
13 Meins et al. (2002).
14 Tronick (1998).
15 "Rather than seeing our results as contradicting these earlier findings, we would instead suggest that our account is fully consistent with the work of Dunn and others relating to exposure to mental state language during the preschool years." (p. 1726). In Meins et al. (2002).
16 Dunn (1994).
17 Lewis et al. (1996).
18 Ruffman et al. (1998). Also see Perner et al. (1994).
19 Lewis et al. (1996).

20 Brown et al. (1996).
21 Dunn & Munn (1985).
22 Dunn, Bretherton & Munn (1987).
23 Zahn-Waxler et al. (1992).
24 Dunn et al. (1991A).
25 Dunn et al. (1991B).
26 Dunn (1995).
27 Pons & Harris (2005).
28 Brown et al. (1996).
29 Much & Shweder (1978). Also see Slomkowski & Killen (1992).
30 Dunn, Cutting & Demetriou (2000).
31 Flavell (2004).
32 Lecce, Caputi & Pagnin (2014).
33 Hughes & Dunn (1998).
34 Peterson & Siegal (2002).
35 Bagwell, Newcomb & Bukowski (1998).
36 Parker et al. (1999). Also Laursen et al. (2007).
37 Numerous studies link being 'friendless' with adverse mental health outcomes including low self-worth, social anxiety, depression, loneliness, and suicide. The negative impact of friendlessness on loneliness, depression, and psychological well-being prevail through childhood, and adulthood. A 12-year longitudinal study found that a young adult having no mutual childhood friends uniquely predicted adult psychopathology (especially depression). See Bagwell, Newcomb & Bukowski (1998). Also Pedersen, et al. (2007).
38 Fink et al. (2014).
39 Hare, Call & Tomasell (2001).
40 Byrne & Whiten (1992).
41 See http://www.dailymail.co.uk/news/article-2574863/Family-law-judge-caught-beating-16-year-old-daughter-video-posted-online-2011-losses-election-bid.html.
42 Theory of Mind tests used in brain imaging studies:
   1. Judging objects from different people's point of view: Goel,V. (1995).
   2. Stories, still cartoons, cartoon strips: Mellet, et al. (1988). Also Dolan, et al. (1996).

3. Judging emotion from eye expression: Baron-Cohen, et al. (1999).
4. Interactive games: human vs. computer: Gallagher, et al. (2002).
5. Social transgressions: stories: Berthoz, et al. (2002).
6. Watching silent animations: random movement vs. scripted movement: Castelli, et al. (2002).
7. Watching gestures: instrumental vs. expressive: Gallagher & Frith (2003).

43 Wallace (2014).

44 Howlin, Baron-Cohen & Hadwin (1998).

45 See Christensen (2016) for estimated statistics. Also see "Study Suggests Autism Is Being Over-diagnosed." Autism may be over-diagnosed in as many as 9 percent of children, U.S. government researchers report. It might be because autism covers such a broad range of symptoms and behaviors and is difficult to diagnose, and it may also be because increasing awareness about autism according to the team at the Centers for Disease Control and Prevention and the University of Washington. http://www.nbcnews.com/health/kids-health/study-suggests-autism-being-overdiagnosed-n450671

46 Lorena and John Bobbit became a famous case of spouse abuse in 1993, wherein John had multiple affairs, the couple fought constantly, and Lorena eventually cut off his penis. https://en.wikipedia.org/wiki/John_and_Lorena_Bobbitt.

47 For example, "People with autism have an underdeveloped TOM ability" See Arden (2015).

48 Hughes & Leekam (2004).

49 Hughes & Dunn (1998); Youngblade & Dunn (1995).

50 Slomkowski & Dunn (1996).

51 Cutting & Dunn (1999; Dunn 2002); (1995).

52 Baron-Cohen et al. (1985). Baron-Cohen et al. (1997).

53 Leslie (1987).

54 Leslie & Frith (1988).

55 Phillips, Baron-Cohen & Rutter (1998).

56 Baron-Cohen, Spitz & Cross (1993).

57 Sodian (1991).
58 Leekam & Prior (1994).
59 Happé (1994).
60 Jarrold, Boucher & Smith (1996); Lewis & Boucher (1988).
61 Wing & Gould (1979).
62 Repacholi & Slaughter (2003).

Chapter 5

1 This applies to every other decision in a relationship as well, whether it's having a baby, disciplining your kids, spending money, or having your mother-in-law move in. LDP/HDP positions usually change from issue to issue.

2 In fact, in many cases (particularly when the LDP is female), the LDP knows more about sex, likes sex more, and has more sexual experience than the HDP. These LDPs know that the sex they're having isn't worth wanting, which explains why they're the LDPs. To handle the HDP, they mask their mind so successfully, the HDP believe the LDP is a sexual dud.

3 This rule only holds where rape is not tolerated and women control access to their bodies. The reason why the LDP controls sex is because the HDP makes most of the initiations, and the LDP decides which initiations to accept.

4 See my book, *Resurrecting Sex: Resolving Sexual Problems and Rejuvenating Your Relationship*.

5 "Emergency lie" is actually a well-known term in many European countries.

Chapter 6

1 Savitsky & Gilovich (2003).
2 Gilovich, Savitsky & Medvec (1998); Miller & McFarland (1987, 1991); Vorauer & Ross (1999).
3 Van Boven, Gilovich & Medvec (2003); Vorauer & Claude (1998); Kassin & Fong (1999).
4 Gilovich et al. (1998).
5 Gilovich & Savitsky (1999).
6 Gilovich et al. (1998).

Chapter 7

1 People are able to maintain a sense of self that is supported by semantic memory of personal facts in the absence of direct access to the memories that describe the episodes on which the knowledge is based. Research indicates patients with severe amnesia can have accurate and detailed semantic knowledge of what they are like as a person. For example, they can know their particular personality traits and characteristics. Self-knowledge about your traits can be accessed without the need for episodic memory retrieval.

2 Descartes (1683/2007).

3 Locke (1690/1979).

4 Shoemaker (1990)

5 Descartes (1641/1911).

6 I understand what I'm reporting here runs contrary to most people's expectations–including many therapists. For an overview of issues about introspection, see the Internet Encyclopedia of Philosophy. The existential aspect of infallible introspection bothers many philosophers: If introspective mistakes are impossible, then the notion of gaining knowledge by introspection doesn't make sense. Gaining self-knowledge only makes sense if it's possible to think wrongly. If failure is logically impossible, then gaining knowledge through introspection is meaningless. See Wittgenstein (1958) and Armstrong (1963).

7 Dennett (1991).

8 According to some, introspection is simply a process of perceptual replay, calling to mind things we did and said in a given situation. It is not a privileged executive monitoring process, over and above perception, memory, and imagination. Introspection involves these three processes being put to a particular use. See Lyons (1986).

9 In one study, subjects were asked to indicate which of four pairs of stockings had the highest quality. Although all four pairs of stockings were identical, the leftmost pair was preferred by a factor of four to one. Though positional effects clearly were a factor, participants explicitly denied position having played a role. See Churchland (1988).

10 Your memory is organized into "mental schemas," clusters of knowledge within a conceptual framework that help quickly organize and simplify information about the world around us. By cataloging what you already know, believe and expect to see, mental schemas expedite recognizing and interpreting events, which is helpful during times of stress, overload, or threat—as long as the pictures you've created are accurate. If they're not, your mental schemas interfere with acquiring accurate self-knowledge, dealing with other people, and breaking out of your past experiences and expectations. If your mental schemas lack relevant mind-mapping data due to traumatic mind-mapping, you're going to be at a distinct disadvantage.

11 Gopnik & Astington (1988).

12 Introspective judgments are prone to gross error even about your current visual experience. We typically assume our ability to know what we're seeing is excellent, except perhaps for imprecision at the borders of our visual field. However, introspective experiments that force you to direct your attention away from the focal center of your vision reveals you have clarity and precision in a surprisingly small portion of your visual field. See Schwitzgebel (2007) and Dennett (1991).

13 James (1890/1950).

14 Nisbett & Wilson (1977).

15 Here are other ways introspection may not yield accurate self-knowledge:

1. *Perception set:* A predisposition to see things a particular way. Quickly drawing conclusions without sufficient checking.
2. *Being right (denial):* Dogmatically holding onto an opinion or belief, or defending an action. A failure to acknowledge evidence. Being wrong is unthinkable. Going to any length to demonstrate your rightness.
3. *Unwarranted blaming:* Holding other people responsible for your difficulties. Two common subtle versions are disproportionate responsibility (inappropriate distribution of accountability) and out-

ward causes (blaming things outside yourself). Conversely, this also includes blaming yourself unjustly.
4. *Polarized thinking (false choice, "black and white" dichotomy):* Compressing complex information into simplistic categories, often aimed at rapid decision-making during times of stress, conflict, or threat. A way of manipulating decision-making. Everything is perceived in extremes with little room for middle ground. Things are black or white, good or bad. Your way is the good way.
5. *Over-generalization:* Coming to conclusions based on insufficient or unrepresentative evidence, or extrapolating beyond what the available data warrants.
6. *Over-personalization (egocentric bias):* The tendency to relate everything to yourself. Incorrectly thinking everything people say or do is a reaction to you, or is aimed at you, or a reflection on you.
7. *Asch Effect:* Changing your opinions to agree with group affiliations, despite clear evidence to the contrary. Named after the researcher who demonstrated the effects of group pressure on individual judgment modification and distortion.
8. *Catastrophizing:* Unreasonable anticipation of disaster based on a small problem. Turning bad news into an inevitable tragedy. A pervasive explanatory style or personality characteristic.

16 Ryle (1949).

17 This explains how therapists, who don't really have their own act together, can come up with interesting insights and be somewhat helpful to their clients.

Chapter 8

1 Research by neuroscientist Simone Shamay-Tsoory on the neuroanatomical basis of sarcasm indicates: (1) Your language-based left hemisphere interprets the literal meaning of

the speaker's utterances; and (2) your right hemisphere and frontal lobes process the intentional, social, and emotional context, and use this information to identify contradictions between literal meanings and social/emotional context. (3) The right rear parts of your frontal lobes identify sarcasm by integrating emotion processing with perspective taking. This area is the inferior (rear) part of the prefrontal cortex (vMPFC). See Shamay-Tsoory, et al. (2005).

2 Sutton, Smith & Swettenham (1999). It turns out children who have less mind mapping ability are more likely to be victims of bullying. Also see Waterman, et al, (1981).

3 Workshops for therapists based on traumatic mind-mapping have been available since 2009. Schnarch (2009); Morehouse & Schnarch (2010).

4 Walker et al. (2016).

5 Porges et al. (1996); Porges et al. (1994).

6 Bishop et al. (2016).

7 Segerstrom & Miller (2004).

8 National Sexual Violence Resource Center (2015). Also see CDC Sexual violence data sheet (2012).

9 Gradus (2016). Also http://www.ptsd.va.gov/professional/PTSD-overview/epidemiological-facts-ptsd.asp.

10 Ford (2013).

11 Rhodes & Chan (2010).

12 de Fabrique et al. (2007). Also see Namnyak et al. (2007).

13 Fuselier (1999).

14 Dutton & Painter (1981).

Chapter 9

1 Cartwright (2013).

2 Allen (2013).

3 Cartwright (2013).

4 Both ABC NEWS and the Huffington Post provide copious documentation in news stories and videos of reported child abuse.

5 See http://my.clevelandclinic.org/health/articles/munchausen-syndrome.

6 Sutton, Smith & Swettenham (1999, 2001). Also Caravita, Di Blasio & Salmivalli (2010); Carreras et al. (2014); and Renouf et al. (2010).

7 Dolan & Fullam (2004); Ellis (1982)

8 Singer (2008A).

9 You can read more about "normal marital sadism" in my books, *Passionate Marriage* and *Intimacy & Desire*.

10 Di Pellegrino et al. (1992). Also Rizzolatti et al. (1996).

11 The insular cortex, particularly its most anterior portion, is considered a limbic-related cortex. The insula has increasingly become the focus of attention for its role in body representation and subjective emotional experience. Antonio Damasio proposes this region plays a role in mapping visceral states associated with emotional experience, giving rise to conscious feelings.

12 Jabbi, Swart & Keysers (2007). Also Jabbi, Bastiaansen & Keysers (2008).

13 Olatunji & McKay (2009).

14 Schnarch (2012).

Chapter 10

1 Houston (2013). "Spending To Piss Off My Father." p. 60 *More* Magazine, February 2013.

2 Surguladze (2003).

3 Fox (2016).

4 For information on resolving troublesome orgasm trigger fantasies, see *Resurrecting Sex*.

5 I only conduct therapy via teleconference with people whom I've had face-to-face sessions. I do not support or recommend therapy via teleconference without this prerequisite.

6 Escher & Romme (2012); Corstens et al. (2014).

7 Posey & Losch (1983); Beavan, Read & Cartwright (2011); Pearson et al. (2008); Tien (1991).

8 Escher (2009).

9 Myelin is a fatty white substance surrounding the axons of some nerve cells, that forms an electrically insulating layer. It is essential for the proper functioning of the nervous system.

In humans, myelination naturally begins during fetal development, and rapidly increases during infancy, and continues through adolescence. The myelin sheath increase the speed of impulse transmission along a nerve.

10 Caspi et al. (2002).
11 Kulis & Esteller (2010).
12 Lund et al. (2004).
13 McKay (2006).
14 Finkelhor (1994a); (1994b).
15 Video of Donald Trump's "Access Hollywood" interview.
16 Donald and Ivanka Trump on "The View" TV show.

Chapter 11

1 Frank, Menasco & O'Sullivan (2008).
2 Geiselman et al. (2011).
3 Hauch et al. (2016). This meta-analysis of 30 studies examined whether training improves detection of deception. Results indicated a small to medium training effect for lie accuracy, but not for truth accuracy. Training based on verbal content cues produced larger effects than nonverbal cue training.
4 Ekman (1957).
5 Ekman (1990).
6 Keltner & Ekman (2000). Micro expressions are very brief facial expressions, lasting only a fraction of a second. They occur when a person either deliberately or unconsciously conceals a feeling.
7 Hontz et al. (2009). This study examined a micro-expression training program called Screening Passengers by Observation Technique (SPOT), designed to identify people who could pose a threat to airline travel. A 2007 report on SPOT stated that "simply put, people (including professional lie-catchers with extensive experience of assessing veracity) would achieve similar hit rates if they flipped a coin."
8 Parsimony, the basic principle underlying all modern science (also known as Occam's razor), advises choosing the simplest explanation that accounts for the widest amount of evidence.
9 Webster (2012).
10 Horowitz (2013).

11 Georgetown University Medical Center (2012). Reported in a study by P. E. Turkeltaub and colleagues at the Center for Brain Plasticity and Recovery, presented at Neuroscience 2012, the annual meeting of the Society for Neuroscience.

12 It's long been known blind people have extraordinary hearing. Research now explains why. Their brain's visual cortex becomes part of their auditory cortex. (See Doidge, 2007.) They literally have more of their brain devoted to hearing than sighted people do. I took this to heart early in my therapy training, when I had the good fortune to have a blind supervisor who had the most amazing mind mapping ability with clients. When I sat in on his sessions, he'd often start talking to the client about things I hadn't picked up on when they spoke. After that I began doing therapy sometimes with my eyes closed to improve my hearing acuity. It really helped.

Chapter 12

1 Doidge (2007).

2 There are ways to handle things you've been told about your past that you don't remember. For instance, let's say your mother told you that your father often hit you with a belt, but you don't remember this. In this case, you would visualize the conversation with your mother, and you would map your mother's mind while she's telling you this. What is she thinking and feeling? Why is she telling you this? What picture of your father's mind is she proposing? Then you can study your mother's descriptions of how your father went about beating you, what he said, how often he did this, how he explained the beatings, and what he did afterwards. In other words, you not only map the mind of your antagonist in the event you don't remember, you do the same with the person reporting the incident(s). This allows you to better assess the potential accuracy of what you're being told.

3 When clients report they can't see their antagonist's face in first-person or third-person perspectives, I'll suggest alternatives to work around this. For example, I might ask which direction the antagonist's faces are pointed, or what he is looking at.

Or if they can't see the antagonist's head at all, I might ask what he is doing with his hands. There is always a way to get around traumatic mind mapping blockages. Don't give up. Get creative.

Chapter 13

1 Stephen certainly had his own unflattering problems, which contributed to Rachael's sexual passivity. I'm only highlighting one side of their marriage, so you can track Rachael's processes in depth. If you want to see the interactive aspects of couple's sexual difficulties, read *Passionate Marriage* or *Intimacy & Desire*.

Chapter 14

1 You can read more about the Crucible Four Points of Balance in *Intimacy and Desire*.
2 Updegraff & Taylor (2000).
3 Tedeschi (1999).

Chapter 15

1 For an interesting article on this subject, see Don't be a stupid Cupid!
2 This doesn't include making moves on your partner, like sharing your sexual fantasies to pressure her into doing things you know she doesn't want to do.
3 Stern (2004) highlights the importance of intense and profound intersubjective moments of meeting. Cozolino (2006) lists 6 other conditions:
   1. A strong and resilient collaborative alliance.
   2. Moderate stress and emotional arousal, followed with calm.
   3. Gathering information and experiences across multiple dimensions of cognition, emotion, sensation, and behavior.
   4. Activity in brain neural networks processing and regulating thoughts, feelings, sensations and behaviors.
   5. New conceptual knowledge integrating emotional

and bodily experiences.
6. Organizing experiences in ways that foster continued neural growth and integration.
4 Ruiz (2014).
5 Dunn (1999a).
6 Lilienfeld & Arkowitz (2014).
7 Jaenisch & Bird (2003).
8 Paul McCartney, *"Let It Be."*

Appendix A

1 Abu-Akel & Shamay-Tsoory (2011).
2 Kim (2010).
3 Hare Psychopath Checklist-Revised (1991).
4 Baron-Cohen et al. (2001).
5 Richell et al. (2003).
6 Shamay-Tsoory et al. (2010).

Appendix B

1 See http://www.nimh.nih.gov/health/topics/posttraumatic-stress-disorder-ptsd.
2 Finkelhor (2008), p. 23.
3 Kessler et al. (1995).
4 Cloitre et al (2001).
5 Breslau et al. (1997); Kessler et al. (1995).
6 Acierno et al. (1999).
7 Ibid.
8 Breslau et al. (1998).
9 See http://www.ptsd.va.gov/professional/PTSD-overview/epidemiological-facts-ptsd.asp.
10 Kulka et al. (1990).
11 Kang et al. (2003).
12 Tanielian & Jaycox (2008).
13 Brett, Spitzer, & Williams (1988); Briere (1988); Cole & Putnam (1992); Scheeringa et. al (1995, 2003).
14 Burgess & Holstrom (1974).
15 Rollstin & Kern (1998); Walker (1984).

16 Krystal (1968).
17 Putnam (2003).
18 Ibid.
19 Ackerman, et al. (1998).
20 van der Kolk et al. (2005).
21 Ibid.
22 Ford & Kidd (1998); McDonagh-Coyle et al. (2005); Zlotnick (1999).
23 See https:www.psychiatry.org/psychiatrists/practice/dsm/history-of-the-dsm.
24 See http://www.dsm5.org/Documents/PTSD%20Fact%20Sheet.pdf.
25 D'Andrea et al. (2012).
26 Butler & Wolfner (2000).
27 Brown & Kulik (1977).
28 Conway & Dewhurst (1995); Brewin (2005); Pillemer (1998).
29 Pillemer (1998).
30 Cahill, Gorski & Le (2003).
31 Strange, Hurlemann & Dolan (2003).
32 McCloskey & Cohen (1989).
33 Damasio (2000), p. 28.
34 van der Kolk (2006).
35 Ibid.
36 Ibid.
37 Bremner et al. (1997).
38 Cailan et al. (2007).
39 Bremner et al. (1995);
40 Woon & Hedge (2008).
41 Gilbertson, et al. (2002).
42 DeBellis, Hooper, & Sapia (2005).
43 Vyas, Bernal, & Chattarji (2003).
44 Sapolsky (2003).
45 van der Kolk (2006).
46 Ledoux, Romanski & Xagoraris (1991).
47 Porges (1991).
48 Porges et al. (1996).

49 Porges & Doussard–Roosevelt (1997); Porges et al. (1994).
50 Porges (1991, 1995).
51 See Porges et al. (1994) for a review.
52 For a review of heart rate variability research, see Beauchaine (2001).
53 Eisenberg, Fabes & Guthrie (1997).
54 Katz & Gottman (1995).
55 Sahar, Shalev & Porges (2001).
56 Hopper et al. (2005).
57 Porges et al. (1996).
58 Cohen et al. (2002).
59 van der Kolk (2006).
60 Field et al. (1982); Stifter et al. (1989).
61 Rohleder et al. (2004).
62 Pervanidou (2008)
63 Chrousos (1996).
64 Koopman et al. (2016).

Appendix C

1 Augustine (1996).

2 Your insula sends and receives input from the amygdala, lateral orbital cortex, olfactory cortex, anterior cingulate cortex (ACC), and STS. See Mesulam & Mufson (1982); also Mufson & Mesulam, (1982).

3 Craig (2002, 2008). For instance, posterior insula activity correlates with actual changes in temperature, whereas anterior activation tracks perceived temperature changes. See Craig et al. (2000).

4 Singer et al. (2004).

5 Craig (2009).

6 Bartels & Zeki (2004).

7 Von Economo or "spindle" neurons are thought to be unique to higher primates. They are found in humans, bonobos, chimpanzees, gorillas, and orangutans, but in no other primate species See Nimchinsky, et al. (1999).

8 Allman, Hakeem & Watson (2002).

9 Allman et al. (2005).

10 See Craig (2002) for review.
11 Allman, Hakeem & Watson (2002).
12 Singer et al. (2004).
13 Jabbi, Swart & Keysers (2007).
14 Abnormal spindle neuron development may cause social disabilities characteristic of autism spectrum disorders, particularly problems evaluating social situations (Frith, 2001; Mundy, 2003), however empirical support is lacking. One study assessed emotional awareness of self and others in people with high functioning autism and healthy adults. Difficulties in emotional awareness were related to low AI activity in autistic individuals and controls alike, predominately in the right AI. Poorer awareness of one's own and others' emotions was related to weaker AI activity, regardless of whether people had ASD or not (Silani, et al., 2008). Another study showed no difference in spindle neuron density between autistic and normal brains (Kennedy, Semendeferi, & Courchesne, 2007). However, short- and long-range neuron connectivity may be disrupted in people with ASD, rather than the number of spindle cells per se (Uddin & Menon, 2009).
15 Baron-Cohen & Wheelwright (2004).
16 Arden (2015).
17 Jabbi, Swart & Keysers (2007).
18 Singer (2008a).
19 Singer (2008b).
20 Elster (1989)
21 van Dijk et al. (2006).
22 Smith et al. (1999).
23 Feather (1989, 1991)
24 Shamay-Tsoory, bi-Elhanany & Aharon-Peretz (2007).
25 Decety & Jackson (2004).
26 Shamay-Tsoory, Tibi-Elhanany & Aharon-Peretz (2007).
27 Preston & de Waal (2002); Decety & Jackson (2004); Leiberg & Anders, (2006).
28 Shamay-Tsoory, Aharon-Peretz & Perry (2009).
29 Jabbi et al. (2007); Schulte-Ruther et al. (2007).

30 Kaplan & Iacoboni (2006).
31 Shamay-Tsoory, Aharon-Peretz & Perry (2009).
32 D'Argembeau et al. (2008).
33 Shamay-Tsoory, Aharon-Peretz & Perry (2009).
34 Mitchell et al. (2006); Gallagher & Frith (2003).
35 De Waal (2008).
36 Shamay-Tsoory, Aharon-Peretz & Perry (2009), p. 618.
37 Reiman et al. (1997); Carr et al. (2003); Wicker et al. (2003); Singer et al. (2004).
38 Samson, et al. (2004); Gallagher & Frith (2003).
39 Hughes & Leekam (2004).
40 Chamorro-Premuzic et al. (2011); Paulhus & Williams (2002); Horowitz & Strack (2010).
41 Paulhus & Williams (2002).
42 Jones & Paulhus (2010b).
43 Byrne & Whiten (1988); Christie & Geis (1970); Linton & Wiener (2001); Wilson, Near, & Miller (1996).
44 Christie & Geis (1970).
45 Fehr, Samsom & Paulhus (1992); McHoskey, Worzel & Szyarto (1998); McHoskey (1995).
46 Jakobwitz & Egan (2006).
47 Kohut (1977).
48 Vernon et al. (2008).
49 Skeem et al. (2011).
50 Gunnthorsdottir, McCabe & Smith (2002); McIllwain (2003).
51 Repacholi et al. (2003); Wilson, Near & Miller (1998).
52 Wilson et al. (1998).
53 Repacholi & Slaughter (2003); Repacholi et al. (2003).
54 Davis & Stone (2003).
55 Sutton et al. (2001).
56 Paal & Bereczkei (2007).
57 Buckels, Jones & Paulhus (2013).
58 See *Passionate Marriage* for discussion of normal marital sadism.
59 Jones & Paulhus (2010a).
60 Woodworth & Porter (2002).
61 Jones & Paulhus (2011).

62 Campbell et al. (2004).
63 Jones & Paulhus (2010a).
64 Baumeister & Campbell (1999); Nell (2006).
65 Taylor (2009).
66 Bushman & Whitaker (2010).
67 Malamuth (2003); Woodworth & Porter (2002).
68 Baumeister, Catanese & Wallace (2002); Campbell et al. (2004).
69 Baumeister & Campbell (1999); Nell (2006); Taylor (2009).
70 Reidy, Zeichner & Seibert (2011).
71 Baumeister & Campbell (1999)
72 Taylor (2009).
73 Reidy et al. (2011).
74 Jonason et al. (2009).
75 Jonason et al. (2011).
76 Jonason & Kavanagh (2010).
77 Jonason, Li & Buss (2010).
78 Jonason, Koenig & Tost (2010).
79 Jonason & Tost (2010).
80 Jonason & Webster (2012); Jonason, Slomski & Partyka (2012).
81 Streep (2016).
82 Ibid.
83 Ibid.
84 Darwin (1872/1965).
85 Craig (2009).
86 Critchley et al. (2002).
87 Critchley et al. (2004) ; Critchley (2005); Schandry (1981); Schandry et al. (1986).
88 Jabbi, Swart & Keysers (2007).
89 Jackson, Meltzoff, & Decety (2005); Lamm, Batson & Decety (2007); Saarela et al. (2007); Singer et, al. (2004).
90 Carr et al. (2003); Wicker et al. (2003).
91 Adolphs, Tranel & Damasio (2003); Calder et al. (2000).
92 Naqvi et al. (2007); Craig (2002); Critchley et al. (2004); Critchley (2005).
93 Damasio et al. (2000); Preston et al. (2007).
94 Kikuchi et al. (2005).

95 Craig (2008).
96 Damasio et al. (2000).
97 Damasio (2003).
98 Jabbi, Bastiaansen & Keysers (2008).
99 Aertsen & Preißl (1991).
100 Friston (2011).
101 Rekkas & Constable (2005).
102 van der Gaag, Mindera & Keysers (2007).
103 Adolphs et al. (2003)
104 Jabbi, Swart & Keysers (2007). Also see Wicker et al. (2003).
105 Craig (2009, 2011).
106 Craig (2009).
107 Ibid.
108 Ibid.
109 Ibid.

Appendix D

1 Buonomano & Merzenich (1998).
2 Grawe (2007).
3 DeAngelis (2008).
4 Resick & Schnicke (1992, 1993).
5 Resick et al. (2002).
6 Schnurr et al. (2007).
7 Van der Kolk (2006).
8 Ibid.
9 Ibid.
10 In one study, adult veterans with PTSD demonstrated semantic memory deficits compared to controls (Yehuda et al., 1995). In other research, inpatient adolescents who experienced trauma reported reduced autobiographical memory, where the degree of memory loss correlated with frequency and severity of traumatic events (de Decker, et al., 2003).
11 Harvey & Byant (1999).
12 de Decker et al. (2003).
13 Clark et al. (2003).
14 Brown & Kulik (1977).

15 Pillemer & White (1989).
16 Brewin (2001, 2003); Brewin, Dalgleish & Joseph (1996).
17 Brewin (2005).
18 Mack & Rock (1998).
19 Grey, Holmes & Brewin (2001).
20 Brewin (2001, 2003).
21 Ehlers & Clark (2000).
22 Nitschke, et al. (2006).
23 Craig (2009).
24 Pessoa (2008); Keysers & Gazzola (2007); Sun et al. (2016).
25 Jabbi, Bastiaansen & Keysers (2008).
26 Ibid.
27 Singer et al. (2004).
28 Butler & Wolfner (2000).
29 Amaral, et al. (1992).
30 Brewin, C. R. (2005).
31 DeAngelis (2008).
32 Hackmann (1998); Layden et al. (1993); Smucker et al. (1995).
33 Ruby & Decety (2004).
34 Singer (2008b).
35 Singer et al. (2004).
36 Sacco et al. (2006).
37 Ploghaus et al. (1999).
38 Keysers and Gazzola (2007).
39 Vasterling et al. (1998).
40 Singer (2008b).
41 Ehlers & Clark (2000).
42 Lehrer et al. (2000).
43 Gevirtz (2013).
44 Ibid.
45 Volz et al. (1990).
46 Friedman & Thayer (1998); Gorman & Sloan (2000); Carney & Freedland (2009); Kemp et al. (2010).
47 De Meersman (1993); Hautala et al. (2003); McNarry & Lewis (2012)

48 Kudaiberdieva et al. (2007); Politano et al. (2008); Ranpuria et al. (2008); Thayer et al. (2010); Huikuri & Stein (2013).
49 Strack & Deutsch, (2004); Linton & Shaw (2011); Paul & Garg (2012).
50 Lehrer et al. (2010).
51 Lehrer & Gervitz (2014).
52 Ibid.
53 Grundy (2002).
54 Klarer et al. (2014).
55 Kok et al. (2012).
56 Borovikova et al. (2000);
57 Koopman et al. (2016); Nahas et al. (2005); Daban et al. (2008); George et al. (2008.
58 Fang et al. (2016); Sackeim et al. (2001).
59 Brown & Gerbarg (2005a, 2005b).
60 Huang et al. (2014).
61 Tan, Wang & Ginsberg (2011).
62 Gevirtz & Dalenberg (2008).
63 Karavidas et al. (2007); Reiner (2008); Siepmann et al. (2008); McCraty et al. (2009); Tan et al. (2011); Patron et al. (2013).
64 Perls, Hefferine & Goodwin (1951).
65 Schnarch (2003).
66 Vyas, Bernal & Chattarji (2003); Vyas et al.(2002).
67 Alderson & Novack (2002); Kim & Diamond (2002); Sapolsky (2003).
68 Ehlers & Clark (2000).
69 Yerkes & Dodson (1908).
70 Arden (2015), p. 171-172.
71 Updegraff & Taylor (2000).
72 Linley & Joseph (2004).
73 Affleck et al. (1987).
74 Tedeschi (1999).
75 Tennen & Affleck (1998).
76 Calhoun & Tedeschi (2006).

REFERENCES

(3:1 = Chapter 3, end note1)
(**B:2** = Appendix B, end note 2)

A:1
Abu-Akel, A., & Shamay-Tsoory, S. G. (2011). Neuroanatomical and neurochemical bases of theory of mind. *Neuropsychologia*, 49, 2971–2984.

B:6 B:7
Acierno, R., Brady, K. L., Gray, M., Kilpatrick, D. G., Resnick, H. S., & Best, C. L. (2002). Psychopathology following interpersonal violence: A comparison of risk factors in older and younger adults. *J. Clinical Geropsychology*, 8, 13–23.

B:19
Ackerman, P.T., Newton, J. E. O., McPherson, W. B., Jones, J. G., & Dykman, R. A. (1998). Prevalence of post traumatic stress disorder and other psychiatric diagnoses in three groups of abused children (sexual, physical, and both). *Child Abuse and Neglect*, 22(8), 759–774.

C:91 C:103
Adolphs, R., Tranel, D., & Damasio, A. R. (2003). Dissociable neural systems for recognizing emotions. *Brain Cognition*, 52, 61–69.

C:99
Aertsen A. & Preißl, H. (1991). Dynamics of activity and connectivity in physiological neuronal networks. In H. G. Schuster (ed.) *Nonlinear dynamics and neuronal networks. Proceedings of the 63rd W. E. Heraeus Simonar Friedrichsdorf 1990.*

D:73
Affleck, G., Tennen, H., Croog, S., & Levine, S. (1987). Causal attribution, perceived benefits, and morbidity after a heart attack: An 8-year study. *J. Consulting Clinical Psychology*, 55(1), 29–35.

REFERENCES

(3:1 = Chapter 3, end note1)
(B:2 = Appendix B, end note 2)

A:1
Abu-Akel, A., & Shamay-Tsoory, S. G. (2011). Neuroanatomical and neurochemical bases of theory of mind. *Neuropsychologia,* 49, 2971-2984.

B:6 B:7
Acierno, R., Brady, K. L., Gray, M., Kilpatrick, D. G., Resnick, H. S., & Best, C. L. (2002). Psychopathology following interpersonal violence: A comparison of risk factors in older and younger adults. *J. Clinical Geropsychology,* 8, 13-23.

B:19
Ackerman, P.T., Newton, J. E. O., McPherson, W. B., Jones, J. G., & Dykman, R. A. (1998). Prevalence of post traumatic stress disorder and other psychiatric diagnoses in three groups of abused children (sexual, physical, and both). *Child Abuse and Neglect,* 22(8), 759-774.

C:91 C:103
Adolphs, R., Tranel, D., & Damasio, A. R. (2003). Dissociable neural systems for recognizing emotions. *Brain Cognition,* 52, 61-69.

C:99
Aertsen A. & Preißl, H. (1991). Dynamics of activity and connectivity in physiological neuronal networks. In H. G. Schuster (ed.) *Nonlinear dynamics and neuronal networks. Proceedings of the 63rd W. E. Heraeus Simonar Friedrichsdorf 1990.*

D:73
Affleck, G., Tennen, H., Croog, S., & Levine, S. (1987). Causal attribution, perceived benefits, and morbidity after a heart attack: An 8-year study. *J. Consulting Clinical Psychology,* 55(1), 29-35.

D:67
Alderson A. L., & Novack T. A. (2002). Neurophysiological and clinical aspects of glucocorticoids and memory: A review. *J. Clinical Experimental Neuropsychology, 24*, 335-355.

9:2
Allen, J. G. (interviewee) (2013). Bringing mentalizing to mind: Interview with Jon G. Allen. *New Therapist, 86*, 6-10.

C:9
Allman, J. M., Watson, K. K., Tetreault, N. A., & Hakeem, A. Y. (2005). Intuition and autism: a possible role for Von Economo neurons. *Trends Cognitive Science, 9*(8), 367-373.

C:8 C:11
Allman, J., Hakeem, A., & Watson K. (2002). Two phylogenetic specializations in the human brain. *Neuroscientist, 8*(4), 335-346.

4:56
Alphons, R., Damasio, H., Tranel, D., Cooper, G., & Damasio AR (2000). A role for somatosensory cortices in the visual recognition of emotion as revealed by three-dimensional lesion mapping. *J. Neuroscience, 20*, 2683-2690.

D:29
Amaral, D. G., Price, J.L., Pitkanen, A. & Carmichael, S.T. (1992). Anatomical organization of the primate amygdaloid complex. In J. P. Aggleton (ed.) *The Amygdala: Neurobiological aspects of emotion, memory, and mental dysfunction.* (pp 1-66.) New York: Wiley-Liss.

4:47 C:16 D:70
Arden, J. B. (2015). *Brain to brain: Enacting client change through the persuasive power of neuroscience.* New York: Wiley.

7:6
Armstrong, D. M. (1963). Is introspective knowledge incorrigible? *Philosophical Review, 62*(4), 417-432.

C:1
Augustine, J. R. (1996). Circuitry and functional aspects of the insular lobe in primates including humans. *Brain Research Review, 22*(3), 229-244.

4:35 4:37
Bagwell, C. L., Newcomb, A. F., & Bukowski, W. M. (1998). Friendship and peer rejection as predictors of adult adjustment. *Child Development, 69*(1), 140-153.

C:15
Baron-Cohen, S. & Wheelwright, S. (2004). The empathy quotient: an investigation of adults with Asperger's syndrome or high functioning autism, and normal sex differences. *J. Autism Developmental Disorder, 34*(2), 163-175.

4:52
Baron-Cohen, S., Jolliffe, T., Mortimore, C., & Robertson, M. (1997). Another advanced test of theory of mind: Evidence from very high functioning adults with autism or Asperger's syndrome. *J. Child Psychology Psychiatry, 38*(7), 813-822.

3:4 4:52
Baron-Cohen, S., Leslie, A., M., & Frith, U. (1985). Does the autistic child have a Theory of Mind? *Cognition, 21*(1), 37-46.

4:42
Baron-Cohen, S., Ring, H. D., Wheelwright, S., Bullmore, E. T., Brammer, M. J., Simmons, A., & Williams, S. C. R. (1999). Social intelligence in the normal and autistic brain: an fMRI study. *European J. Neuroscience, 11*(6), 1891-1898.

4:56
Baron-Cohen, S., Spitz, A., & Cross, P. (1993). Do children with autism recognize surprise? A research note. *Cognition Emotion, 7*(6), 507-516.

A:4
Baron-Cohen, S., Wheelwright, S., Hill, J., Raste, Y., & Plumb, I. (2001). The "Reading the Mind in the Eyes" Test Revised Version: A Study with Normal Adults, and Adults with Asperger Syndrome or High-functioning Autism. J. Child Psychology Psychiatry, 42, 241-251.

C:6
Bartels, A., & Zeki, S. (2004). The neural correlates of maternal and romantic love. *Neuroimage, 1*(3), 1155-1166.

C:64 C:69 C:71
Baumeister, R. F., & Campbell, W. K. (1999). The intrinsic ap-

peal of evil: Sadism, sensational thrills, and threatened egotism. *Personality Social Psychology Review, 3*(3), 210-221.

C:68

Baumeister, R. F., Catanese, K. R., & Wallace, H. M. (2002). Conquest by force: A narcissistic reactance theory of rape and sexual coercion. *Review General Psychology, 6,* 92-135.

B:52

Beauchaine, T. (2001). Vagal tone, development, and Gray's motivational theory: Toward an integrated model of autonomic nervous system functioning in psychopathology. *Development Psychopathology, 13,* 183-214

10:7

Beavan, V., Read, J., & Cartwright, C. (2011). The prevalence of voice-hearers in the general population: A literature review. *J. Mental Health, 20*(3), 281-292.

2:11

Bednya, M., Pascual-Leonea, A., & Saxe, R. R. (2009). Growing up blind does not change the neural bases of Theory of Mind. *PNAS, 106*(27), 11312-11317.

4:42

Berthoz, S., Armony, J. L., Blair, R. J. R., & Dolan R. J. (2002). An fMRI study of intentional and unintentional (embarrassing) violations of social norms. *Brain, 125,* 1696-1708.

8:6

Bishop, D. G., Wise, R. D., Lee, C., von Rahden, R. P., & Rodseth, R. N. (2016). Heart rate variability predicts 30-day all-cause mortality in intensive care units. *S. African J. Anesthesia Analgesia, 22*(4), 125-128.

D:56

Borovikova, L. V., Ivanova, S., Zhang, M., Yang, H., Botchkina, G. I., Watkins, L. R., Wang, H., et al. (2000). Vagus nerve stimulation attenuates the systemic inflammatory response to endotoxin. *Nature, 405*(6785), 458-462.

B:37

Bremner J. D., Randall, P., Vermetten, E., Staib, L., Bronen, R. A., Mazure, C., Capelli, S., et al. (1997). Magnetic resonance imaging-based measurement of hippocampal volume in

posttraumatic stress disorder related to childhood physical and sexual abuse: A preliminary report. *Biology Psychiatry, 41*(1), 23-32.
B:39
Bremner, J. D., Krystal, J. H., Southwick, S. M., & Charney, D. S. (1995). Functional neuroanatomical correlates of the effects of stress on memory. *J. Trauma Stress, 8*(4), 527-553.
B:5
Breslau, N., Davis, G. C., Andreski, P., Peterson, E. L., & Schultz, L. R. (1997). Sex differences in posttraumatic stress disorder. *Arch. General Psychiatry, 54*(11), 1044-1048.
B:8
Breslau, N., Davis, G., Andreski, P., Federman, B., & Anthony, J. C., (1998). *Epidemiological findings on posttraumatic stress disorder and co-morbid disorders in the general population.* New York: Oxford University Press.
B:13
Brett, E. A., Spitzer, R. L., & Williams, J. B. (1988). DSM-III-R criteria for posttraumatic stress disorder. *Am. J. Psychiatry, 145,* 1232-1236.
B:28 D:17 D:30
Brewin, C. R. (2005). Encoding and retrieval of traumatic memories. In Vasterling, J. J., & Brewing, C. R. (Eds.), *Neuropsychology of PTSD: Biological, Cognitive, and Clinical Perspectives* (Kindle Locations 3917-3926, Kindle Edition). New York: Guilford Press.
D:16 D:20
Brewin, C. R., (2001). A cognitive neuroscience account of posttraumatic stress disorder and its treatment. *Behaviour Research Therapy, 39,* 373-393.
D:16 D:20
Brewin, C. R., (2003). *Posttraumatic stress disorder: Malady or myth?* New Haven, CT: Yale University Press.
D:16
Brewin, C. R., Dalgleish, T., & Joseph, S. (1996). A dual representation theory of posttraumatic stress disorder. *Psychology Review, 103*(4), 670-686.

B:13

Briere, J. (1988). The long-term clinical correlates of childhood sexual victimization. *Annals New York Academy Science, 528*, 327-334.

D:59

Brown R. P., & Gerbarg P. L. (2005a). Sudarshan Kriya Yogic breathing in the treatment of stress, anxiety, and depression. Part II – clinical applications and guidelines. *J. Altern. Complement. Med., 11*, 711-717.

D:59

Brown R. P., & Gerbarg P. L. (2005b). Sudarshan Kriya yogic breathing in the treatment of stress, anxiety, and depression: Part I-neurophysiologic model. *J. Altern. Complement. Med., 11*, 189-201.

4:20 4:28

Brown, J. R., Donelan-McCall, N., & Dunn J. (1996). Why talk about mental states? The significance of children's conversations with friends, siblings, and mothers. *Child Development, 67*(3), 836-849.

B:27 D:14

Brown, R., & Kulik, J. (1977). Flashbulb memories. *Cognition, 5*(1), 73-99.

C:57

Buckels, E. E., Jones D. N., & Paulhus D. L. (2013). Behavioral confirmation of everyday sadism. *Psychological Science, 24*(11), 2201-2209.

D:1

Buonomano, D. V., & Merzenich, M. M. (1998). Cortical plasticity: From synapses to maps. *Annual Review Neurosci., 21*, 149-186.

B:14

Burgess, A. W., & Holmstrom, L. L. (1974). Rape trauma syndrome. *Am. J. Psychiatry, 131*(9), 981-986.

C:66

Bushman, B. J., & Whitaker, J. L. (2010). Like a magnet: Catharsis beliefs attract angry people to violent video games. *Psychological Science, 21*(6), 790-792.

B:26 D:28
Butler, L. D., & Wolfner, A. L. (2000). Some characteristics of positive and negative ("most traumatic") event memories in a college sample. *J. Trauma Dissociation, 1,* 45–68.

C:43
Byrne, R. W., & Whiten, A. (1988). *Machiavellian intelligence: Social expertise and the evolution of intellect in monkeys, apes, and humans.* New York: Oxford University Press.

4:40
Byrne, R. W., & Whiten, A. (1992). Cognitive evolution in primates: Evidence from tactical deception. *Man, 27*(3), 609–627.

B:30
Cahill, L., Gorski, L., & Le, K. (2003). Enhanced human memory consolidation with post-learning stress: Interaction with the degree of arousal at encoding. *Learn Memory, 10*(4), 270–4.

B:38
Cailan, H., Jun, L., Kun, W., Lingjiang, L., Meng, L., Zhong, H., Yong, L., et al. (2007). Brain responses to symptom provocation and trauma-related short-term memory recall in coal mining accident survivors with acute severe PTSD. *Brain Research, 1144,* 165–174.

C:91
Calder, A. J., Keane, J., Manes, F., Antoun, N., & Young, A.W. (2000). Impaired recognition and experience of disgust following brain injury. *Nature Neuroscience, 3,* 1077–1078.

D:76
Calhoun, L.G., & Tedeschi, R.G. (2006). *Handbook of Posttraumatic Growth.* New York: Psychology Press.

C:68
Campbell, W. K., Bonacci, A. M., Shelton, J., Exline, J. J., & Bushman, B. J. (2004). Psychological entitlement: Interpersonal consequences and validation of a self-report measure. *J. Personality Assessment, 83*(1), 29–45.

C:62
Campbell, W. K., Goodie, A. S., & Foster, J. D. (2004). Narcissism, overconfidence, and risk attitude. *J. Behavioral Decision Making, 17,* 297–311.

9:6
Caravita, S. C. S., Di Blasio, P, & Salmivalli, C. (2010). Early adolescents' participation in bullying: Is ToM involved? *J. Early Adolescence*, *30*(1), 138–170.

D:46
Carney, R. M., & Freedland, K. E. (2009). Depression and heart rate variability in patients with coronary heart disease. *Cleveland. Clinic. J. Medicine*, *76*(2), 13–17.

C:37 C:90
Carr, L., Iacoboni, M., Dubeau, M. C., Mazziotta, J. C., & Lenzi, G. L. (2003). Neural mechanisms of empathy in humans: A relay from neural systems for imitation to limbic areas. *Proceedings National Academy Science. USA*, *100*, 5497–5502.

9:6
Carreras, M. R., Braza, P., Muñoz, J. M., Braza, F., Azurmendi, A., Pascual-Sagastizabal, E., Cardas, J., et al. (2014). Aggression and prosocial behaviors in social conflicts mediating the influence of cold social intelligence and affective empathy on children's social preference. *Scand. J. Psychol.*, *55(4)*, 371–379.

9:1 9:3
Cartwright, D. (2013). Mentalizing: Feeling thoughts and thinking feelings. *New Therapist*, *86*, 11–15.

10:10
Caspi, A., McClay, J., Moffitt, T. E., Mill, J., Martin, J., Craig, I. W., Taylor, A., & Poulton, R. (2002.) Role of genotype in the cycle of violence in maltreated children. *Science*, *297*(5582), 851–854.

4:42
Castelli, F., Frith, C., Happe, F., & Frith, U., (2002). Autism, Asperger syndrome and brain mechanisms for the attribution of mental states to animated shapes. *Brain*, *125*, 1839–1849.

8:8
Centers for Disease Control and Prevention (2012). *CDC Sexual violence data sheet (pdf)*.

C:40
Chamorro-Premuzic, T., von Stumm, S., & Furnham, A. (2011). *The*

Wiley-Blackwell Handbook of Individual Differences (pp. 527). New York: John Wiley.

4:45

Christensen, D. L., Baio, J., Braun, K. V., Bilder, D., Charles, J., Constantino, J. N., Daniels, J., et al. (2016). Prevalence and characteristics of autism spectrum disorder among children aged 8 years–Autism and developmental disabilities monitoring network, 11 Sites, United States, 2012. *MMWR Surveillance Summaries, 65*(3), 1–23.

C:43 C:44

Christie, R., & Geis, F. L. (1970). *Studies in Machiavellianism.* New York: Academic Press.

B:63

Chrousos, G. P. (Ed.) (1996). Stress: Basic mechanisms and clinical implications. *Annals New York Acad. Sci., Vol. 771.* New York: New York Academy of Sciences.

7:9

Churchland, P. S. (1988). Can neurobiology teach us anything about consciousness? In N. J. Block, O. J. Flanagan, & G., Guzeldere (Eds.). *The nature of consciousness: Philosophical debates* (pp. 127–141). Cambridge, MT: MIT Press.

D:13

Clark, D. M., McManus, E. A., Hackmann, A., Fennell, M., Campbell, H., Flower, T., & Louis, B. (2003). Cognitive therapy versus fluoxetine in generalized social phobia: A randomized placebo-controlled trial. *J. Consulting Clinical Psychology, 71*(6), 1058–1067.

B:4

Cloitre, M., Cohen, L. R., Edelman, R. E., & Han, H. (2001). Posttraumatic stress disorder and extent of trauma exposure as correlates of medical problems and perceived health among women with childhood abuse. *Women Health, 34*(3), 1–17.

B:58

Cohen, J. A., Perel, J. M., DeBellis, M. D., Friedman, M. J., & Putnam, F. W. (2002). Treating traumatized children: Clinical implications of the psychobiology of posttraumatic stress disorder. *Trauma Violence Abuse, 2*(3), 91–108.

B:13
Cole, P., & Putnum, F. W. (1992). Effect of incest on self and social functioning: A developmental psychopathology perspective. *J. Consulting Clinical Psychology, 60,* 174-184.

B:28
Conway, M. A. & Dewhurst, S. A. (1995). The self and recollective experience. *Applied Cognitive Psychology, 9*(1), 1-19.

10:6
Corstens, D., Hayward, E., McCarthy-Jones, S., Waddingham, R., & Thomas, N. (2014). Emerging perspectives from the Hearing Voices Movement: Implications for research and practice. *Schizophrenia Bulletin, 40*(S4), S285-S294.

15:3
Cozolino, L. (2006). The neuroscience of psychotherapy: Healing the social brain (2nd ed.). New York: W. W. Norton.

C:3 C:10 C:92
Craig, A. D. (2002). How do you feel? Interoception: The sense of the physiological condition of the body. *Nature Review Neuroscience, 3,* 655-666.

C:3 C:95
Craig, A. D. (2008). Interoception and emotion: A neuroanatomical perspective. In: M. Lewis (Ed.) *Handbook of Emotions* (pp. 272-288). New York: Guilford Press.

C:5 C:85 C:105 C:106 C:107 C:108 C:109 D:23
Craig, A. D. (2009). How do you feel now? The anterior insula and human awareness. *Nature Review Neuroscience, 10,* 59-70.

C:105
Craig, A. D. (2011). A. D. (Bud) Craig on the Anterior Insula and Human Awareness. *Science Watch Fast Moving Front Commentary May 2011.*

C:3
Craig, A. D., Chen, K., Bandy, D., & Reiman, E. M. (2000). Thermosensory activation of insular cortex. *Nature Neuroscience, 3*(2), 184-190.

C:94
Cristancho, P., Cristancho, M. A., Baltuch, G. H., Thase, M. E.,

& O'Reardon, J. P. (2011). Effectiveness and Safety of Vagus Nerve Stimulation for Severe Treatment-Resistant Major Depression in Clinical Practice After FDA Approval: Outcomes at 1 Year *J. Clinical Psychiatry, 72*, 1376-1382.
C:87 C:92

Critchley, H. D. (2005). Neural mechanisms of autonomic, affective, and cognitive integration. *J. Comparative Neurology, 493*, 154-166.
C86

Critchley, H. D., Mathias, C. J., & Dolan, R. J. (2002). Fear conditioning in humans: The influence of awareness and arousal on functional neuroanatomy. *Neuron, 33*, 653-663.
C:87 C:92

Critchley, H. D., Wiens, S., Roshtein, P., Ohman, A., & Dolan, R. J. (2004). Neural systems supporting interoceptive awareness. *Nature Neuroscience, 7*, 189-195.
4:51

Cutting, A. L. and Dunn, J. (1999) Theory of mind, emotion understanding, language, and family background: Individual differences and interrelations. *Child Development, 70*, 853-865.
4:51

Cutting, A. L., & Dunn, J. (2002). The cost of understanding other people: social cognition predicts young children's sensitivity to criticism. *J. Child Psychology Psychiatry, 43*(7), 849-860.
B:25

D'Andrea, W., Ford. J., Stolbach, B., Spinazzola, J. & van der Kolk, B. A. (2012). Understanding interpersonal trauma in children: Why we need a developmentally appropriate trauma diagnosis. *J. Orthopsychiatry, 82*(2), 187-200.
C:32

D'Argembeau, A., Feyers, D., Majerus, S., Collette, F., Van der Linden, M., Maquet, P., et al. (2008). Self-reflection across time: Cortical midline structures differentiate between present and past selves. *Social Cognitive Affective Neuroscience. 3*, 244-252.

D:57

Daban, C., Martinez-Aran, A., Cruz, N., & Vieta, E. (2008). Safety and efficacy of vagus nerve stimulation in treatment-resistant depression: A systematic review. *J. Affect Disorders,* *110*(1-2), 1-15.

C:97

Damasio, A. R. (2003). *Looking for Spinoza: Joy, sorrow and the feeling brain.* New York: Harcourt Brace.

9:11 B:33 C:93 C:97

Damasio, A. R., (2000). *The Feeling of what happens: Body and emotion in the making of consciousness.* New York: Harcourt Brace.

C:93

Damasio, A. R., Grabowski, T. J., Bechara, A., Damasio, H., Ponto, L. L., Parvizi, J., et al. (2000). Subcortical and cortical brain activity during the feeling of self-generated emotions. *Nature Neurosci., 3,* 1049-1056.

C:84

Darwin, C. (1965). *The expression of the emotions in man and animals.* Chicago: University of Chicago Press. (Original work published 1872).

C:54

Davis, M., & Stone, T. (2003). Synthesis: Psychological understanding and social skills. In B. Repacholi & V. Slaughter (Eds.) *Individual differences in theory of mind. Macquarie monographs in cognitive science* (pp. 305-352). Hove, E. Sussex: Psychology Press.

B:42

De Bellis, M. D., Hooper, S. R., & Sapia, J. L. (2005). Early trauma exposure and the brain. In J. Vasterling and C. Brewin (Eds.) *Neuropsychology of PTSD: Biological, cognitive, and clinical perspectives* (pp. 271-291). New York: The Guilford Press.

D:10 D:12

de Decker, A., Hermans, D., Raes, F., & Eelen, P. (2003). Autobiographical memory specificity and trauma in inpatient adolescents. *J. Clinical Child Adolescent Psychology, 32*(1), 22-31.

8:12
de Fabrique, N., Romano, S. J., Vecchi, G. M., & van Hasselt, V. B. (2007). Understanding Stockholm Syndrome. *FBI Law Enforcement Bulletin (Law Enforcement Communication Unit)*, *76*(7), 10–15.

D:47
de Meersman, R. E. (1993). Heart rate variability and aerobic fitness. *Amer. Heart Journal*, *125*(3), 726–731.

C:35
de Waal, F. B. M. (2008). Putting the altruism back into altruism: The evolution of empathy. *Annual Review Psychology*, *59*, 279–300.

D:3 D:31
DeAngelis, T. (2008). PTSD treatments grow in evidence, effectiveness. *Monitor Psychology*, *39*(1), 40.

C:25 C:27
Decety, J., & Jackson, P. L. (2004). The functional architecture of human empathy. *Behavioral Cognitive Neuroscience Reviews*, *3*(2), 71–100.

7:7 7:12
Dennett, D. C. (1991). *Consciousness explained*. Boston: Little, Brown & Company.

7:5
Descartes, R. (1641/1911). Meditations on first philosophy. In E. S. Haldane (translator), *The Philosophical Works of Descartes (1911 edition)*. New York: Cambridge University Press.

7:2
Descartes, R. (1683/2007). Discourse on the Method. In P. Kraus and F. Hunt (Eds.) *Rene Descartes Discourse on Method*. Indianapolis, IN: Focus Publishing.

9:10
Di Pellegrino, G., Fadiga, L., Fogassi, L., Gallese, V., & Rizzolatti, G. (1992). Understanding motor events: A neurophysiological study. *Experimental Brain Research*, *91*, 176–180.

3:2
Doherty, M. (2009). *Theory of mind: How children understand other's thoughts and feelings*. New York: Psychology Press.

11:12 12:1
Doidge, N. (2007). *The brain the changes itself: Stories of personal triumph from the frontiers of brain science.* New York: Penguin.
9:7
Dolan, M., & Fullam, R. (2004). Theory of mind and mentalizing ability in antisocial personality disorders with and without psychopathy. *Psychology Medicine, 34*(6), 1093–1102.
4:42
Dolan, R. J., Fletcher, P., Morrisa, J., Kapurc, N., Deakind, J. F. W., & Frithe, C. D. (1996). Neural activation during covert processing of positive emotional facial expressions. *NeuroImage, 4*(3), 194–200.
2:10
Domes, G., Heinrichs, M., Michel, A., Berger, C. & Herpertz, S. C. (2007). Oxytocin improves "mind-reading" in humans. *Biological Psychiatry, 61*(6), 731–733.
4:16
Dunn, J. (1994). Changing mind and changing relationships. In C. Lewis & P. Mitchell (Eds.) *Children's early understanding of mind: Origins and development.* Hillsdale (USA): Erlbaum.
4:26 4:51
Dunn, J. (1995). Children as psychologists: The later correlates of individual differences in understanding of emotions and other minds. *Cognition Emotion, 9*(23), 187–201.
4:21
Dunn, J., & Munn, P. (1985). Becoming a family member: Family conflict and the development of social understanding in the second year. *Child Development, 56*, 480–492.
4:6 4:24 15:5
Dunn, J., Brown, J., & Beardsall, L. (1991a). Family talk about feeling states and children's later understanding of others' emotions. *Developmental Psychology, 27*(3), 448–455.
4:25
Dunn, J., Brown, J., Slomkowski, C., Tesla, C., & Youngblade, L. (1991b). Young children's understanding of other people's feelings and beliefs: Individual differences and their antecedents. *Child Development, 62*(6), 1352–1366.

4:30

Dunn, J., Cutting, A. L., & Demetriou, H. (2000). Moral sensibility, understanding others, and children's friendship interactions in the preschool period. *British J. Development Psychology, 18*(2), 159–177.

4:22

Dunn, J., Bretherton, I., & Munn, P. (1987). Conversations about feeling states between mothers and their young children. *Developmental Psychology, 23*(1), 132–139.

8:14

Dutton, D. G., & Painter, S. L. (1981). Traumatic bonding: The development of emotional attachments in battered women and other relationships of intermittent abuse. *Victimology, 1*(4), 139–155.

D:21 D:41 D:68

Ehlers, A., & Clark, D. M. (2000). Cognitive model of posttraumatic stress disorder (PTSD). *Behavior Research & Therapy, 38*, 319–345.

B:53

Eisenberg, N., Fabes, R. A., & Guthrie, I. K. (1997). Coping with stress: The roles of regulation and development. In S. A. Wolchik, & I. Sandler (Eds.), *Handbook of children's coping: Linking theory and intervention* (pp. 41–70). New York: Plenum.

11:4

Ekman, P. (1957). A methodological discussion of nonverbal behavior. *J. Psychology, 43*, 141–149.

11:5

Ekman, P. (1990). Duchenne and facial expression of emotion. In Cuthbertson, R. A. (Ed. and Transl.), *The Mechanism of Human Facial Expression* (pp. 270–284). Cambridge: Cambridge University Press.

11:6

Keltner, D., & Ekman, P. (2000). Facial expression and emotion. In M. Lewis and J. Haviland-Jones (Eds.) *Handbook of emotions, 2nd edition.* New York: Guilford Publications

9:7

Ellis, P. L. (1982). Empathy: A factor in antisocial behavior. *J. Abnormal Child Psychology, 10*(1), 123-133.

C:20

Elster, J. (1989). *Solomonic Judgements: Studies in the Limitation of Rationality.* Cambridge, UK: Cambridge University Press.

10:8

Escher, S. (2009). Accepting voices and finding a way out. In M. Romme, S. Escher, J. Dillon, D. Corstens, & M. Morris (Eds.) *Living with Voices: 50 stories of recovery* (pp. 48-54). Herefordshire, UK: PCCS Books.

10:6

Escher, S., & Romme, M. A. J. (2012), The Hearing Voices Movement. In J. D. Blom, & I. E. C. Sommer (Eds.) *Hallucinations: Research and practice* (pp. 385-393). New York: Springer.

D:58

Fang, J., Rong, P., Hong, Y., Fan, Y., Liu, J., Wang, H., Zhang, G., et al. (2016). Transcutaneous vagus nerve stimulation modulates default mode network in major depressive disorder. *Biological Psychiatry, 79*(4), 266-273.

C:23

Feather, N. T. (1989). Attitudes towards the high achiever: The fall of the tall poppy. *Australian J. Psychology, 1,* 239-267.

C:23

Feather, N. T. (1991). Attitudes towards the high achiever: Effects of perceiver's own level of competence. *Australian J. Psychology, 43,* 121-124.

C:45

Fehr, B., Samsom, D., & Paulhus, D. L. (1992). The construct of Machiavellianism: Twenty years later. In C. D. Spielberger & J. N. Butcher (Eds.), *Advances in personality assessment (vol. 9)* (pp. 77-116). Hillsdale, NJ: Erlbaum.

B:60

Field, T., Widmayer, S., Greenberg, R., & Stoller, S. (1982). Effects of parent training on teenage mothers and their infants. *Pediatrics, 69,* 703-707.

4:38

Fink, E., Beger, S., Peterson, C. C., Slaughter, V., & Rosnay, M. (2014). Friendlessness and Theory of Mind: A prospective longitudinal study. *British J. Developmental Psychology, 33*(1), 1037–1051.

10:14

Finkelhor, D. (1994a). Current information on the scope and nature of child sexual abuse. *Future of Children, 4,* 31–53.

10:14

Finkelhor, D. (1994b). The international epidemiology of child sexual abuse. *Child Abuse & Neglect, 18*(5), 409–417.

B:2

Finkelhor, D. (2008). *Childhood victimization: Violence, crime, and abuse in the lives of young people.* New York: Oxford University Press.

4:31

Flavell, J. H. (2004). Theory-of-Mind development: Retrospect and prospect. *Merrill-Palmer Quarterly, 50*(3), 274–290.

8:10

Ford, J. D. (2013). Trauma exposure and posttraumatic stress disorder in the lives of adolescents. *J. Amer. Academy Child Adolescent Psychiatry, 52*(8), 780–783.

B:22

Ford, J. D., & Kidd, P. (1998). Early childhood trauma and disorders of extreme stress as predictors of treatment outcome with chronic posttraumatic stress disorder. *J. Traumatic Stress, 11*(4), 743–761.

10:3

Fox, M. (2016). *Poor parenting can be passed from generation to generation: Study.* Research paper by A. M. Conn presented at Pediatric Academic Societies meeting in Baltimore on 5/3/2016. (http://www.nbcnews.com/health/kids-health/poor-parenting-can-be-passed-generation-generation-study-n566036)

11:1

Frank, M. G., Menasco, M. A., & O'Sullivan, M. (2008). Human behavior and deception detection. In J. G. Voeller (Ed.), *Handbook of science and technology for homeland security (vol. 5).* New York: John Wiley & Sons.

D:46

Friedman, B. H., & Thayer, J. F. (1998). Autonomic balance revisited: Panic anxiety and heart rate variability. *J. Psychosomatic Research, 44*(1), 133-151.

C:100

Friston, K. J. (2011). Functional and effective connectivity: A review. *Brain Connectivity, 1*(1), 13-36.

C:14

Frith, U. (2001). Mind blindness and the brain in autism. *Neuron, 32*(6), 969-979.

8:13

Fuselier, G. D. (1999). Placing the Stockholm Syndrome in perspective. *FBI Law Enforcement Bulletin, July,* 22-25.

C:34 C:38

Gallagher, H. L., & Frith, C. D. (2003). Functional imaging of 'theory of mind.' *Trends Cognitive Science, 7*(2), 77-83.

4:42

Gallagher, H. L., Jack, A. I., Roepstorff, A., & Frith, C. D. (2002). Imaging the intentional stance in a competitive game. *NeuroImage, 16*(3), 814-821.

11:2

Geiselman, R. E., Elmgren, S., Green, C. & Rystad, I. (2011). Training laypersons to detect deception in oral narratives and exchanges. *Amer. J. Forensic Psychiatry, 32*(2):43-61.

D:57

George, S. Z., Wallace, M. R., Wright, T. W., Moser, M. W., Greenfield III, W. H., Sack, B. K., Herbstman, D. M., et al. (2008). Evidence for a biopsychosocial influence on shoulder pain: Pain catastrophizing and catechol-O-methyltransferase (COMT) diplotype predict clinical pain ratings. *Pain, 136*(12), 53-61.

11:11

Georgetown University Medical Center. (2012). What you hear could depend on what your hands are doing. *Science Daily,* October 14, 2012.

2:2

Gervais, R., & Robinson, M. (producers) (2009). *The Invention of Lying.* (movie.) Warner Bros.

D:43 D:44
Gevirtz, R. (2013). The promise of heart rate variability biofeedback: Evidence-based applications. *Biofeedback, 41,* 110–120.

D:62
Gevirtz, R., & Dalenberg, C. (2008). Heart rate variability biofeedback in the treatment of trauma symptoms. *Biofeedback, 36*(1), 22–23.

B:41
Gilbertson, M. W., Shenton, M. E., Ciszewski, A., Kasai, K., Lasko, N. B., Orr, S. P., Pitman, R. K. (2002). Smaller hippocampal volume predicts pathologic vulnerability to psychological trauma. *Nature Neuroscience, 5*(11), 1242–1247.

6:5
Gilovich, T., & Savitsky, K. (1999). The spotlight effect and the illusion of transparency: Egocentric assessments of how we are seen by others. *Current Directions Psychological Science, 8*(6), 165–168.

6:2 6:4 6:6
Gilovich, T., Savitsky, K., & Medvec, V. H. (1998). The illusion of transparency: Biased assessments of others ability to read our emotional states. *J. Personality Social Psychology, 75,* 332–346.

4:42
Goel, V. (1995). *Sketches of Thought.* Harvard: MIT Press.

7:11
Gopnik, A. & Astington, J. W. (1988). Children's understanding of representational change and its relation to the understanding of false belief and the appearance-reality distinction. *Child Development, 59*(1), 26–37.

D:46
Gorman, J. M., & Sloan, R. P. (2000). Heart rate variability in depressive and anxiety disorders. *American Heart J., 40*(4), 77–83.

8:9
Gradus, J. L. (2016). *Epidemiology of PTSD.* National Center for PTSD.

D:2
Grawe, K. (2007). *Neuropsychotherapy: How the neurosciences inform effective psychotherapy.* New York: Taylor & Francis.

D:19
Grey, N., Holmes, E., & Brewin, C. R. (2001). Peritraumatic emotional "hotspots" in traumatic memory. *Behavioural Cognitive Psychotherapy, 29,* 367–372.

D:53
Grundy, D. (2002). Neuroanatomy of visceral nociception: vagal and splanchnic afferent. *Gut, 29,*(1), 12–15.

C:50
Gunnthorsdottir, A., McCabe, K., & Smith, V. (2002). Using the Machiavellianism instrument to predict trustworthiness in a bargaining game. *J. Economic Psychology, 23*(1), 49–66.

D:32
Hackmann, A., Clark, D. M., & McManus, F. (2000). Recurrent images and early memories in social phobia. *Behavior Research Therapy, 38*(6), 601–610.

4:59
Happé, F. G. E. (1994). An advanced test of theory of mind: Understanding of story characters' thoughts and feelings by able autistic, mentally handicapped, and normal children and adults. *J. Autism Developmental Disorders, 24*(2), 129–154.

4:39
Hare, B., Call, J., & Tomasell, M. (2001). Do chimpanzees know what conspecifics know? *Animal Behaviour, 61*(1), 139–151.

A:3
Hare, R. D. (1991). *The Hare Psychopathy Checklist – Revised.* Toronto, Ontario: Multi-Health Systems.

D:11
Harvey, A. G., & Bryant, R. A. (1999). The relationship between acute stress disorder and posttraumatic stress disorder: A 2-year prospective evaluation. *J. Consulting Clinical Psychology, 67*(6), 985–988.

11:3
Hauch, V., Sporer, S. L., Michael, S. W., & Meissner, C. A. (2016). Does training improve the detection of deception? A meta-analysis. *Communication Research, 43*(3) 283–343.

D:47
Hautala, A. J., Mäkikallio, T. H., Kiviniemi, A., Laukkanen, R. T., Nissilä, S., Huikuri, H. V., & Tulppo, M. P. (2003) Cardiovascular autonomic function correlates with the response to aerobic training in healthy sedentary subjects. *Amer. J. Physiology Heart Circulatory Physiology, 285*(4), 1747-1752.

11:7
Hontz, C. R., Hartwig, M., Kleinman, S. M. & Meissner, C. A. (2009). Credibility Assessment at Portals, *Portals Committee Report* (2009).

11:7 B:56
Hopper, J. W., Spinazzola, J., Simpson, W. B., & van der Kolk, B. A. (2006). Preliminary evidence of parasympathetic influence on basal heart rate in posttraumatic stress disorder. *J. Psychosomatic Research. 60*(1), 83-90.

C:40
Horowitz, L. M., & Strack, S. (2010). *Handbook of interpersonal psychology: Theory, research, assessment and therapeutic interventions (pp. 252-55)*. New York: John Wiley.

11:10
Horowitz, S. (2013). *The universal sense: How hearing shapes the mind.* New York: Bloomsbury USA.

10:1
Houston, P. (2013). Spending to piss off my father. *More Magazine*, February, 60-62.

4:44
Howlin, P., Baron-Cohen, S., & Hadwin, J. (1998). *Teaching children with autism to mind-read: A practical guide for teachers and parents.* New York: Wiley.

D:60
Huang, Y., Kendrick, K. M., & Yu. R. (2014). Conformity to the opinions of other people lasts for no more than 3 days. *Psychological Science, 25*(7), 1388-1393.

4:48 C:39
Hughes, C. & Leekam, S. (2004). What are the Links Between Theory of Mind and Social Relations? Review, reflections and new directions for studies of typical and atypical development. *Social Development, 13*(4), 590-619.

4:33 4:49
Hughes, C., & Dunn, J. (1988). Understanding mind and emotions: Longitudinal associations with mental-state talk between young friends. *Developmental Psychology, 34*(5), 1026–1037.

4:33 D:48
Huikuri, H. V., & Stein, P. K. (2013). Heart rate variability in risk stratification of cardiac patients. *Progressive Cardiovascular Disease, 56*(2), 153–159.

9:12 C:98 D:25 D:26
Jabbi, M., Bastiaansen, J., & Keysers, C. (2008). A common anterior insula representation of disgust observation, experience and imagination shows divergent functional connectivity pathways. *PLoS ONE, 3*(8) e2939.

9:12 C:13 C:17 C:29 C:88 C:104
Jabbi, M., Swart, M., & Keysers, C. (2007). Empathy for positive and negative emotions in the gustatory cortex. *NeuroImage, 34*(4), 1744–1753.

C:89
Jackson, P. L., Meltzoff, A. N., & Decety, J. (2005). How do we perceive the pain of others? A window into the neural processes involved in empathy. *Neuroimage, 24*, 771–779.

15:7
Jaenisch, R. & Bird, A. (2003). Epigenetic regulation of gene expression: How the genome integrates intrinsic and environmental signals. *Nature Genetics, 3*, 245–254.

C:46
Jakobwitz, S., & Egan, V. (2006). The 'dark triad' and normal personality traits. *Personality Individual Differences, 40*(2), 331–339.

7:13
James, W. (1890/1950). *The principles of psychology, vol. 2.* New York: Dover Publications.

4:60
Jarrold, C., Boucher, J., & Smith, P. K. (1996). Generativity deficits in pretend play in autism. *British J. Developmental Psychology, 14*(3), 275–300.

7:13 C:76

Jonason, P. K., & Kavanagh, P. (2010). The dark side of love: The Dark Triad and love styles. *Personality Individual Differences, 49*, 606–610.

C:79

Jonason, P. K., & Tost, J. (2010). I just cannot control myself: The Dark Triad and self-control. *Personality Individual Differences, 49*, 611–615.

C:80

Jonason, P. K., & Webster, G. D. (2012). A protean approach to social influence: Dark Triad personalities and social influence tactics. *Personality Individual Differences, 52*(4), 521–526.

C:78

Jonason, P. K., Koenig, B., & Tost, J. (2010). Living a fast life: The Dark Triad and Life History Theory. *Human Nature, 21*(4), 428–442.

C:77

Jonason, P. K., Li, N. P., & Buss, D. M. (2010). The costs and benefits of the Dark Triad: Implications for mate poaching and mate retention tactics. *Personality Individual Differences, 48*(4), 373–378.

C:74

Jonason, P. K., Li, N. P., Webster, G. W., & Schmitt, D. P. (2009). The Dark Triad: Facilitating short-term mating in men. *European J. Personality. 23*, 5–18.

C:80

Jonason, P. K., Slomski, S., & Partyka, J. (2012). The Dark Triad at work: How toxic employees get their way. *Personality Individual Differences, 52*(3), 449–453.

C:75

Jonason, P. K., Valentine, K. A., Li, N. P., & Harbeson, C. L. (2011). Mate-selection and the Dark Triad: Facilitating a short-term mating strategy and creating a volatile environment. *Personality Individual Differences, 51*(6), 759–763.

C:59 C:63

Jones D. N., & Paulhus D. L. (2010a). Different provocations trigger aggression in narcissists and psychopaths. *Social Personality Psychology Science, 1*, 12–18.

C:42

Jones, D. N., & Paulhus, D. L. (2010b). Differentiating the Dark Triad within the interpersonal circumplex. In L. M. Horowitz & S. Strack, (Eds.), *Handbook of interpersonal theory and research* (pp. 249-267). New York: Guilford.

C:61

Jones, D. N., & Paulhus, D. (2011). The role of impulsivity in the Dark Triad of personality. *Personality Individual Differences, 51*(5), 679-682. ·

B:11

Kang, H. K., Natelson, B. H., Mahan, C. M., Lee, K. Y., & Murphy, F. M. (2003). Posttraumatic stress disorder and chronic fatigue syndrome-like illness among Gulf War veterans: A population-based survey of 30,000 Veterans. *Amer. J. Epidemiology, 157*(2), 141-148.

C:30

Kaplan, J. T., & Iacoboni, M. (2006). Getting a grip on other minds: Mirror neurons, intention understanding, and cognitive empathy. *Social Neuroscience, 1*(34), 175-183.

D:63

Karavidas, M. K., Lehrer, P. M., Vaschillo, E., Vaschillo, B., Marin, H., Buyske, S., et al. (2007). Preliminary results of an open label study of heart rate variability biofeedback for the treatment of major depression. *Applied Psychophysiology Biofeedback, 32,* 19-30.

6:3

Kassin, S. M., & Fong, C. T. (1999). "I'm innocent!" Effects of training on judgments of truth and deception in the interrogation room. *Law Human Behavior, 23,* 499-516.

B:54

Katz, L. F., & Gottman, J. M. (1995). Marital interaction and child outcomes: A longitudinal study of mediating and moderating processes. In D. Cicchetti & S. L. Toth (Eds.), *Rochester symposium on developmental psychopathology: Vol. 6. Emotion, Cognition, and Representation* (pp. 301-342). Rochester, NY: University of Rochester Press.

D:46
Kemp, A. H., Quintana, D. S., Gray, M. A., Felmingham, K. L., Brown, K., & Gatt, J. M. (2010). Impact of depression and antidepressant treatment on heart rate variability: A review and meta-analysis. *Biological Psychiatry, 67*(11), 1067–1074.

C:14
Kennedy, D. P., Semendeferi, K., & Courchesne, E. (2007). No reduction of spindle neuron number in frontoinsular cortex in autism. *Brain Cognition, 64*(2), 124–129.

B:3 B:5
Kessler, R. C., Sonnega, A., Bromet, E., Hughes, M., & Nelson, C. B. (1995). Posttraumatic stress disorder in the National Comorbidity Survey. *Archives General. Psychiatry, 52,* 1048–1060.

D:24 D:38
Keysers, C., & Gazzola, V. (2007). Integrating simulation and theory of mind: From self to social cognition. *Trends Cognitive Science, 11,* 194–196.

C:94
Kikuchi, S., Kubota, F., Nisijima, K., Washiya, S., & Kato, S. (2005). Cerebral activation focusing on strong tasting food: A functional magnetic resonance imaging study. *Neuroreport, 16,* 281–283.

A:2
Kim, H. (2010). Dissociating the roles of the default mode, dorsal, and ventral networks in episodic memory retrieval. *NeuroImage, 50,* 1648–1657.

D:67
Kim, J. J., & Diamond, D. M. (2002). The stressed hippocampus, synaptic plasticity and lost memories. *Nature Reviews Neuroscience, 3,* 453–462.

D:54
Klarer, M., Arnold, M., Günther, L., Winter, C., Langhans, L. & Meyer, U. (2014). Gut vagal afferents differentially modulate innate anxiety and learned fear. *J. Neuroscience, 34*(21), 7067–7076.

C:47

Kohut, H. (1977). *The Restoration of the Self.* New York: International Universities Press.

D:55

Kok, B. E., Coffey,K. A., Cohn, M. A., Catalino, L. I., Vacharkulksemsuk, T., Algoe, S. B., Brantley, M., et al. (2012). How positive emotions build physical health: Perceived positive social connections account for the upward spiral between positive emotions and vagal tone.*Psychological Science, 24*(7) 1123-1132.

B:64 D:57

Koopman, F. A., Chavan, S. S., Miljko, S., Grazio, S., Sokolovic, S., Schuurman, P. R., Mehta, A. D., et al. (2016). Vagus nerve stimulation inhibits cytokine production and attenuates disease severity in rheumatoid arthritis. *PNAS, 113*(29), 8284-8289.

B:16

Krystal, H. (1968). *Integration and self-healing: Affect, trauma, alexithymia.* Hillsdale, NJ: Analytic Press.

D:48

Kudaiberdieva, G., Gorenek, B., & Timuralp, B. (2007). Heart rate variability as a predictor of sudden cardiac death. *Anadolu. Kardiyol. Derg. 7*(1), 68-70.

10:11

Kulis, M., & Esteller, M. (2010). DNA methylation and cancer. *Advances Genetics, 70,* 27-56.

B:10

Kulka, R. A., Schlenger, W. A., Fairbanks, J. A., Hough, R. L., Jordan, B. K., Marmar, C. R., & Cranston, A. S. (1990). *Trauma and the Vietnam war generation: Report of findings from the National Vietnam Veterans Readjustment Study.* New York: Brunner/Mazel.

C:89

Lamm, C., Batson, C. D., & Decety, J. (2007). The neural substrate of human empathy: Effects of perspective-taking and cognitive appraisal. *J. Cognitive Neuroscience, 19,* 42-58.

4:36
Laursen, B., Bukowski, W. M., Aunola, K., & Nurmi, J. E. (2007). Friendship moderates prospective associations between social isolation and adjustment problems in young children. *Child Development, 78*(4), 1395-1404.

D:32
Layden, M. A., Newman, C. F., Freeman, A., & Morse, S. B. (1993). *Cognitive therapy of borderline personality disorder.* Boston: Allyn and Bacon.

4:32
Lecce, S., Caputi, M., & Pagnin, A. (2014). Long-term effect of Theory of Mind on school achievement: The role of sensitivity to criticism. *European J. Developmental Psychology, 11*(3), 305-318.

B:46
Ledoux, J. E., Romanski L., & Xagoraris, A. (1991). Indelibility of subcortical emotional memories. *J. Cognitive Neuroscience, 1,* 238-243.

4:58
Leekam, S. R., & Prior, M. (1994). Can autistic children distinguish lies from jokes? A second look at second-order belief attribution. *J. Child Psychol. Psychiat., 35*(5), 901-915.

D:51 D:52
Lehrer, P. M., & Gervitz, R. (2014). Heart rate variability biofeedback: How and why does it work? *Frontiers Psychology, 5,* 1-9.

D:42
Lehrer, P. M., Vaschillo, E., & Vaschillo, B. (2000). Resonant frequency biofeedback training to increase cardiac variability: Rationale and manual for training. *Applied. Psychophsiol. Biofeedback, 25*(3), 177-191.

D:50
Lehrer, P., Karavidas, M. K., Lu, S. E., Coyle, S. M., Oikawa, L. O., Macor, M., et al. (2010). Voluntarily produced increases in heart rate variability modulate autonomic effects of endotoxin induced systemic inflammation: An exploratory study. *Applied Psychophysiol. Biofeedback, 35,* 303-315.

C:27

Leiberg, S., & Anders, S. (2006). The multiple facets of empathy: A survey of theory and evidence. *Progress Brain Research, 156,* 419–440.

4:53

Leslie, A. M. (1987). Pretense and representation: The origins of "Theory of Mind." *Psychological Review, 94*(4), 412–426.

4:54

Leslie, A. M., & Frith, U. (1988). Autistic children's understanding of seeing, knowing and believing. *British J. Developmental Psychology, 6,* 315–324.

4:17 4:19

Lewis, C., Freeman, N. H., Kyriakidou, C., Maridaki-Kassotaki, K., & Berridge, D. M. (1996). Social influences on false belief access: Specific sibling influences or general apprenticeship? *Child Development, 67*(6), 2930–2947.

4:60

Lewis, V., & Boucher, J. (1988). Spontaneous, instructed, and elicited play in relative able autistic children. *British J. Developmental Psychology, 6*(4), 325–339. ·

15:6

Lilienfeld, S. O., & Arkowitz, H. (2014.) Why "just say no" doesn't work. *Scientific American.* January.

D:72

Linley, P. A., & Joseph, S. (2004). Positive change following trauma and adversity: A review. *J. Traumatic Stress, 17*(1), 11–21.

C:43

Linton, D. K., & Wiener, N. I. (2001). Personality and potential conceptions: Mating success in a modern Western male sample. *Personality Individual Differences, 31*(5), 675–688.

D:49

Linton, S. J., & Shaw, W. S. (2011). Impact of psychological factors in the experience of pain. *Physical Therapy, 91,* 700–711.

7:3

Locke, J. (1690/1979). An essay concerning human understanding. In P. H. Nidditch (ed.), *The Claredon edition of the works of John Locke.* New York: Oxford University Press.

10:12

Lund, G. L., Andersson, L., Lauria, M., Lindholm, M., Fraga, M. F., Villar-Garea, A., Ballestar, E., et al. (2004). DNA methylation polymorphisms precede any histological sign of atherosclerosis in mice lacking Apolipoprotein E. *J. Biology Chemisty, 279*(28), 29147–29154.

7:8

Lyons, W. (1986). *The disappearance of introspection.* Chicago: University of Illinois Press.

D:18

Mack, A., & Rock, I. (1998). *Inattentional blindness.* Cambridge, MA: MIT Press.

2:1

MacLean, P. D. (1990). *The triune brain in evolution.* New York: Plenum.

C:67

Malamuth, N. M. (2003). Criminal and noncriminal sexual aggressors: Integrating psychopathy in a hierarchical-mediational confluence model. *Annals New York Academy Sciences. 989,* 33–58.

15:8

McCartney, P. & Lennon, J. (1970). *Let It Be.* (song). Apple Music.

B:32

McCloskey, M. and Cohen, N. J. (1989). Catastrophic interference in connectionist networks: The sequential learning problem. In G. H. Bower, (Ed.), *The psychology of learning and motivation, Vol. 24* (pp. 109–164). San Diego, CA: Academic Press.

D:63

McCraty, R., Atkinson, M., Tomasino, D., & Bradley, R. T. (2009). The coherent heart: Heart-brain interactions, psychophysiological coherence, and the emergence of system-wide order. *Integral Review, 5*(2), 1–115.

B:22

McDonagh, A., Friedman, M., McHugo, G., Ford, J. D., Mueser, K., Sengupta, A., et al. (2005). Randomized trial of cognitive-behavioral therapy for chronic posttraumatic stress disorder in adult female survivors of childhood sexual abuse. *J Consulting Clinical Psychology, 73*(3), 515–524.

C:45

McHoskey, J. (1995). Narcissism and Machiavellianism. *Psych Reports, 77,* 755-759.

C:45

McHoskey, J. W., Worzel, W., & Szyarto, C. (1998). Machiavellianism and psychopathy. *J. Personality Social Psychology, 74,* 192-210.

C:50

McIllwain, D. (2003). Bypassing empathy: A Machiavellian theory of mind and sneaky power. In B. Repacholi & V. Slaughter (Eds.), *Individual differences in theory of mind. Macquarie monographs in cognitive science* (pp. 39-66). Hove, E. Sussex: Psychology Press.

13:13

McKay D. (2006). Treating disgust reactions in contamination-based obsessive-compulsive disorder. *J. Behavior Therapy Experimental Psychiatry, 37*(1), 53-59.

D:47

McNarry, M. A., & Lewis, M. J. (2012). Heart rate variability reproducibility during exercise. *Physiology Measurment. 33*(7), 1123-1233.

4:8

Meins, E. (1998). The effects of security of attachment and material attribution of meaning on children's linguistic acquisitional style. *Infant Behavior Development, 21*(2), 237-252.

4:11 4:12

Meins, E., Fernyhough, C., Fradley, E., & Tuckey, M. (2001). Rethinking maternal sensitivity: Mothers' comments on infants' mental processes predict security of attachment at 12 months. *J. Child Psychology Psychiatry, 42*(5), 637-648.

4:7 4:9 4:10

Meins, E., Fernyhough, C., Russell, J., & Clark-Carter, D. (1998). Security of attachment as a predictor of symbolic and mentalizing abilities: A longitudinal study. *Social Development, 7*(1), 1-24.

4:13 4:15
Meins, E., Fernyhough, C., Wainwright, R., Gupta, M. D., Fradley, E., & Tuckey, M. (2002). Maternal mind-mindedness and attachment security as predictors of Theory of Mind understanding. *Child Development, 73*(6), 1715–1726.

4:42
Mellet, E., Petit, L., Mazoyer, B., Denis, M., & Tzourio, N. (1988). Reopening the mental imagery debate: Lessons from functional anatomy. *NeuroImage, 8*(2), 129–139.

C:2
Mesulam, M. M., & Mufson, E. J. (1982). Insula of the old world monkey. III: Efferent cortical output and comments on function. *J. Comparative Neurology, 212,* 38–52.

6:2
Miller, T., & McFarland, C. (1987). Pluralistic ignorance: When similarity is interpreted as dissimilarity. *J. Personality Social Psychology, 53,* 298–305.

6:2
Miller, T., & McFarland, C. (1991). When social comparison goes awry: The case of pluralistic ignorance. In J. Suls & T.A. Wills (Eds.), *Social comparison: Contemporary theory and research* (pp. 287–313). Hillsdale, NJ: Erlbaum.

C:34
Mitchell, J. P., Macrae, C. N., & Banaji, M. R. (2006). Dissociable medial prefrontal contributions to judgments of similar and dissimilar others. *Neuron, 50*(4), 655–663.

8:3
Morehouse, R. K. & Schnarch, D. M. (2010). *Sexual desire, intimacy, interpersonal neurobiology and family psychology.* Annual Convention of the American Psychological Association, Division 43 Presidential Address. August 13, 2010, San Diego, CA.

4:29
Much, N. C., & Shweder, R. . (1978). Speaking of rules: The analysis of culture in breach. *Child Adolescent Development, 2,* 19–39.

C:2
Mufson, E. J., & Mesulam, M. M. (1982). Insula of the old world monkey. II: Afferent cortical input and comments on the claustrum. *J. Comparative Neurology, 212,* 23–37.

C:14
Mundy, P. (2003). The neural basis of social impairments in autism: The role of the dorsal medial-frontal cortex and anterior cingulate system. *J. Child Psychology Psychiatry Allied Disciplines, 44,* 793–809.

D:57
Nahas, Z., Marangell, L.B., Husain, M.M., Rush, A.J., Sackeim, H.A., Lisanby, S.H., Martinez, J. M., et al. (2005) Two-year outcome of vagus nerve stimulation (VNS) for treatment of major depressive episodes. *J. Clinical Psychiatry, 66,* 1097–1104.

8:12
Namnyak, M., Tufton, N., Szekely, R., Toal, M., Worboys, S., & Sampson, E. L. (2007). 'Stockholm syndrome': Psychiatric diagnosis or urban myth? *Acta Psychiatrica Scandinavia, 117*(1), 4–11.

C:92
Naqvi, N. H., Rudrauf, D., Damasio, H., & Bechara, A. (2007). Damage to the insula disrupts addiction to cigarette smoking. *Science, 315*(5811), 531–534.

8:8
National Sexual Violence Resource Center (2015). *Statistics: Understanding sexual violence.*

C:64 C:69
Nell, V. (2006). Cruelty's rewards: The gratifications of perpetrators and spectators. *Behavior Brain Science. 29*(3), 211–224.

C:7
Nimchinsky, E. A., Gilissen, E., Allman, J. M., Perl, D. P., Erwin, J. M., & Hof, P. R. (1999). A neuronal morphologic type unique to humans and great apes. *PNAS, 96*(9), 5268–5273.

7:14
Nisbett, R. E., & Wilson, T. D. (1977). Telling more than we can know: Verbal reports on mental processes. *Psychological Review, 84*(3), 231–259.

D:22
Nitschke, J. B., Dixon, G. E., Sarinopoulos, I., Short, S. J., Cohen, J. D., Smith, E. E., Kosslyn, S. M., et al. (2006). Altering expectancy dampens neural response to aversive taste in primary taste cortex. *Nature Neuroscience, 9,* 435–442.

9:13
Olatunji, B., & McKay, D. (Eds.) (2009). *Disgust and its disorders: Theory, assessment, and treatment implications.* Washington, DC: American Psychological Assn.

C:56
Paal, T., & Bereczkei, T. (2007). Adult theory of mind, cooperation, Machiavellianism: The effect of mindreading on social relations. *Personality Individual Differences, 43,* 541–551.

4:36
Parker, J. G., Saxon, J. L., Asher, S. R., & Kovacs, D. M. (1999). Dimensions of children's friendship adjustment: Implications for understanding loneliness. In K. J. Rotenberg & S. Hymel (Eds.), *Loneliness in Childhood and Adolescence.* Cambridge: Cambridge U. Press.

D:63
Patron, E., Messerotti Benvenuti, S., Favretto, G., Valfre, C., Bonfa, C., Gasparotto, R., et al. (2013). Biofeedback assisted control of respiratory sinus arrhythmia as a biobehavioral intervention for depressive symptoms in patients after cardiac surgery: A preliminary study. *Applied Psychophysiol. Biofeedback, 38,* 1–9.

D:49
Paul, M., & Garg, K. (2012). The effect of heart rate variability biofeedback on performance psychology of basketball players. *Applied Psychophysiol. Biofeedback, 37*(2), 131–144.

C:40 C:41
Paulhus, D. L. & Williams, K. M. (2002). The Dark Triad of personality: narcissism, Machiavellianism, and psychopathy. *J. Research Personality, 36*(6), 556–563.

10:7
Pearson, D., Smalley, M., Ainsworth, C., Cook, M., Boyle, J., & Flury, S. (2008). Auditory hallucinations in adolescent and adult students: Implications for continuums and adult pa-

thology following child abuse. *J. Nervous Mental Disease, 196*(8), 634–638.
4:37
Pedersen, S., Vitaro, F., Barker, E. D., & Borge, A. I. (2007). The timing of middle-childhood peer rejection and friendship: Linking early behavior to early adolescent adjustment. *Child Development, 78*(4), 1037–1051.
D:64
Perls, F., Hefferine, R. F., & Goodwin, P. (1951). *Gestalt therapy: Excitement and growth in the human personality.* Alexandria: VA: Julian Press.
4:18
Perner, J., Ruffman, T., & Leekam, S. R. (1994). Theory of Mind is contagious: You catch it from your sibs. *Child Development, 65*(4), 1228–1238.
B:62
Pervanidou, P. (2008). Biology of posttraumatic stress disorder in childhood and adolescence. *J. Neuroendocrinology, 20,* 632–638.
D:24
Pessoa, L. (2008). On the relationship between emotion and cognition. *Nat. Review Neuroscience, 2,* 148–158.
4:34
Peterson, C. C., & Siegal, M. (2002). Mindreading and moral awareness in popular and rejected preschoolers. *British J. Developmental Psychology, 20*(2), 205–224.
4:55
Phillips, W., Baron-Cohen, S., & Rutter, M. (1998). Understanding intention in normal development and in autism. *British J. Developmental Psychology, 16*(3), 337–348.
B:28 B:29
Pillemer, D. (1998). *Momentous events, vivid memories.* Cambridge, MA: Harvard University Press.
D:15
Pillemer, D. B., & White, S. H. (1989). Childhood events recalled by children and adults. In H. W. Reese (Ed.), *Advances in child development and behavior, volume 21* (pp. 297–340). Orlando, FL: Academic Press.

D:37

Ploghaus, A., Tracey, I., Gati, J. S., Clare, S., Menon, R.S., Matthews, P.M., & Rawlins, J.N. (1999). Dissociating pain from its anticipation in the human brain. *Science, 284*(5422), 1979–1981.

D:48

Politano, L., Palladino, A., Nigro, G., Scutifero, M., & Cozza, V. (2008). Usefulness of heart rate variability as a predictor of sudden cardiac death in muscular dystrophies. *Acta Myologica, 27*, 114–122.

4:27

Pons, F., & Harris, P. (2005). Longitudinal change and longitudinal stability of individual differences in children's emotion understanding. *Cognition Emotion, 19*(8), 1158–1174.

B:49

Porges, S. & Doussard-Roosevelt, J. (1997). The psychophysiology of temperament. In J. D. Noshpitz (Ed.), *The Handbook of Child and Adolescent Psychiatry* (p. 250–268). New York: Wiley Press.

B:47 B:50

Porges, S. (1991). Vagal tone: An autonomic mediator of affect. In J. A. Garber & K. A. Dodge (Eds.), *The Development of Affect Regulation and Dysregulation* (pp.11–128). New York: Cambridge University Press.

B:50

Porges, S. (1995). Orienting in a defensive world: Mammalian modifications of our evolutionary heritage. A Polyvagal theory. *Psychophysiology 32*(4), 301–318.

8:5 B:48 B:57

Porges, S., Doussard-Roosevelt, J. A., Portales, A. L., & Greenspan, S. I. (1996). Infant regulation of the vagal brake predicts child behavior problems: A psychobiological model of social behavior. *Developmental Psychobiology, 29*, 697–712.

8:5 B:49 B:51

Porges, S., Doussard-Roosevelt, J., Portales, A. & Suess, P. E. (1994). Cardiac vagal tone: Stability and relation to difficultness in infants and three-year-old children. *Developmental Psychobiology, 27*, 289–300.

10:7
Posey, T. B., & Losch, M. E. (1983). Auditory hallucinations of hearing voices in 375 normal subjects. *Imagination Cognition Personality, 3*(2), 99-113.

C:27
Preston, S. D., & de Waal, F. B. M. (2002). Empathy: Its ultimate and proximate bases. *Behavioral Brain Sciences, 25*, 1-72.

C:93
Preston, S. D., Bechara, A., Damasio, H., Grabowski, T. J., Stansfield, R. B., & Sonya, M. (2007). The neural substrates of cognitive empathy. *Social Neuroscience 2*, 254-275.

B:17 B:18
Putnam, F. W. (2003). Ten-year research update review: Child sexual abuse. *J. Amer. Academy Child Adolescent Psychiatry, 42*(3), 269-278.

D:48
Ranpuria, R., Hall, M., Chan, C. T., & Unruh, M. (2008). Heart rate variability (HRV) in kidney failure: Measurement and consequences of reduced HRV. *Nephrology Dial. Trans. 23*, 444-449.

C:70 C:73
Reidy, D. E., Zeichner, A., & Seibert, L. A. (2011). Unprovoked aggression: Effects of psychopathic traits and sadism. *J. Personality, 79*(1), 75-100.

C:37
Reiman, E. M., Lane, R. D., Ahern, G. L., Schwartz, G. E., Davidson, R. J., Friston, K. J., et al. (1997). Neuroanatomical correlates of externally and internally generated human emotion. *Am. J. Psychiatry, 154*, 918-925.

D:63
Reiner, R. (2008). Integrating a portable biofeedback device into clinical practice for patients with anxiety disorders: Results of a pilot study. *Applied Psychophysiol. Biofeedback, 33*(1), 55-61.

C:101
Rekkas, P. V., & Constable, R. T. (2005). Evidence that autobiographic memory retrieval does not become independent

of the hippocampus: An fMRI study contrasting very recent with remote events. *J. Cognitive Neuroscience, 17,* 1950-1961.

9:6

Renouf, A., Brendgen, M., Séguin, J. R., Vitaro, F., Boivin, M., Dionne, G., Tremblay, R. E., et al. (2010). Interactive links between theory of mind, peer victimization, and reactive and proactive aggression. *J. Abnormal Child Psychology, 38,* 1109-1123.

4:62 C:53

Repacholi, B. & Slaughter, V. (Eds). (2003). *Individual differences in theory of mind: Implications for typical and atypical development.* New York: Psychology Press.

C:51 C:53

Repacholi, B., Slaughter, V., Pritchard, M., & Gibbs, V. (2003). Theory of mind, Machiavellianism, and social functioning in childhood. In B. Repacholi & V. Slaughter (Eds.), *Individual differences in theory of mind* (pp 99-120). New York: Psychology Press.

D:4

Resick, P. A., & Schnicke, M. K. (1992). Cognitive processing therapy for sexual assault victims. *J. Consulting Clinical Psychology, 60,* 748-756.

D:4

Resick, P. A., & Schnicke, M. K. (1993). *Cognitive processing therapy for rape victims: A treatment manual.* Newbury Park, CA: Sage.

D:5

Resick, P. A., Nishith, P., Weaver, T. L., Astin, M. C., & Feuer, C. A. (2002). A comparison of cognitive-processing therapy with prolonged exposure and a waiting condition for treatment of chronic posttraumatic stress disorder in female rape victims. *J. Consulting Clinical Psychology, 70*(4), 867-879.

8:11

Rhodes, J. & Chan, C. (2010). The impact of Hurricane Katrina on the mental and physical health of low-income parents in New Orleans. *Amer. J. Orthopsychiatry, 80*(2), 237-247.

A:5
Richell, R. A., Mitchell, D. G. V., Newman, C., Leonard, A., Baron-Cohen, S., & Blair, R. J. R. (2003). Theory of mind and psychopathy: Can psychopathic individuals read the 'language of the eyes'? *Neuropsychologia, 41*(5), 523-526.

9:10
Rizzolatti, G., Fadiga, L., Gallese, V., & Fogassi, L. (1996). Premotor cortex and the recognition of motor actions. *Cognitive Brain Research, 3*(2), 131-141.

B:61
Rohleder, N., Joksimovic, L., Wolfe, J. M., & Kirschbaum, C. (2004). Hypocortisolism and increased glucocorticold sensitivity of pro-inflammatory cytokine production in Bosnian war refugees with posttraumatic stress disorder. *Biological Psychiatry, 55*, 745-751.

B:15
Rollstin, A. O., & Kern, J. M. (1998). Correlates of battered women's psychological distress: Severity of abuse and duration of the post-abuse period. *Psychology Report, 82*(2), 387-394.

D:33
Ruby, P., & Decety, J. (2004). How would you feel versus how do you think she would feel? A neuroimaging study of perspective-taking with social emotions. *J. Cognitive Neuroscience, 16*(6), 988-999.

4:18
Ruffman, T., Perner, J., Naito, M., Parkin, L., & Clements, W. A. (1998). Older (but not younger) siblings facilitate false belief understanding. *Developmental Psychology, 34*(1), 161-174.

15:4
Ruiz, R. (2014). How childhood trauma could be mistaken for ADHD. *The Atlantic magazine, July 7.*.

7:16
Ryle, G. (1949). *The concept of mind.* Chicago: University of Chicago Press.

C:89
Saarela, M. V., Hlushchuk, Y., Williams, A. C., Schurmann, M.,

Kalso, E., & Hari, R. (2007). The compassionate brain: Humans detect intensity of pain from another's face. *Cerebral Cortex, 17*, 230-237.

D:36

Sacco, K., Cauda, F., Cerliani, L., Mate, D., Duca, S., & Geminiani, G.C. (2006). Motor imagery of walking following training in locomotor attention: The effect of "the tango lesson." *Neuroimage, 32*(3), 1441-1449.

D:58

Sackeim, H. A., Rush, A. J., George, M. S., Marangell, L. B., Husain, M. M., Nahas, Z., Johnson, C. R., et al. (2001). Vagus nerve stimulation (VNS™) for treatment-resistant depression: Efficacy, side effects, and predictors of outcome. *Neuropsychopharmacology, 25*, 713-728.

B:55

Sahar, T., Shalev, A., & Porges, S. (2001). Vagal modulation of responses to mental challenge in posttraumatic stress disorder. *Biological Psychiatry, 49*, 637-643.

C:38

Samson, D., Apperly, I. A., Chiavarino, C., & Humphreys, G. W. (2004). The left temporo-parietal junction is necessary for representing someone else's belief. *Nature Neuroscience, 7*, 449-500.

B:44 D:67

Sapolsky, R. M. (2003). Stress and plasticity in the limbic system. *Neurochemisty Research, 28*(11), 1735-1742.

6:1

Savitsky, K., & Gilovich, T. (2003). The illusion of transparency and the alleviation of speech anxiety. *J. Experimental Social Psychology, 39*, 618-625.

C:87

Schandry, R. (1981). Heart beat perception and emotional experience. *Psychophysiology, 18*(4), 483-488.

C:87

Schandry, R., Sparrer, B., & Weitkunat, R. (1986). From the heart to the brain: A study of heartbeat contingent scalp potentials. *International J. Neuroscience, 30*, 261-275.

B:13

Scheeringa, M. S., Zeanah, C. H., Drell, M. J., & Larrieu, J. A. (1995). Two approaches to the diagnosis of posttraumatic stress disorder in infancy and early childhood. *J. Amer. Academy. Child Adolescent Psychiatry, 34*(2), 191–200.

B:13 D:72

Scheeringa, M. S., Zeanah, C. H., Meyers, L., & Putnam, F. W. (2003). New findings on alternative criteria for PTSD in preschool children. *J. Amer. Academy Child Adolescent Psychiatry, 42*(5), 561–570.

9:9 13:1 C:58

Schnarch, D. M. (1991). *Passionate Marriage: Keeping Love and Intimacy Alive in Emotionally Committed Relationships.* New York: W. W. Norton.

5:4 10:4

Schnarch, D. M. (2002). *Resurrecting sex: Resolving sexual problems and rejuvenating your relationship.* New York: HarperCollins.

D:65

Schnarch, D. M. (2003). *Outside the comfort zone: Growing through relationship conflict.* Workshop presented at the Annual Conference of the American Association for Marriage & Family Therapy. October: Long Beach, CA.

9:9 13:1 14:1

Schnarch, D. M. (2009). *Intimacy & Desire: Awaken the passion in your relationship.* New York: Beaufort Books.

8:3

Schnarch, D. M. (2009). *Sexual Crucible Clinical Workshops for Therapists.* Evergreen Colorado: Marriage & Family Health Center.

9:14

Schnarch, D. M. (2012). *Interpersonal Neurobiology of Disgusting Parenting.* Presentation delivered at the Annual Convention of the American Psychological Association. August, 2012: Orlando, FL. Part of symposium: *Do parents always do the best they can?* (D. M. Schnarch, J. Thoburn, & K. Berry, presenters, R. Morehouse, Chair.)

D:6
Schnurr, P. P., Friedman, M. J., Engel, C. C., Foa, E. B., Shea, M. T., Chow, B. K., Resick, P. A., et al. (2007). Cognitive behavioral therapy for posttraumatic stress disorder in women: A randomized controlled trial. *JAMA, 297*(8), 820–830.

C:29
Schulte-Rüther, M., Markowitsch, H. J., Fink, G. R., & Piefke, M. (2007). Mirror neuron and theory of mind mechanisms involved in face-to-face interactions: A functional magnetic resonance imaging approach to empathy. *J. Cognitive Neuroscience, 19,* 1354–1372.

7:12
Schwitzgebel, E. (2007). No unchallengeable epistemic authority, of any sort, regarding our own conscious experience–contra Dennett? *Phenomenology Cognitive Science, 6,* 107–113.

8:7
Segerstrom S. C., & Miller, G. E. (2004). Psychological stress and the human immune system: A meta-analytic study of 30 years of inquiry. *Psychology Bulletin, 130*(4), 601–630.

C:28 C:31 C:33 C:36
Shamay-Tsoory, S. G., Aharon-Peretz, J., & Perry, D. (2009). Two systems for empathy: A double dissociation between emotional and cognitive empathy in inferior frontal gyrus versus ventromedial prefrontal lesions. *Brain, 132*(3), 617–27.

A:6
Shamay-Tsoory, S. G., Harari, H., Aharon-Peretz, J., & Levkovitz, Y. (2010). The role of the orbitofrontal cortex in affective theory of mind deficits in criminal offenders with psychopathic tendencies. *Cortex, 46*(5), 668–677.

9:8 C:26
Shamay-Tsoory, S. G., Tibi-Elhanany, Y., & Aharon-Peretz, J. (2007). The green-eyed monster and malicious joy: The neuroanatomical bases of envy and gloating (schadenfreude). *Brain, 30*(6), 1663–1678.

8:1
Shamay-Tsoory, S. G., Tomer, R., & Aharon-Peretz, J. (2005). The neuroanatomical basis of understanding sarcasm and

its relationship to social cognition. *Neuropsychology, 19*(3), 288-300.
7:4
Shoemaker, S. (1996). *The first-person perspective and other essays*. New York: Cambridge University Press.
D:63
Siepmann, M., Aykac, V., Unterdörfer, J., Petrowski, K., & Mueck-Weymann, M. (2008). A pilot study on the effects of heart rate variability biofeedback in patients with depression and in healthy subjects. *Applied Psychophysiol. Biofeedback, 33*(4), 195-201.
C:14
Silani, G., Bird, G., Brindley, R., Singer, T., Frith, C., & Frith, U. (2008). Levels of emotional awareness and autism: An fMRI study. *Social Neuroscience, 3*(2), 97-112.
9:8 C:18
Singer, T. (2008a). I feel how you feel but not always: The empathic brain and its modulation. *Current Opinion Neurobiology, 18*, 153-158.
C:19 D:34 D:40
Singer, T. (2008b). Understanding others: Brain mechanisms of theory of mind and empathy. In P. W. Glimcher, C. F. Camerer, E. Fehr, & R. A. Poldrack (Eds.), *Neuroeconomics: Decision making and the brain* (pp. 233-250). Amsterdam: Elsevier.
C:4 C:12 C:37 C:89 D:27 D:35
Singer, T., Seymour, B., O'Doherty, J., Kaube, H., Dolan, R. J., & Frith, C. D. (2004). Empathy for pain involves the affective but not sensory components of pain. *Science, 303*, 1157-1162.
C:49
Skeem, J. L., Polaschek, D. L. L., Patrick, C. J., & Lilienfeld, S. O. (2011). Psychopathic personality: Bridging the gap between scientific evidence and public policy. *Psych. Science Public Interest, 12*(3), 95-162.
4:29
Slomkowski, C. L., & Killen, M. (1992). Young children's conceptions of transgressions with friends and nonfriends. *Intl. J. Behavior Development, 15*(2), 247-258.

4:50

Slomkowski, C. L, & Dunn, J. (1996). Young children's understanding of other people's beliefs and feelings and their connected communication with friends. *Developmental Psychology, 32*(3), 442–447.

C:22

Smith, R. H., Parrott, G. W., Deiner, E. F., Hoyle, R. H., & Kim, S. H. (1999). Dispositional envy. *Personality Social Psychology. Bulletin, 25*(8), 1007–1020.

D:32

Smucker, M. R., Dancu, C. V., Foa, E. B., & Niederee, J. L. (1995). Imagery rescripting: A new treatment for survivors of childhood sexual abuse suffering from posttraumatic stress. *J. Cognitive Psychotherapy, 9*, 3–17.

APPENDIX C

Sobotta, J. (1909). *Atlas and text-book of human anatomy.* Philadelphia: W. B. Saunders.

4:57

Sodian, B. (1991). The development of deception in young children. *British J. Developmental Psychology, 9*(1), 173–188.

15:3

Stern, D. N. (2004). The present moment in psychotherapy and everyday life. New York: W.W. Norton.

B:60

Stifter, C. A., Fox, N. A., & Porges, S. W. (1989). Facial expressivity and vagal tone in 5- and 10-month-old infants. *Infant Behavior Development, 12*, 127–137.

D:49

Strack, F. & Deutsch, R. (2004). Reflective and impulsive determinants of social behavior. *Personality Social Psychology Review, 8*(3), 220–247.

B:31

Strange, B. A., Hurlemann, R., & Dolan, R. J. (2003). An emotion-induced retrograde amnesia in humans is amygdala- and beta-adrenergic-dependent. *Proceedings Natl. Academy Sciences. USA., 100*(23), 13626–13631.

C:81 C:82 C:83
Streep, P. (2016). *Why it's so hard to see this form of childhood abuse* PsychologyToday.com

D:24
Sun, F. W., Stepanovic, M. R., Andreano, J., Feldman Barrett, L., Touroutoglou, A., & Dickerson, B. C. (2016). Youthful brains in older adults: Preserved neuroanatomy in the default mode and salience networks contributes to youthful memory in Superaging. *J. Neuroscience, 36*(37), 9659–9668.

10:2
Surguladze, S., Keedwell, P., & Phillips, M. (2003). Neural systems underlying affective disorders. *Advances Psychiatric Treatment,* 9(6), 446–455.

8:2 9:6
Sutton, J., Smith, P. K., & Swettenham, J. (1999). Socially undesirable need not be incompetent: A response to Crick and Dodge. *Social Development, 8,* 132–134.

9:6 C:55
Sutton, J., Smith, P. K., & Swettenham, J. (2001). Bullying and Theory of Mind: A critique of the 'social skills deficit' view of anti-social behaviour. *Social Development, 8*(1), 117–127.

D:61 D:63
Tan, G., Wang, P., & Ginsberg, J. (2011). Heart rate variability (HRV) and posttraumatic stress disorder. *Assn. Applied Psychophysiology Biofeedback, 41*(3), 131–135.

B:12
Tanielian, T., & Jaycox, L. (Eds.) (2008). *Invisible wounds of war: Psychological and cognitive injuries, their consequences, and services to assist recovery.* Santa Monica, CA: RAND Corporation.

C:65 C:69 C:72
Taylor, K. (2009). *Cruelty: Human evil and the human brain.* New York: Oxford University Press.

14:3 D:74
Tedeschi, R, G. (1999). Violence transformed: Posttraumatic growth in survivors and societies. *Aggression Violent Behavior, 4,* 319–341.

D:75

Tennen, H., & Affleck, G. (1998). Personality and transformation in the face of adversity. In R. G. Tedeschi, C. L. Park, & L. G. Calhoun (Eds.), *Posttraumatic growth: Positive changes in the aftermath of crisis* (pp. 65-98). Mahwah, New Jersey: Lawrence Erlbaum.

D:48

Thayer, J. F., Yamamoto, S. S., & Brosschot, J. F. (2010). The relationship of autonomic imbalance, heart rate variability and cardiovascular disease risk factors. *Intl. J. Cardiology, 141,* 122-131.

10:7

Tien, A. Y. (1991). Distributions of hallucinations in the population. *Social Psychiatry Psychiatric Epidemiology, 26*(6), 287-292.

4:14

Tronick, E. Z. (1998). Dyadically expanded states of consciousness and the process of therapeutic change. *Infant Mental Health, 19*(3), 290-299.

C:14

Uddin, L.Q., & Menon, V., (2009). The anterior insula in autism: Under-connected and under-examined. *Neuroscience Biobehavior Review, 33*(8), 1198-1203.

14:2 D:71

Updegraff, J. A., & Taylor, S. E. (2000). From vulnerability to growth: Positive and negative effects of stressful life events. In J. Harvey & E. Miller (Eds.), *Loss and trauma: General and close relationship perspectives* (pp. 3-28). Philadelphia, PA: Brunner-Routledge.

6:3

Van Boven, L., Gilovich, T., & Medvec, V. H. (2003). The illusion of transparency in negotiations. *Negotiation Journal, 19*(2), 117-131.

C:102

Van der Gaag, C., Minderaa, R., & Keysers, C. (2007). Facial expressions: What the mirror neuron system can and cannot tell us. *Social Neuroscience, 2,* 179-222.

B:34 B:35 B:36 B:45 B:59 D:7 D:8 D:9
Van der Kolk, B. A. (2006). Clinical implications of neuroscience research in PTSD. *Annals New York Academy Sciences, 1071*, 277–293.

B:20 B:21
Van der Kolk, B. A., Roth, S., Pelcovitz, D., Sunday, S., & Spinazzola, J. (2005). Disorders of extreme stress: The empirical foundation of a complex adaptation to trauma. *J. Traumatic Stress, 18*(5), 389–399.

C:21
Van Dijk, W. W., Ouwerkerk, J. W., Goslinga, S., Nieweg, M., & Gallucci, M. (2006). When people fall from grace: Reconsidering the role of envy in schadenfreude. *Emotion, 6*, 156–160.

D:39
Vasterling, J. J., Brailey, K., Constans, J. I. & Sutker, P.B. (1998). Attention and memory dysfunction in posttraumatic stress disorder. *Neuropsychology, 12*, 121–133.

C:48
Vernon, P. A., Villani, V. C., Vickers, L. C., & Harris, J. A. (2008). A behavioral genetic investigation of the Dark Triad and the Big 5. *Personality Individual Differences, 44*(2), 445–452.

D:45
Volz, H. P., Rehbein, G., Triepel, J., Knuepfer, M. M., Stumpf, H., & Stock, G. (1990). Afferent connections of the nucleus centralis amygdalae. A horseradish peroxidase study and literature survey. *Anatomy Embryology (Berl.), 181*, 177–94.

6:3
Vorauer, J. D., & Claude, S. D. (1998). Perceived versus actual transparency of goals in negotiation. *Personality Social Psychology Bulletin, 24*(4), 371–385.

6:2
Vorauer, J. D., & Ross, M. (1999). Self-awareness and feeling transparent: Failing to suppress one's self. *J. Experimental Social Psychology, 35*(5), 415–440.

B:43 D:66
Vyas, A., Bernal, S., & Chattarji, S. (2003). Effects of chronic stress on dendritic arborization in the central and extended amygdala. *Brain Research, 965*(1-2), 290-294.

D:66
Vyas, A., Mitra, R., Rao, B. B. S., & Chattarji, S. (2002). Chronic stress induces contrasting patters of dendritic remodeling in hippocampal and amygdaloid neurons. *J. Neuroscience, 22*(15), 6810-6818.

8:4
Walker, A., McKune, A., Ferguson, S., Pyne, D. B., & Rattray, B. (2016). Chronic occupational exposures can influence the rate of PTSD and depressive disorders in first responders and military personnel. *Extreme Physiol. Medicine, 5,* 8.

B:15
Walker, L. J. (1984). Sex differences in the development of moral reasoning: A critical review. *Child Development, 55*(3), 677-691.

4:43
Wallace, B. (2014). Autism spectrum: Are you on it? *New York Magazine, May 12,* http://nymag.com/news/features/autism-spectrum-2012-11/.

8:2
Waterman, J., Sobesky, W., Silvern, L., Aoki, B., & McCauley, M. (1981). Social perspective-taking and adjustment in emotionally disturbed, learning disabled and normal children. *J. Abnormal Child Psychology, 9,* 133-148.

11:9
Webster, M. (2012, December 28). Ears don't lie. *Radiolab, Podcast.*

3:1
Wellman, H. M., Cross, D., & Watson, J. (2001). Meta-analysis of Theory of Mind development: The truth about false beliefs. *Child Development, 72*(3), 655-684.

C:37 C:90 C:104
Wicker, B., Keysers, C., Plailly, J., Royet, J. P., Gallese, V., & Rizzolatti, G. (2003). Both of us disgusted in my insula: The common neural basis of seeing and feeling disgust. *Neuron, 40*(3), 655-664.

C:51 C:52
Wilson, D. S., Near, D. C., & Miller, R. R. (1998). Individual differences in Machiavellianism as a mix of cooperative and exploitative strategies. *Evolution Human Behavior, 19,* 203–212.

C:43
Wilson, D. S., Near, D., & Miller, R. R. (1996). Machiavellianism: A synthesis of the evolutionary and psychological literatures. *Psychology Bulletin, 119*(2), 285–299.

3:3
Wimmer, H., & Perner, J. (1983). Beliefs about beliefs: Representation and constraining function of wrong beliefs in young children's understanding of deception. *Cognition, 13,* 103–128.

4:61
Wing, L., & Gould, J. (1979). Severe impairments of social interaction and associated abnormalities in children: epidemiology and classification. *J. Autism Developmental Disorders, 9*(1), 11–29.

7:6
Wittgenstein, L. (1958). *Philosophical investigations (3rd edition).* New York: Macmillan.

C:60 C:67
Woodworth, M., & Porter, S. (2002). In cold blood: Characteristics of criminal homicides as a function of psychopathy. *J. Abnormal Psychology, 3*(3), 436–445.

B:40
Woon, F. L., & Hedge, D. W. (2008). Hippocampal and amygdala volumes in children and adults with childhood maltreatment-related posttraumatic stress disorder: A meta-analysis. *Hippocampus, 18*(8), 729–36.

D:10
Yehuda, R., Boisoneau, D., Lowy, M. T., & Giller, E. L. (1995). Dose-response changes in plasma cortisol and lymphocyte glucocorticoid receptors following dexamethasone administration in combat veterans with and without posttraumatic stress disorder. *Archives General Psychiatry, 52*(7), 583–593.

D:69
Yerkes, R. M., & Dodson, J. D. (1908). The relation of strength of stimulus to rapidity of habit-formation. *J. Comparative Neurology Psychology, 18,* 459–482.

4:49
Youngblade, L. M., & Dunn, J. (1995). Individual differences in young children's pretend play with mother and sibling: Links to relationships and understanding of other people's feelings and beliefs. *Child Development, 66*(5), 1472–1492.

4:23
Zahn-Waxler, C., Radke-Yarrow, M., Wagner, E., & Chapman, M. (1992). Development of concern for others. *Developmental Psychology, 28*(1), 126–136.

B:22
Zlotnick, C. (1999). Antisocial personality disorder, affect dysregulation and childhood abuse among incarcerated women. *J. Personality Disorders, 13,* 90–95.

Index

A

Abu-Akel, Ahmed, 25, 351–52, 356
abuse, 83, 129, 133, 139, 145, 171, 227, 363–64, 366, 401, 436. *See also domestic violence, sexual abuse.*
abuser, 125, 133, 216, 303, 390
accidents, 151, 277, 306, 362, 366, 446
accountability, 50, 213
accuracy, 70, 98, 131, 139, 160, 199, 208, 244, 270, 315, 429. *See also mind mapping.*
accusations, 126, 186, 188, 201, 205, 306
activation, brain, 386–87, 393, 419, 432. *See also brain.*
adolescents, 42, 55–56, 67, 117, 125, 143, 216, 255, 274, 281, 329, 375, 380, 394, 396, 418. *See also children.*
adult love relationships, 23, 57, 67, 70, 79, 83, 87, 103, 141, 167, 183, 444. *See also marriage.* adultery, 76, 323. *See also affairs extramarital, infidelity.*
adults, 38, 43, 52, 57–59, 63, 67, 95, 99, 330, 332, 362–63, 380–381
adversarial, 32, 204, 307, 436. *See also antagonist.*
affairs, extramarital, 24, 45, 75–77, 79, 89, 111–13, 131, 170, 178, 198–200, 221, 225–26, 239, 247, 323–24. *See also infidelity.*
age, mind-mapping development, 20, 34, 36–38, 40, 50–51, 53–56, 58, 82, 114, 246–47, 254, 274–75, 283, 362, 368. *See also mind mapping.*
aggression, 138, 194, 366, 380, 396, 399, 403
alliances, 248, 333
    collaborative, 287
    collusive, 183, 323, 340, 397
    combative, 15, 176, 204

allostatic load, 379, 381. *See also heart-rate variability.*
Alzheimer's disease, 341, 356
amnesia, 29, 365
amusia, 372, 374
amygdala, 350, 353, 377, 379, 391, 395, 405, 410–411, 415, 419, 422, 426, 441, 445
anger, 42, 149, 153–54, 163–65, 184, 188, 213, 231, 244, 257, 265, 298, 301, 363, 365, 372, 386, 390
anhedonia, 167
anorexia nervosa, 356
anosognosia, 373–74
antagonist, 130–131, 147, 162, 166, 176–79, 205–7, 215, 217–21, 224–25, 227–28, 233, 266, 271, 288–91, 295–98, 300, 308, 358–59, 424–31, 435–36. *See also Crucible Neurobiological Therapy methods.*
    handling antagonist
        6 points for dealing with antagonist, 435
        10 points for dealing with moves, 288-291
        dealing with people who do disgusting things, 273
        decipher your antagonist's moves, 286
        don't let your antagonist get around you, 297
        don't try to pin your antagonist down, 298
        email interactions, 304
        fear of showing antagonist you can see him, 291
        focus on observable behavior, 312
        handling people who make moves on you, 284
        increasing accuracy in mind map-

ping antagonist, 203-04, 431
keep people's mind–mapping abilities in mind, 206, 244-45, 248
keep up with your antagonist in real time, 308
look for signs of impact, 311
make gold-standard responses, 295
show your antagonist you can see him, 290
written mental dialogs, 266-70, 285-86, 427-430
antagonist inflicts
exploitation, 395-404
negative feelings and emotions, 388
odors, 337, 432
traumatic mind mapping, 109-134
antagonist plays three moves ahead, 17, 204-06, 239, 336
detecting, 204
anterior cingulate cortex (ACC), 350, 353-54, 418
anterior insula (AI), 154-55, 162, 172, 373-74, 384-86, 388, 391, 403-7, 409-10, 415, 417, 422, 430. *See also disgust.*
AI-generated interoceptive map, 410
and adjacent frontal operculum (IFO), 376, 385, 403-6, 420-421
anticipatory traumatic mind mapping, 132-34, 184, 213, 267, 289, 291, 300, 302, 335, 363, 378, 409 *See also traumatic mind mapping.*
antisocial, 82, 109, 135, 138, 154, 187, 197, 244, 285, 358, 387, 411
behavior, 177, 397-98, 402, 404
mind-mapping ability, 110
personality disorder, 138, 354, 356, 358
antisocial empathy, 137–43, 145–51, 153, 155–57, 159, 182–83, 245, 247–48, 270–271, 312–13, 349, 383, 387–92, 398–99, 401–3, 408–10. *See also compassion, empathy.*
common characteristics, 388
enjoyment of other people's suffering, 42, 147, 188, 387
reality of, 139, 141, 145
anxiety, 84, 88, 102, 113, 119, 165, 186, 206, 213, 325–26, 329, 331, 356, 364, 378, 380, 397, 415, 419–20, 425–28, 433, 441–45
anxiety attacks, 244, 369
anxiety disorders, 127, 185
apology, 204, 263, 282, 298, 301, 306, 308, 335–36, 435 *See also forgive, remorse.*
arousal, 27, 367, 376-7, 380, 417, 420-21, 431, 440, 443, 445
Asperger's syndrome, 49, 60-64, 113
where do schizophrenia, autism, and Asperger's syndrome fit in, 60
attachment, 21, 51, 137, 141, 145, 385, 438. *See also differentiation.*
attention, 23
attentional networks, 25
dorsal attentional system, 25
ventral attentional system, 25
joint attention, 33
ventral-dorsal switching process, 27
attention-deficit hyperactivity disorder (ADHD), 61, 156, 185, 326, 354, 365, 371
attributions, 12, 26, 210, 357, 437. *See also mind mapping.*
absent, 428
errors, 357
unwarranted, 130, 210, 212
autism, 20, 49, 60-64, 356-58, 386. *See also Asperger's syndrome.*
alternative aspects of false-belief comprehension, 63
autobiographical memory, 29–32, 179, 181, 213, 218–19, 221, 235, 237, 243–47, 249, 251–55, 259–61, 263–67, 271, 340–341
memory gaps and distortions, 244, 368, 370, 418
autonomic hyperarousal, 380

B

babies, 33–34, 51, 117, 119, 122, 257, 317. *See also children.*
Baron-Cohen, Simon, 35
basal ganglia, 356, 407
battering, domestic, 130, 365. *See also abuse, domestic violence.*
batting away response, reflexive, 176. *See interpersonal neurobiological problems.*
bed-wetting, 185, 368
Beers, Katie, 120
behaviors, bad, 24, 62, 84, 130, 139, 148, 150, 184, 206, 213, 248, 435. *See also Dark Tetrad.*
beliefs, 14, 19, 26, 29, 36, 57, 62, 101, 122, 243, 247, 275, 318, 354, 429
betrayal, 229, 362
bias, 100, 141, 398
binomial thinking, 232
blaming, 206
blindness, emotional, 4, 12, 28, 90, 129–30, 177–78, 187, 231, 233, 239–40, 301, 310, 313–14, 320, 328. *See also interpersonal neurobiological problems.*
 blind spots, 104, 178, 182
 go blind, 189
bluffing, 15, 64, 72, 196, 291
blunting of emotion, 372, 374. *See also interpersonal neurobiological problems.*
Bobbitt, Lorena, 63
body, 73, 75, 94, 121–22, 302–3, 324–25, 375–76, 386, 388, 405–6, 408–10, 422–24, 425, 432–34, 442
 body cues, 418. *See also interoception.*
 body learning, 325, 431, 433
 body states, 376, 417
bottom-up process, 23
Bowen, Murray,157, 394
bowel (intestinal) difficulties, 341, 405. *See also disgust.*
brain, 3–6, 19–29, 31–32, 109–13, 154–56, 161–65, 167–70, 174–76, 216–20, 324–27, 351–54, 376–78, 383–84, 386–88, 404
 anatomy, 351, 384, 410
 children's, 135, 156, 319, 326
 circuitry, 356, 378, 433

 conditions facilitating brain-change, 293, 325
 damage, 21,60, 125, 219, 363
 hemispheres
  left, 6–7, 28, 30, 110, 209, 20, 224, 238
  right, 7, 28, 29, 31, 224, 238, 252
 maps, 28, 325
 plasticity, 331
 reactions, involuntary, 284, 362
 regressions, 169
 science, 8, 18–21, 23, 25, 27, 29, 31, 64, 89, 102–3, 144, 344, 350, 387–88
 studies of disgust, 155
 studies of empathy, 388, 425
Brain Talk, 4–8, 109, 273, 349, 357, 368, 390, 413, 435–36
brain-based psychotherapies, 369, 381. *See also Crucible® Neurobiological Therapy.*
Brewin, Chris, 419–21, 433
brightening, 316. *See also steady-state regressions.*
bullying, 111, 138, 184, 187, 396, 398. *See also antisocial empathy.*
business, 16, 123, 164, 216, 221, 240, 300, 309, 374

C

Calhoun, Larry, 447
callousness, 61, 140, 237, 397, 404. *See also antisocial empathy.*
calming, 325, 415, 433
Calm Heart,Quiet Mind: Second Point of Balance, 294. *See also Crucible Four Points of Balance.*
care, taking, 33, 57, 114, 128, 259, 305, 314, 341 343, 363, 399
catecholamine, 378
cells, brain, 22–23, 144, 386. *See also brain.*
 growing new, 378
 increased glial, 414
 spindle, 386, 405
central nervous system (CNS), 376–77
cerebellum, 388, 433–34
child abuse, 250, 362, 365, 446. *See also*

abuse, sexual abuse.
  child molester, 305
children, 33–43, 48–54, 58–59, 63–64, 146–47, 154, 156–58, 169–72, 178, 183–86, 326–29, 331–36, 389–90, 394–96, 402–4
  false-belief comprehension, 63
  gaslighting, 403
  hyperactive, 326
  mind-mapping abilities, 32, 35, 39, 40-4, 43, 45–46, 50, 51, 64, 67, 329
    precocious development, 41
    proof of mind-mapping ability, 38
    rudimentary abilities, 34, 50, 54
  traumatized, 363, 365
childhood, 39, 42–43, 127, 130, 168, 171, 175–77, 179–80, 219, 222–23, 263–64, 313, 316
  disgust reactions, 421
  PTSD, 368
  traumatic mind mapping, 169
clients, 29–31, 40–42, 90–91, 112–16, 118–20, 149–50, 174–77, 217–18, 231–33, 314–16, 368–70, 416–17, 426–30, 435–40, 443–46
clinicians, 7, 61, 137, 365. *See also therapists.*
cognitions, 367, 377, 421–22
cognitive, 37, 156, 215, 415
  and emotional empathy, 392
  complexity, 317
  decoupling, 64
  empathy, 383, 392–95,
  flexibility, 394, 436
  mind mapping, 354, 358
cognitive behavior therapy (CBT), 414-16, 442
cognitive processing therapy (CPT), 414-16, 429, 442
collaborative alliance, 75, 77, 176, 276, 325, 336, 397
collusive alliance, 183, 323, 340
combat veterans, 362, 364, 378
combative alliance, 15, 176, 204
common denominator, lowest, 393
compassion, 21, 136–38, 140, 187, 251, 313, 319, 337, 344, 387, 395. *See also empathy.*
confabulation, 99, 101, 311, 332. *See also deception, lying.*
conflicts, 13, 52–53, 71, 88, 178, 333
confrontations, 121, 231
connections
  bidirectional, 352
  brain-to-brain and mind-to-mind, 438
  emotional, 17, 136–37
  new neural, 28, 219
  positive emotional, 21
  synaptic, 414
connectivity, effective, 407–8
consciousness, 365, 388, 405, 421. *See also interoception, self-awareness.*
contagion, emotional, 393–95, 420. *See also empathy.*
contempt, 72, 154, 439
context, 53, 99, 129, 199, 260, 399, 421, 427
contradictions, 90, 104–5, 113, 205
control, 49–51, 84–85, 121, 126, 128, 164, 204, 215, 218, 265, 282, 286, 298, 300, 395, 397, 417–18. *See also power.*
  anxiety, 434
  decisions, 280
  impulse, 363
  sex, 71
  usurping, 377
conversations, 131–32, 197–99, 202–7, 255–56, 262, 264, 286, 288, 290, 293, 295, 301, 303, 308–11, 339
cooperation, 212, 320, 395, 397
correspondence, written, 173, 260–261, 264
corticotrophin-releasing hormone (CRH), 378
cortisol, 376, 378
couples, 1–2, 71–72, 74, 77–78, 88, 137, 140, 143, 149, 186, 196, 323, 328
  fail-to-thrive, 1–2, 179–80. *See also disgusting parenting.*
courage, 171–72, 316, 447
co-workers, 13, 17, 22, 31, 45, 48, 114, 146–47, 158, 190, 207, 211, 285, 296, 337–38
Craig, Bud, 404, 410–411, 422
crisis, national health, 158, 186

Crucible® Four Points of Balance, 373
Crucible® Neurobiological Therapy (CNT)
(mind-mapping-based), 2, 5, 189-91, 423-448
   considerations for more effective therapy, 416-24
      brain-to-brain contact, 438
      exploit the IFO (anterior insula), 422-24
      focus on inner experience, 418
      insight is not enough, 417
      synchronize implicit and explicit memory, 418-22
   treating traumatic mind mapping and disgust reactions, 404-11, 413-448
Crucible® Neurobiological Therapy (CNT) methods, 412–44. *See also antagonist (dealing with)*.
   detecting, 213, 436, 438
      antisocial empathy, 146-49
      deception and lying, 201-206
      gaps in mind-mapping radar, 249
      holes in autobiographical memory, 243-249
      mind-mapping ability, 193-213
      mind-masking ability, 197-204
   dialogs, 285-311
      aspects of CNT written dialogs, 431
      written mental dialogs with antagonist, 266-71, 427-430
   increase differentiation, 436
   modify current relationship with antagonist, 435
   intensity of sessions
      high arousal and "safe emergencies", 443
      post-traumatic growth, 446
      window of tolerance, 444
   interpersonal domain
      6 points for dealing with antagonist, 435
      10 points for dealing with moves, 288-291
      analyze audio and video recordings, 264-65
      analyze correspondence, 260-263
      cross-reference information, 259
      cross-reference memories of traumatic events, 258-59
   memory reintegration
      facilitate SAM to VAM processing, 418-22
   neuroplastic activities
      aspects of CNT conjoint neuroplastic activities, 434
      conjoint neuroplastic activities, 430-34
      Eyes-Open Sex and Orgasms, 327, 432
      Heads on Pillows, 326, 430-34
      Hugging 'till Relaxed, 324-26, 430-34
      self-regulation during contact with partner, 324-27, 433
   psychophysiology
      heart-rate variability biofeedback, 441-42
      in-session psychophysiological monitoring, 440-41
   role of the therapist in CNT, 436-40
      brain-to-brain contact, 438
      CNT therapist activities, 437 commentary, written, 437
      use your left brain to corner your right brain, 237
   visualization techniques, 218-242, 425-27
      aspects of CNT visualization activities, 428
      guided imagery format, 427
Crucible® Therapy (differentiation-based), 3, 417, 436, 444
   outside the comfort zone, 444
cruel mental "voice", 174–75. *See also voice hearers*.
cruelty, 49, 61, 111–12, 134, 137-38, 141, 145, 157–58, 163, 166, 174–75, 183, 208, 215, 247, 288, 296–97, 358, 399-400. *See also antisocial empathy*.
   craving, 399
   identifying, 62
cues, interoceptive bodily, 359, 408, 414, 424, 428, 432, 434, 442, 445
culture, 136, 141, 155, 157, 163

cytokines, inflammatory, 441–42
  immune-system-damaging IL-6, 381

D

damage, 5, 74, 111–12, 125, 156, 183, 300, 332, 357, 359, 373, 391
Damasio, Anthony, 376
danger, 133, 363, 378, 435
dark personality, 23, 74, 77, 90, 134, 189, 208, 327, 377, 396, 398
Dark Tetrad, 187, 399–400, 403, 409
Dark Triad, 396–97, 399, 401
dead spot, emotional, 166, 371. *See also interpersonal neurobiological problems.*
death, 127–28, 266, 275, 314, 338, 341–42, 367, 369
decency, basic (moral), 154, 157–58, 189. *See also moral disgust.*
deception, 23–24, 75, 77, 82, 84, 198–99, 201, 205–6, 209, 212, 222, 249, 396–97. *See also lying.*
  detecting, 23, 75, 132, 201-03
decoupling, cognitive, 64, 394
deductive logic, 6, 23, 28–30, 98, 110, 145, 160, 178, 237, 252, 416
deficits, 64, 168, 207, 219, 373–74, 408
defying, 84, 235, 265, 282, 298
dendritic atrophy, 445
depression, 20, 163, 371, 375, 380–381, 416, 441–42
Descartes, Rene, 97–98
DESNOS symptoms, 365–67, 370. *See also Disorders of Extreme Stress Not Otherwise Specified.*
destructive
  people, 273. *See also antagonist.*
  relationships, 232. *See also marriage.*
detecting
  mind-mapping abilities, 193-95, 197, 199, 201, 203, 205, 207, 209, 211, 213. *See also mind mapping.*
  mind-masking abilities, 197. *See also mind masking*
diagnoses, 61, 171, 365
  personality disorder, 187

diagnostic system, brain-based, 369
dialogs, written mental, 266, 270, 408, 415, 421, 424-27, 429–31, 435–37, 439, 444, 446. See also *Crucible® Neurobiological Therapy methods.*
differentiation, 97, 157, 184, 293, 295, 317, 383, 393, 395, 417, 433, 436, 439, 444. *See also attachment.*
  intergenerational, 284
  operationalizing, 395
  therapist's, 440
differentiation framework, 437
digest (mentally process information), 120, 127, 227, 231, 349–50, 376
dilemma, two-choice, 147
discomfort, 114, 147, 232, 388, 390, 444
disgust, 84–85, 154–55, 163, 168, 174, 184, 229, 284, 383, 386–87, 400, 403–9, 415–16, 421–23
  emotion, 164, 390, 423
  facial expression, 226
  instilling, 184
disgusting, 154–56, 158, 163, 166, 170, 175, 177–79, 181, 183–85, 212–13, 254, 284, 308, 314, 404, 408–11
  behavior, 154, 160, 183, 186, 246, 293, 313, 319, 374, 400, 409, 411, 413
  dynamics, 183
  imagining, 170, 406
  observing, 407
  tasting, 406
  traumatic mind mapping, 171
disgusting parenting, 154–58, 178, 183, 185, 284, 314, 340, 362, 371, 409
  doctors overlook, 156
disgust brain reactions, 154, 156, 159–162, 168, 170–173, 178, 183–85, 189, 251, 359, 384, 406, 408–10, 413, 421–22, 425
  automatic, involuntary, 155, 158, 170, 172, 363, 383
  impaired, 154, 170, 373–74
    absent, 172, 372
    displaced, 186
    eroticized, 170–172, 372
    reduced, 170
  moral disgust, 372, 409
  repeated, 157, 159, 189, 296, 374, 409, 411

disorders, 20, 60–61, 63–64, 185, 357, 362, 365, 373, 413, 441–42. *See also interpersonal neurobiological problems.*
  attachment, 365
  attention-deficit hyperactivity, 61, 356, 365
  autism-spectrum, 115
  bipolar, 356
  character, 366
  impacting mind mapping, 356-57
  obsessive-compulsive, 185
  oppositional defiant, 365
  phobic, 365
  post-traumatic stress, 127, 361, 364
Disorders of Extreme Stress Not Otherwise Specified (DESNOS), 365–66, 368–70
  beyond DSM and DESNOS, 368–70
  list of DESNOS symptoms, 365-66
dissociation, 363, 366–68
  neural network, 172, 238
distortions, mind-mapping, 100, 129, 179, 213, 244, 437. *See also mind mapping.*
  in mapping your own mind, 103-4. *See also introspection.*
divorce, 24, 72, 77, 88, 153, 178, 216, 222, 273. *See also marriage.*
DNA methylation, 176–77
Dodson, John, 445
dogs, 21, 37, 45–49
  emotional contagion, 46
  endearing connection, 46
  mind-mapping ability, 47
Doidge, Norman, 220
domestic violence, 443. *See also abuse.*
dominance, 194, 285, 396 *See also control, power.*
dopamine, 26, 167, 354–56
  disruption in Asperger's syndrome, 358
  dopamine and serotonin system (DS), 354-57
drinking, alcohol, excessive, 42, 59, 84, 91, 127, 151, 160, 280, 332, 336
DSM-3 PTSD criteria, 364. *See also post-traumatic stress disorder (PTSD).*
DSM-4 PTSD criteria, 364–66, 376
DSM-5 PTSD criteria, 174, 367–69
dual representation model of memory, 419-22. *See also situationally-accessible memory (SAM), verbally-accessible memory (VAM).*
Dunn, Judy, 50–54, 58
dysfunctional relationships, 131, 318, 430. *See also adult love relationships.*
dysregulation, emotional, 365–66, 438

E

eating disorders, 185. *See also disgust.*
effectiveness, 3, 312, 416
ego, 289, 399
egocentrism, 208, 396. *See also narcissism.*
Ekman, Paul, 202
elicitation window, 437. *See also Crucible® Neurobiological Memory methods.*
embarrassment, 48, 148, 102, 130, 228, 257, 421
embodied knowing, 145. *See also body learning, empathy, mirror neurons, von Economo neurons.*
emergency, 76, 443
emergency lies, 23, 75, 197, 285, 453. *See also lying.*
emotional abuse, 8, 170, 183, 319, 363, 369, 443. *See also abuse, sexual abuse.*
emotional
  arousal, 380, 417–18, 433
  blocks, 437
  crashes, 173. *See also regressions.*
  empathy, 392–95
  fusion, 150, 274
  gridlock, 88, 444
  intelligence, 101, 385
  maps, 26–27
  reactions, 26, 189, 313
  regulation, 244, 317, 380, 419
emotional contagion, 392, 445. *See also empathy.*
emotional depth, 164, 375
emotional super-glue, 184, 284, 296, 314-15, 403. *See also disgust reaction, disgusting parenting.*
emotions, 27–30, 53–54, 64, 353–54, 371–74, 376–77, 381, 385–88, 404–5,

407, 410, 421, 423–24
  competitive, 389, 391
  negative, 385, 387, 389, 406
  positive, 385, 389, 442
  recognition, 53, 393
empathic, 109, 135, 380, 387, 393–94, 420, 433
  abilities, 138
  contagion information, 421
  responses, 393
empathy, 15, 135–42, 145, 187, 383–88, 390–392, 394–95, 398–99, 404–06, 408, 425, 434, 436. *See also antisocial empathy.*
  and mind mapping, 387, 411
  conflated with compassion, 136–37, 187
  differentiation-based view, 394
  scales, 386-87, 393
empathy system, two-part, 394
employees, 32, 285, 296
employers, 31, 285
environment, 120, 377, 385. *See also home.*
  family, 325
  hostile, 70
  stable, 85
  unstable, 58
envisioning, *7, 56, 68, 133, 135, 182, 231, 251, 260, 300, 393. See also imagery, visualization.*
envy, 389–90. *See also fortune of others emotions, jealousy.*
epigenetics, 176, 335
epinephrine, 376
equilibrium, emotional, 313, 371. *See also Crucible Four Points of Balance.*
erectile difficulties, 73. *See also sexual problems.*
eroticism, 17, 73, 241
erroneous conclusions, 89, 415
errors, detecting, 205
evil, 139. *See also antisocial empathy, cruelty.*
evolution, 132, 293, 385, 411, 430
executive functioning, 377
experience, 29, 41, 43, 47, 133, 165–66, 168, 178–80, 342, 344, 368–69, 390, 417–18, 423, 446

body-learning, 326, 390
confabulated, 315
interoceptive, 400, 418, 431
traumatic mind-mapping, 114
explicit memory, 378, 418–19, 422, 431, 445. *See also memory, implicit memory.*
explicit mind mapping, 34, 386. *See also mind mapping.*
exposure therapy, prolonged, 414-415, 422, 425-26, 429, 442
exposure to trauma, *52, 58, 293, 367, 415-16, 435, 445. See also traumatic mind mapping.*
  repeated, 170, 183, 367, 381
extramarital affairs, 24, 40, 88, 148, 153, 216, 240, 246, 263, 323, 368
eyes, 13, 16–17, 22, 31, 34, 45, 47, 78, 132–33, 224–25, 227–29, 231, 233, 309–10, 326-28
  eye contact, 33, 227–28, 231, 233, 276, 282, 430
  eye tracking, 227
Eyes-Open Sex and Orgasms, 327, 417, 432, 434, 437. *See also Crucible® Neurobiological Therapy methods.*

F

facade, 71, 83
facial expressions, ·13, 163, 175, 209, 233, 264, 381, 408
  disgusted, 387, 406, 408
  fixed, 372
fail-to-thrive clients, 149–50, 154–55, 163, 178, 181, 200, 212, 296
failure, predicting treatment, 307, 395, 431, 435. *See also disgusting parenting.*
false beliefs, 4, 35–37, 39, 54–55, 58, 62, 64, 72, 76, 88, 99, 195
  implanting, 39, 56, 65, 88–89, 196, 204, 330, 401. *See also lying.*
  tests, 35, 63-64
false memories, 298, 310, 315
false narratives, creating, 187. *See also confabulation, lying.*

INDEX

families, 49–50, 52–53, 62–63, 131–33, 179, 249–50, 255, 275, 278, 311, 315, 334, 342–43, 373–75, 403. *See also home.*
　dynamics, 126, 259
　dysfunctional, 6, 23, 57, 60, 67, 86, 109, 186, 314–15, 340, 373
　influence children's mind-mapping abilities, 50
　meals, 246, 255-56, 259, 276
　problems, 41, 52, 436
　rules, 53, 302
family therapy, 117–18
fantasies, sexual, 17, 62, 170, 172–73, 246, 330, 372
feeling states, 53, 405, 425. *See also interoception, mental states.*
　bodily, 406, 408, 423
　emotional, 423
fight or flight response, 27, 116, 367, 377, 381. *See also anticipatory traumatic mind mapping.*
fights, 16, 40, 160–161, 195, 205, 217, 281, 308, 332
first level mind-mapping ability: mind mapping, 22. *See also mind mapping.*
First Point of Balance: Solid Flexible Self, 294. *See also Crucible Four Points of Balance.*
flashbacks, 364, 366–67, 369, 419–21
flashbulb memories, 375–76, 419
fMRI studies, 20, 406, 432
fog, mental, 156, 178, 316, 370. *See also spaghetti brain.*
forgive, 63, 152, 263, 297, 306, 323, 335. *See also apology, remorse.*
fortitude, 172, 448
"fortune of others" emotions, 391
fourth level mind-mapping ability: mind twisting, 90. *See also mind twisting.*
Fourth Point of Balance: Meaningful Endurance, 294. *See also Crucible Four Points of Balance.*
freedom, 218, 234, 328, 340, 438
friendships, 6, 55, 333, 336–37
Frith, Uta, 36
functional
　connectivity research, 407
　control, 162
　neural networks, 352

G

gaps, autobiographical, 29–30, 179, 181, 203, 205, 209, 213, 219, 221, 226, 243–46, 252, 374, 418, 420, 429, 431, 438. *See also autobiographical memory.*
gaslighting, 187, 373, 401–3. *See also antisocial empathy, mind twisting.*
gauging people's mind-mapping abilities, 194. *See also mind mapping.*
Geiselman, Edward, 201–3
gestures, 16, 33, 68, 152, 202, 342, 397
gift of mind, 321
give your dilemma meaning, 447. *See also self-soothing.*
gloating, 391
gold-standard responses, 271, 286, 295–97, 431, 434. *See also antagonist.*
grandchildren, 274, 279, 311, 318
grandparents, 52, 216, 264, 274–76, 311, 317. *See also parents.*
Grounded Responding, Third Point of Balance, 294. *See also Crucible Four Points of Balance.*
growth, 102–3, 273, 417, 439, 447–48
　post-traumatic, 447
guilt, 148, 288-89, 291, 293, 312, 366, 416 421
Gulf War veterans, 364
gustatory centers, 405. *See also disgust.*
gut reactions, 384, 420. *See also interoception, vagus nerve.*

H

harming, 359, 399, 446
harm-joy, 140–141. *See also antisocial empathy, schadenfreude.*
hate, 178, 225, 234, 269

Heads on Pillows (HOP), 326, 417, 430, 432–34, 437. *See also Crucible® Neurobio-*

logical Therapy methods.
healing traumatic mind mapping, 182, 312. *See also traumatic mind mapping.*
health, 22, 119, 121, 380
healthy self-doubt, 211
hearing a mental "voice," 174. *See also voice hearers.*
heart, 6, 8, 17, 21, 31, 46, 141, 299, 305, 327, 336, 342, 379, 384, 390
   heart rates, 116–21, 324, 379–80, 420
heart-rate variability (HRV), 116–21, 379, 417, 441–42. *See also vagal tone.*
   biofeedback, 441–42
hemisphere, brain
   left, 47, 209. *See also left hemisphere.*
   right, 28, 47, 209. *See also right hemisphere.*
high ground, moral, 270, 286, 310
Higher Desire Partner (HDP), 71–72. *See also Lower Desire Partner (LDP).*hippocampus, 378, 407, 411, 418–19, 445,
   size, 378
hitting, 38, 114, 148, 152, 216, 238, 240, 281
holes, autobiographical, 34, 76, 101–2, 124, 130–131, 163, 177–78, 180–181, 205, 209, 218–19, 229, 244, 252, 260, 330–331. *See also autobiographical memory.*
   repairing, 86, 206, 244, 258, 260, 340
home, 49–50, 52, 59, 85–87, 122, 124, 126, 182–86, 223, 259, 261, 279, 340–341, 343, 383, 409. *See also families.*
   bad homes raise people with good mind-masking ability, 85
   troubled, 50, 52, 57–60, 85, 122, 139
   what if you come from a troubled home, 57
homeostatic mechanisms, 411, 422
hostage, adult children, 183, 284, 314-315. *See also disgusting parenting, emotional super-glue.*
hostility, 247, 380
House Devil, Street Angel, 59
how the brain
   creates empathy, 392-95. *See also antisocial empathy, empathy.*
   differential routing through the brain,

   dorsal attention network, 25, 27
   ventral attention network, 25
   tracks whose mind you're mapping, 351-57
   dopamine and serotonin system (DS), 354-57
   norepinephrine, 27
hugging, 324–26, 417, 430, 434, 446
Hugging 'till Relaxed (HTR), *326, 430, 432–33, 437*. *See also Crucible® Neurobiological Therapy methods.*
human bonding, 21, 24
human brain, 12, 21–22, 28, 102, 168, 369, 388, 410. *Also see brain.*
human nature, 5–6, 19, 21, 23–24, 104, 109, 141, 288, 383
humor, 302–3, 305
hurt, 72, 76, 101, 112, 128, 137, 139, 148–49, 156, 183, 243, 250, 274, 279, 399–400
hypervigilance, 74, 129, 150, 161, 177–78, 184, 204, 367, 370, 374, 378
hypothalamic-pituitary-adrenal axis (HPA), 116, 120, 377, 379. *See also stress.*
hypothalamus, 377, 379

I

imagery, 68, 123, 151, 220–221, 225, 233, 251, 254, 264, 367, 369 376, 406, 419, 423, 426, 432, 435, 437, 440
   involuntary, 419, 446
imagination, 26, 29, 407, 423, 427,
   imagining disgusting scenarios, 406
   imagining exposure therapy, 426
immune system, 121, 379, 381
impacts of traumatic mind mapping, 159, 161, 163, 165, 167, 169, 171, 173, 175, 177, 179, 181, 183, 185, 187, 189. *See also traumatic mind mapping.*
implanting
   false beliefs, 36, 72, 74, 76, 88–89, 111, 194–95, 203, 211, 270, 339, 342
   uncertainty, 313
implicit mind-mapping ability, 33–34. *See also mind-mapping ability.*

improvement, 121, 203, 231, 241, 371, 442
inaccuracies, 100, 129, 205, 213. *See also mind mapping.*
indicators,
    mind-masking, 198. *See also mind masking ability.*
    antisocial empathy, 147
    deception and lying, 201
infants, 20, 50–51, 380, 394, 410. *See also children.*
infidelity, 181, 200, 226. *See also extramarital affairs.*
inflicting, 337, 359, 388, 397. *See also antisocial empathy.*
    enjoyed inflicting, 148
    misery, 142
    traumatic mind mapping, 125, 149, 188, 273
inner conflict, 105, 146
inner experience, 417-18, 424. *See also interoception.*
    confabulated, 371
innuendo, 196, 207, 281, 291, 315
insula, 384–86, 395, 405, 410–411, 441, 458, 464, 501–2. *See also anterior insula.*
    posterior, 384–85, 407
in-session psychophysiological monitoring, 118, 440
integrity crises, 294
intellectualizing, 7, 31, 220, 237
intensity, 217, 440–441, 444. *See also window of tolerance.*
Intensive Therapy Program, Crucible®, 173, 236. *See also Crucible® Neurobiological Therapy methods.*
intentions, 14, 20, 34, 51, 53, 64, 69, 101, 128, 195, 304, 398
intergenerational transmission of trauma, 318, 363. *See also disgusting parenting, traumatic mind mapping.*
interoception, 405-06, 417-18, 424, 433. *See also self-awareness.*
interpersonal, 194, 284–85, 300, 366, 370, 375, 416, 431, 446
    corrective experience, 164
    manipulation, 396. *See also mind twisting.*
    trauma, 365–67, 413–14, 418, 429–30, 442–43. *See also traumatic mind mapping.*
    victimization, 362, 364, 367
interpersonal neurobiological problems, 159-190, 370-381
    impacts of repeated traumatic mind mapping, 115-34, 159-90, 370-381. *See also traumatic mind mapping.*
    impaired
    acute loss of self-awareness, 373. *See also interoception.*
    disgust reaction, 155, 170-74, 362, 372, 382-390. *See also disgust.*
    emotional functioning, 371–72
    emotional mind mapping, 370. *See also mind mapping.*
    fixed smile, 372
    masked face, 372, 374, 381, 439
    non-psychotic thinking disorders, 370-371
    pattern recognition, 210
    missing
    autobiographical memories, 370, 419. *See also autobiographical memory gaps.*
    mental maps, 370. *See also mental maps.*
    thoughts (thought blockage), 223, 233, 236, 300, 370, 428
    recognizing personal impairments, 215-42
    reflexive
    angry crying, 372
    thought patterns, 371
    uncontrollable,
    behavioral tics, 372
    mental tics, 438
    mind mapping, 61
    noxious touch and ticklishness, 372
    phonic tics, 372
    tangential thinking, 370
    varieties of problems, 370-73
interpersonal neurobiology, 1, 20, 32, 97, 109, 183, 293, 324, 326, 338, 383, 409
interventions, 63, 336, 414, 424. *See also Crucible® Neurobiological Therapy*

methods.
intimacy, 4, 55, 65, 74, 322, 324, 326
Intimacy & Desire, 458, 461, 511
introspection, 93, 95, 97–99, 100, 102, 176, 180, 233-34, 332, 237, 336, 416, 428
    fallibility, 98–99
    perceived specialness of, 97
    privileged access, 97–99, 180
    scientists reject, 102
    sources of self-knowledge, 95-97
intuition, 15, 29, 64, 111, 114, 137, 140, 187, 218, 283, 313
investment, emotional, 16, 153, 327, 336, 340, 441
invisible
    epidemic, 362
    maladies, 178
irritable bowel syndrome, 441. *See also disgust.*

J

Jabbi, Mbembe, 359, 387, 404–9, 422-23, 425, 432, 434
jealousy, 162, 230, 238, 262. *See also fortune of others emotions.*
jokes, 53, 64, 78, 276, 281, 302, 322
joy, 174, 277, 321, 381, 434
judgments, 16, 99, 102, 177, 188, 209, 236, 240, 280, 386

K

kidding, 40, 57, 73, 250, 324–25
kinesthetic cues, 325, 424, 434, 446. *See also situationally-accessible memory (SAM).*
    unprocessed, 415

L

learning, 2, 4–5, 7, 11, 43, 51, 61, 81, 181, 245, 293, 408, 417, 430-431, 434
    learning disabilities, 326. *See also attention deficit disorder.*
left hemisphere. *See right hemisphere, brain.*
    left-brain approach, 223
    left-brain deductive-logic, 28
    left-brain response, 236
Lehrer, Paul, 441
Leslie, Alan, 35
life story, 28, 30, 182, 243–45. *See also autobiographical memory.*
limbic system, 164, 391, 410
long-term relationships, 23, 70, 75, 212. *See also marriage.*
love, 145, 193, 206, 210, 222
    love relationships, 67–80, 83, 87, 103, 144, 167, 183, 320-27
    lover, 131, 152–53
Lower Desire Partner (LDP), 71–72, 149, 222. *See also Higher Desire Partner (HDP).*
lying, third level of mind-mapping ability, 89–90, 92, 111, 201–4, 207, 234, 403. *See also mind mapping, mind masking.*
    detecting, 201
    good therapists need to be good liars, 89-90

M

Machiavellianism, 138, 187, 395–99, 401–3. *See also Dark Triad.*
malevolence, 362, 396
mammals, 21–22, 60, 352
    mammalian brain regions, 353
manipulation, 23–24, 75–76, 89, 92, 114, 187, 198, 207, 265, 287, 398
    mind-twisting, 183
map, mental, 13–17, 25–32, 68–71, 81–82, 89–91, 94–96, 125–26, 193–95, 209–12, 218–20, 233, 288–90, 333–34, 335–37, 353–54, 356, 405. *See also mind mapping.*
marriage, 2, 5–6, 22, 24, 62, 67–68, 70, 77, 88, 141, 148, 150, 153, 178, 185–86, 222, 287, 302, 309–10, 390
mate (partner), 22, 40, 77, 88–89, 131, 137, 158, 195, 198, 368
mealtime memories, 255
Meaningful Endurance, Forth Point of

Balance, 294
meanness, 49. *See also antisocial empathy, cruelty.*
meditation, 98, 381, 417
Meins, Elizabeth, 50–52, 57
memories, 28, 30–31, 178–79, 181–82, 220–221, 248–49, 258, 283–84, 296–98, 320, 376–77, 414–15, 418–20, 422, 428–29
  intrusive, 364
  triggered, 283
  visuospatial, 428
memory
  autobiographical, 179, 245, 249, 419, 421, 429, 431
  retrieval, 178, 219, 415
  situationally accessible (SAM), 283. *See also situationally accessible memory.*
  verbal (VAM), 378, 416, 418. *See also verbally accessible memory.*
mentalizing, 137. *See also mind mapping.*
mental blocks, 223, 233, 236, 300, 428
mental dialogs, written, 266–69, 271, 285, 287, 294, 300, 309, 423, 315, 427–29, 438–39, 444, 446. *See also Crucible Neurobiological Therapy methods.*
mental health, 7, 90, 218
  mental health crisis, 186, 188
mental health professionals, 7, 158, 172, 185, 187, 361, 393
mental imagery, 423, 432. *See also imagery.*
mental maps, 12, 20, 24, 26, 29, 32, 36, 179, 181, 268, 275, 352, 354, 356–57, 429
  affective, 357
  emotional, 27
mental states, 26–27, 29–30, 51, 54, 57, 62, 95, 97–98, 352, 354, 356–58, 432, 434
  affective, 353
  cognitive, 353
mental state language, 50–52, 54, 58. *See also mind mindedness.*
mental tics, 438
mental world, 146, 183, 329, 397
methods
  Crucible® Neurobiological Therapy, 425. *See also Crucible Neurobiological Therapy methods.*

effective self-soothing, 447. *See also Crucible Four Points of Balance.*
manipulation, 310
mind-twisting, 91. *See also mind-twisting.*
perspective-taking, 408
micro expressions, 202
mind blindness, 56 129–30, 150, 162–63, 177, 187, 206–8, 212, 374–75, 385. *See also mind mapping.*
  inability to detect implied meanings, 207
  rule out mind blindness, 206-08
mind mappers, 5, 13, 55, 58, 60, 67, 76, 81-2, 86, 89, 111, 194, 210, 397-98
mind mapping, 21-22 25, 49, 57, 62, 88, 138, 162, 194, 208, 289, 398. *See also traumatic mind mapping.*
  accuracy, 211–12
  and empathic contagion information, 421
  and sexual desire, 70
  brain regions, 26
  deficit hypothesis, 63
  disorders impacting, 356-57
  errors, 129, 209–11
  impairments, 358, 370
  network, 26, 28, 356, 389, 391
  precocious understanding, 51, 54
  system, 23, 27, 168, 184, 392
mind-mapping-based Crucible® Neurobiological Therapy, 418
mind-mapping radar, 4, 14, 41, 129, 131, 177–78, 198, 209, 213, 244, 249, 370, 374, 399, 427, 429, 431, 438
  gaps, 177
mind maps, 13, 25, 29, 31, 72, 181, 203, 431
  missing, 30
mind maskers, 58, 76, 82, 87-88, 90, 268, 397
mind masking, 60, 74, 77, 79, 81–91, 111, 132, 198, 270, 323. *See also mind mapping, mind twisting.*
  abilities, 60, 83, 85, 88–89, 195, 197–98, 201, 218, 244, 246–47, 257, 270, 312–13

behaviors, 198
  deflecting, 114, 176, 199, 268, 312, 322
  derailing topics, 197, 205, 298
  misdirection, 23, 39, 75
  "misspoke", 87, 189, 211
  plausible deniability, 87, 197, 313
  thwarting, repeated, 86, 111, 147, 194, 197, 206
  defeating mind mapping, 81, 83, 85, 87, 89, 91
  flourishes in families, 84
  in adult love relationships, 87-88
  what does it take to mask your mind, 86-88
  why do people mask their minds, 82-85
mind-mindedness, 51-52
mind twisters, 90-92, 147-48, 247, 270. See also gaslighting, mind masking, lying.
mirror neurons, 144-45, 326, 385-86, 394, 407, 410. See also von Economo neurons.
"misfortune of others" emotions, 136, 138, 141, 389, 391
misunderstandings, not understanding, 207, 243, 336. See also mind blindness.
mocking, 109, 111
models
  interoception, 425
  memory dual-representation, 376
  mind mapping neuroanatomy-neurochemistry, 354
  triune brain, 25
moments of meeting, 34, 221, 285, 320-21, 324, 326-27, 330, 335, 341, 441, 438. See also traumatic mind mapping.
  positive moments, 319-46
  traumatic moments, 227-28
money, 3-4, 13, 16, 69, 76, 91-92, 122-24, 162-63, 177, 249, 257, 275, 281, 339, 342
morality, 52-54, 154, 390, 396
moral disgust, 154, 174, 185, 188, 359, 409. See also disgust.
Morehouse, Ruth, 148
motivations, 14, 69, 90, 100, 103-4, 129-30, 169, 188, 194, 210, 220, 248, 275, 313, 399, 410

motor cortex, 144
myelination, 176

N

narcissism, 187, 395-97
narcissists, 398-99, 494
narrow-band mind-blindness, 370. See also mind blindness.
nasty, 183, 222, 294, 388
nausea, 185, 308. See also disgust.
nemesis, 30, 290. See also antagonist.
nerve, 110, 114, 203, 271, 281, 294
  nerve regeneration, 378. See also neurogenesis.
  vagus, 379-81, 441-42. See also vagus nerve.
nervous system
  autonomic, 375, 379-80, 411, 442
  central, 376-77
neural priming, 378. See also anticipatory traumatic mind mapping.
neuroanatomy, 350-351, 354
  basis of antisocial empathy, 387, 392
  basis of mind mapping, 407
neurobiological problems, interpersonal, 170, 284, 300, 370, 375, 416, 436, 438. See also interpersonal neurobiological problems.
Neurobiological Therapy, Crucible®, 8, 425. See also Crucible Neurobiological Therapy (CNT)
  neurobiological underpinnings, 155, 395
neurobiology, 28, 376
neurochemistry of mind mapping, 20, 25-28, 349, 351, 354
neuroendocrine responses, 379
neurogenesis, 378, 413-14
neuroimaging studies, 377, 387, 404-5
neurological disorders, 356
neurons, 26, 344, 385-86, 408, 413. See also mirror neurons, von Economo neurons.
neurophysiology, 20, 26, 349
neuroplasticity, 32, 408, 413-14, 417

INDEX

creating, 349, 413, 425, 444
neuroplastic activities, somatosensory, 324-327, 430-434. *See also Crucible Neurobiological Therapy methods.*
neuropsychology, 376
neurotransmitters, 26-27, 354, 378, 414, 445
nonverbal communication, 29
norepinephrine, 27
"not speaking to each other", 79
noxious touch and ticklishness, 372
numbing, emotional, 367

O

obsessive-compulsive disorder (OCD), 185-86. *See also disgust reaction.*
offenses, reportable, 124 *See also sexual abuse, traumatic mind mapping.*
optimism, 68, 316, 446
orbitofrontal cortex (OFC), 350, 353, 355, 377, 407, 442,
orgasms, 11, 17, 73-74, 88, 170, 172, 222, 327, 372
    orgasm problems, 74, 405
    troublesome orgasm trigger fantasies, 172-73, 372
outbursts, emotional, 184, 244, 312
oxytocin, 28, 47

P

pain, 19, 48, 141, 146, 235, 240, 385, 387-88, 405-6, 432-34
    empathy for, 388
    pain matrix, 388, 433
    recurrent abdominal, 441
    therapist's, 439
panic attacks, 165, 184, 368, 380
paranoia, 198, 200, 238, 261
parenting, 5, 43, 157, 233, 344
    disgusting parenting, 154-58, 178, 183, 185, 284, 314, 340, 362, 371, 409
    influences children's mind-mapping development, 50

parents, 38-42, 58-59, 85, 122-24, 135-39, 149-50, 152-58, 175-78, 182-86, 232-35, 274-79, 284-85, 314-19, 327-29, 331-36, 339-44
    dysfunctional, 133, 150, 173, 178, 185, 189, 221, 314, 396, 435
    effective, 15, 40, 157, 319, 375
    gaslighting children, 403-4
    invisible, 178. *See also disgusting parenting.*
parsimony (Occam's razor), 206, 316
partner (mate, spouse), 2, 17, 23-24, 68-79, 82-83, 87-91, 100-101, 103-5, 136-37, 148-50, 320-327, 331, 395, 399-400, 432-34
    partners map each other, 70-73
    partners mask their minds, 75, 87, 322
    when partners stop masking, 323
Passionate Marriage, 7, 148, 324, 458, 461, 466, 510
Passionate Marriage® Couples Enrichment Weekends, 148
Paulhus, Delroy, 399-400
peace, 69, 129, 134, 175, 323-24, 328
pedophiles, 140, 145-46, 170
perceptions, 36, 96-97, 102, 180, 211, 315, 366-67, 385, 401, 407, 418
Perner, Josef, 35
perpetrators, 249, 366
perseverance, 171, 174, 211, 217, 219, 237
person, prudent, 210, 238, 333
personal
    benefits, 247, 446
    development, 383, 392, 447
    experience, 11, 36, 177, 187, 332, 341, 343, 389, 406, 408, 414, 423, 434
    integrity, 294
personality, 13, 15, 85, 93, 171, 371, 396
    disorders, 356
    multiple, 171-73
    normal personality traits, 187, 371
perspective, 7, 35, 55, 60, 67, 92, 138, 232-33, 254, 330, 391-94, 405, 415, 427-28, 431
    first-person, 427
    third-person (fly on the wall), 233, 332, 424-25, 427

perspective-taking abilities, 393
philandering, chronic, 127, 170, 183
physical contact, 260, 325, 430
physical reactions, 241, 385, 415, 426
physiological monitoring, real-time, 116 120, 379, 440
picture, mental, 7, 12, 28–31, 65, 68, 95–98, 100–101, 115–17, 124–27, 137–41, 175, 178–79, 221, 224–25, 229–33, 234–38, 248–49, 251–57, 333–34, 383–84
    false, 89–91, 105, 180, 194, 203, 260
    implant, 144, 251, 341
    inaccurate, 36, 90, 99, 328
    therapist's, 443
pituitary gland, 379
plausible deniability, 87, 197, 313. See also antagonist.
playing three moves ahead, 17, 206, 204–06, 239, 336
positive moments in love relationships, 320–27
    deeper intimacy in and out of bed, 324
    Eyes-Open Sex and Orgasms, 327
    give "the gift of mind" to those you love, 231–32
    Heads on Pillows, 326
    Hugging 'till Relaxed, 324–26
    resolving extramarital affairs, 323
    stop masking your mind, 322–23
positive moments of meeting, 319–48
positive moments with aging parents, 341–43
positive moments with children, 327–35
    brain-oriented sex education, 331–32
    enter young children's mental worlds, 329–31
    freedom to unmask their minds, 328–29
    give your children permission to see you, 327–28
    three-step repair strategy, 332–34
positive moments with friends, 335–37
positive moments with siblings, 339–40
post-traumatic growth, 316–17, 436, 446–48
post-traumatic stress disorder (PTSD), 127–28, 361–69, 377–78, 380–381, 415–17, 421, 427, 430, 433. See also traumatic mind mapping.
    developing, 381
    diagnosis, 361–62, 364–65, 368
    flashbacks, 420
    incidence, 364
    preschool subtype, 368
    symptoms, 316, 364, 366, 369, 416
    treatments, 414–15, 443, 445
        cognitive behavior therapy (CBT), 414–16, 442
        cognitive processing therapy (CPT), 414–16, 429, 442
        Crucible® Neurobiological Therapy (CNT), 89-91, 423-448
        exposure therapy, 435
    treatment recommendations, 414
power, 29, 174, 259, 300, 338, 375. See also control.
precuneus, 350, 355
prefrontal cortex (PFC), 22, 164, 350, 353–55, 377, 385, 391, 442
Premack, David, 19–20
President Donald Trump, 186–89
pretending, 62–64, 198, 227, 320, 340
primary emotions, 154, 163, 168, 390, 421. See also disgust.
primates, 56, 393
    bonobos, 410
    chimpanzees, 11, 19–20, 34, 45, 47, 394, 410
    gorillas, 410
    monkeys, 94, 385, 410
prisoners, emotional, 77, 183, 201, 307, 342, 361. See also emotional super-glue.
privileged access of introspection, 97–99, 180. See also introspection.
prosocial deception, 329
prosocial empathy, 145, 319. 337, 344, 386, 396, 410. See also antisocial empathy, empathy.
    presumption of prosocial motivation, 138, 387
prosociality, 387
psychopathology, 380, 409
psychopaths, 358, 397–99
    mind-mapping ability, 358

psychopathy, 138, 187, 356, 358, 396–97, 399, 402–3. *See also sociopaths.*
psychopathic behavior, 358
psychotherapists, 12, 62, 98, 110, 413
psychotherapy, 115, 189, 383, 417, 443
psychotherapy sessions, 116, 119–20
punishment, 53–54, 83, 137, 147, 198, 249, 282, 328, 398–99
punitiveness, 148, 361

Q

questions
　trapping, 196
　trick, 157
quicksand reaction, 164–65

R

radar, mind-mapping, 60, 86, 89, 102, 200, 334. *See also mind mapping.*
rage, 162, 168, 233
rape, 126, 361–62, 368–69, 372, 429, 446
　rape victims, 364, 429
rational executive brain, 417
rationalizations, 31, 112, 275
reactivity, 367
　behavioral, 380
　emotional, 121, 363, 436
reality, 141, 292, 397, 401, 403–04, 441
realization, 19, 30, 40, 45, 49, 94, 180, 198, 205, 231, 244, 310, 325, 435
recognizing traumatic mind mapping, 115
recollection, 99, 130, 175, 181, 261
　of traumatic memories, 179
recovering from "Trump Trauma", 186–189
re-experiencing traumatic mind mapping, 431
regressions, emotional
　acute, 164–66, 169, 184, 371
　long-term (steady state), 316
　reactive, 371
regulation, top-down, 433
relationship difficulties, 67, 104, 184, 333
relationships, healthy, 38, 340

relaxation, 69, 75, 133, 320, 324–25, 442
remembering, 29–30, 179, 181, 220, 240, 244, 246–47
remorse, 91, 148, 152
　remorseless, 397, 404
repetition, need for, 160, 175–76, 190, 219, 296, 300, 426, 430–431, 433
representations, mental states, 47, 316, 352–53, 357, 376, 411, 424-25, 432. *See also interoception.*
　bodily, 424
　common neural path, 407
reprocessing SAM memories, 422. *See also situationally accessible memory.*
reptiles, 14, 21
reptilian brain, 22–24, 27
resilience, 169, 173-74, 293, 300, 309, 312, 316-17, 439, 441
Resurrecting Sex, 453, 510
retraumatization, 125, 218, 443
retrieval, memory, 27, 420
　mind-mapping information, 421
　retrieval errors 29, 370
revenge, 366, 398–99
revisualizing traumatic mind mapping, 219–221, 231, 235, 237, 251, 267, 271, 282, 336, 415–16, 418, 421, 424–27, 436–39, 444, 446
revolting behavior, 160
revulsion, 154. *See also disgust.*
right hemisphere (right brain). *See also left hemisphere, brain.*
　abilities, 29
　decouples, 110
　processes, 7, 30-31
　right-brain thinking, 28
　right-brain visualization, 220
romantic relationships, 320, 344

S

sadism, 187, 395, 399–400, 409. *See also antisocial empathy.*
　arousal from, 171
　normal marital, 141, 148, 399
　sadistic mind, 401
sadists, 128, 399–400

sadness, 141, 163, 165, 168, 240
safety, 14, 23, 435, 443
salience, 376, 426
Sally-Anne test, 35, 37
sarcasm, 110–111, 291, 313, 335
Saxe, Rebecca, 22–23
schadenfreude, 138, 140–141, 388–91, 402. *See also antisocial empathy.*
schizophrenia, 20, 60, 174, 356–57
Schnarch, David, 341, 457–58, 470, 502, 510–511, 542
secondary (social) emotions, 389, 421, 423. *See also primary emotions.*
second level mind-mapping ability: mind masking, 81
Second Point of Balance: Quiet Mind and Calm Heart, 294
security, 23, 443
self, human, 83, 93–97, 104, 279, 294, 354, 357, 366, 421, 434
   continuity of, 94
   flexible, 294
   mental maps of, 351
   no accurate knowledge of who you really are, 92
   other people know you better than you know yourself, 102
   solid sense of self, 96–97, 104–5, 166
self-awareness, 95, 155, 371, 373, 375, 385, 388, 405–6, 416, 423, 432–33. *See also interoception.*
   ways of fooling yourself, 100–2
   you can be right about your partner and wrong about yourself, 103–4
self-
   confrontation, 98, 104–5, 333
   control, 188, 401
   deception, 319
   destructive behavior, 165, 365, 367
   efficacy, 312, 316, 446, 448
   esteem, 74, 100
   image, 95–96, 99–100, 102, 279
   knowledge, 95, 97, 100, 102
   perception, 30, 98, 102, 104, 229, 244, 366
   presentation, 59, 204
   regulation, 164, 380, 395, 413, 433

respect, 316
soothing, 380, 395
selfishness, 208, 397
sensations, proprioceptive, 367, 406, 417, 428, 432, 445. *See also interoception.*
sensitivity, 54, 63
serotonin, 26, 354–56
   serotonin receptor binding, reduced, 357. *See also dopamine-serotonin (DS) system.*
sessions, therapy, 115–17, 120, 172–73, 226, 231, 236, 239, 267, 275, 277, 373, 381, 417, 437, 439–40, 442
sex, 4–5, 11–12, 17, 70–74, 125–27, 148, 166–68, 170–172, 222, 245–46, 259–60, 263–66, 280–281, 327–28, 331–32
sex education, brain-oriented, 331
sex life, 40, 258, 276, 302, 319
sex therapy, 430
sexual abuse, 117, 120, 122, 124–25, 127, 133, 260, 263, 361–62, 368, 378. *See also abuse, domestic violence.*
   offenses, reportable, 124
   sexual assault victims, 415
   sexualizing, 126, 246, 287, 390
   sexual violation, 367–68
   via sexual innuendo, 238, 322
   via traumatic mind mapping, 124-27
   vibes, sexual, 223, 226
sexual
   behavior, 299
   disinterest, 72
   innuendo, 322
   interest, 70-71, 125, 188, 194, 241, 369
   problems, 74, 417, 430
   relationships, 56, 170, 222
Shamay-Tsoory, Simone, 25, 350 –52, 356, 358–59, 390–93, 394–95
shame, 148, 421
shared attention, 329
shock, 3, 12, 30, 154, 205, 403
shortcomings, 42, 100, 104, 212, 263, 315, 335
simulation, mental, 406-08, 423, 433
   prospective empathic, 425
   simulation system, 394
Singer, Tanya, 138, 387–88, 425, 432
situationally accessible memory (SAM),

283, 420–421, 424, 433, 445. *See also verbally accessible memory (VAM).*
  and VAM memories, 421
  SAM memories, 359, 420–422, 424, 446
  SAM system, image-based, 420, 422
smells, 154, 334, 404, 406, 420, 426
snap judgments, 386
social intelligence, 23, 138, 285
sociopaths, 88, 111, 140, 177, 187, 372, 395, 401. *See also psychopaths.*
soldiers, 120, 172, 361
Solid Flexible Self, First Point of Balance, 294
somatosensory cortex, 395, 427
soul, human, 318
spaghetti brain, 112–15, 156, 160, 168, 200, 205–6, 215–17, 227, 230–232, 271, 282, 293, 302, 308, 312, 401, 436, 439. *See also traumatic mind mapping.*
spindle cells and mirror neurons, *386. See also von Economo neurons, mirror neurons.*
stability, emotional, 164, 415
status quo, 65, 200, 302
steady-state regressions, 169, 173, 181, 184, 241, 371. *See also regressions.*
Street Angel, House Devil, 59
stress, 115–17, 119–21, 371, 377–80, 425, 428, 431, 440, 442, 445
  cortisol production, 376–77
  healthy HPA response, 165
strictures, brain, 232, 235, 370, 438
stubbornness, 190, 217, 219, 235, 237
super-glue, emotional, 184, 284, 296, 314–15, 403. *See also disgust reaction.*
superior temporal sulcus (STS), 350, 352–55, 405
surprise, 46, 50, 82, 88, 94, 98, 163, 168, 171, 190, 197, 232, 239, 333, 339
survival mechanisms, innate, 14, 21, 24, 56, 85, 132, 155, 185, 385, 387. *See also disgust, empathy, mind mapping.*
survivors, concentration camp, 364
sympathy, 129, 250–251, 280, 387
symptoms, 61, 361–62, 365, 367, 369, 380, 446
synaptic plasticity, reduced, 378

systems,
  attentional, ventral-dorsal, 26-27, 352–54, 357
  behavior-prediction, 21
  cardiovascular, 379
  cognitive perspective-taking, 392
  digestive, 365
  early-developing emotion-matching, 393
  emergency-response, 208
  inflammatory, 442
  limbic-paralimbic, 353
  mirror neuron, 392–93
  neuroendocrine, 379
  parasympathetic, 379–80

T

tactics, 92, 88, 196
Talmud, 100
tangential thinking, 198
taste, 154, 182, 206, 258, 379, 396, 404, 406, 432, 446
teasing, 50, 53–54, 109–10, 393, 395
Tedeschi, Richard, 447
teenagers, 37, 56–57, 84, 117, 127, 299, 310, 331–32, 368
temper, 130, 164, 249–251, 259–60, 269, 332
temporal pole(TP), 350, 353-54, 391, 395
temporoparietal junction (TPJ), 22, 352, 355
tension, 13, 71, 101, 260, 323
terror, 128, 142, 146, 189
Theory of Mind (ToM), 19–20. *See also mind mapping.*
therapists, 3, 5, 14, 61–63, 89, 98, 110–111, 115–16, 131–32, 136–39, 156–57, 180–181, 429, 436–41, 443–44
  experience traumatic mind mapping, 113–14
therapy, 1, 3, 5–6, 113, 115, 117, 149, 171, 173, 190, 414, 416–17, 425–27, 439–40, 443–45
  attachment-based, 62
  brain-based, 120, 190, 408, 413

brain-changing, 416, 444
brain-to-brain, 438
cognitive, 415, 417
emotion-based, 145
sexual, 437
therapy sessions, 173, 217
third level of mind-mapping ability: implanting false beliefs, 88
Third Point of Balance: Grounded Responding, 294
threats, 72, 146, 292, 307, 312, 405, 418, 436
thwarting, repeated, 86, 111, 147, 194, 197, 206. *See also antagonist*.
tolerance, window of, 65, 444–45
tolerating pain for growth, 439. *See also Meaningful Endurance*.
top-down functioning, 377
top end abilities, 165
tormenting, 110, 410
trackers, 81. *See also mind mappers*.
tracking, 22, 24, 27, 34, 76, 78, 88, 137, 195, 202, 217, 220, 227, 283, 289–90. *See also mind mapping*.
training, 11, 132, 180, 202
transparency, 84, 322, 340
trauma, 127–28, 168–69, 316, 318, 366–70, 376, 378–79, 411, 413, 415, 417, 419–22, 424–27, 429, 432, 435–36. *See also traumatic mind mapping*.
   definition, 362
   diagnosing, 369,-70
   disrupts SAM-VAM synchrony, 416, 421
   domestic, peace time, 361, 363, 369
   situational, 366
   treatment, 413, 423, 425
   unprocessed, 376
   war-related, 361
traumatic, 121, 126, 133, 159, 168, 180, 216, 221, 254, 293, 300, 310, 316, 399, 401
   brain injury (TBI), 356
   childhood, 150, 169
   events, 29, 122, 127, 168, 179, 220, 240, 258, 364, 366–67, 376, 378, 414–15, 418–21, 426–27, 446
   revisiting, 218
   witnessing, 364
   memories, 240, 283, 366, 369, 414–15, 419, 421-22, 425, 427–28, 435, 445. *See also memory*.
traumatic mind mapping, 111–15, 117–29, 131–35, 142, 159–65, 171–75, 185–87, 189–90, 201, 216–19, 237, 243–44, 251, 266, 361–64, 368–70, 374–76, 415, 435–37, 445. *See also mind mapping*.
   and PTSD, 367
   impacts
      long-term impacts, 167-184
      ripple through dysfunctional families, 314
      short-term impacts, 160-67
      the closer the relationship, the bigger the impact, 121–22
   resolving, 182, 215–72, 283, 293, 312
   treatment, 89–90, 120, 123, 155, 157, 240–241, 361, 365–67, 408, 414–17, 434–36, 439–41, 445–46
   who fails in therapy, 149–54
Trump, Donald, 186–89
"Trump Trauma", 186-189
trust, 3, 16, 57, 86, 89, 102, 161, 176, 209, 221, 235, 270, 280, 316, 328
truth, 1–2, 23, 76, 86, 99–101, 111–12, 139, 148, 189, 199, 206–7, 222, 229, 231, 308
two-choice dilemmas, 444

U

uncontrollable
   sobbing, 165
   tangential thinking, 370
   ticklishness, 372
unhappiness, 83, 146, 156, 182, 388-89
unmasking, 17, 328
unrepentant, 200
upsetting, 30–31, 41, 97, 102, 113, 150, 185, 189, 205–6, 215, 260, 281, 284, 291, 313
   enjoyed upsetting, 298, 310
Urania auditorium (Germany), 141

V

vagus nerve, 379, 442. *See also heart-rate variability.*
  vagal tone, 380-81, 441–42
validation, 96, 312, 322
van der Kolk, Bessel, 365, 369, 377, 417
vasopressin, 378
von Economo neuron (VEN), spindle cells, 386, 405, 410–411. *See also mirror neurons.*
verbally accessible memory (VAM), 283, 359, 419–22, 424, 429, 433, 446. *See also situationally accessible memory (SAM).*
  VAM memories, 283, 419–21
  making new, 438
  VAM system, 420, 422 VAM-SAM dual-representation systems, 433
veterans, combat, 116, 120, 172, 201, 416
vibes, sexual, 223, 226
victims, 122, 124–25, 128–29, 138, 140, 145–46, 148, 163, 270, 362, 365, 398, 426
video recordings, 20, 31, 157, 264–65, 271, 432
vigilance, 27, 40, 82, 198
violence, 121, 168
  domestic, 122, 142, 368–69, 443
virtual reality exposure techniques, 426
visceral responses, 135, 421
vision, 31, 208, 332
visualization, 29, 31, 35, 73–74, 121, 127, 155, 167–68, 216, 218, 220–221, 223–24, 227, 233, 240, 249, 252, 254, 258, 332, 363, 424-28, 437, 439–40. *See also Crucible® Neurobiological Therapy methods.*
  activities, 428
  memories, 423
  third-person perspective (fly on the wall), 234, 254
  traumatic mind mapping, 425
voice, 128, 174–75, 209, 216, 266, 371, 375
  hostile mental, 174–75
  irritating unmelodic, 373
  tone of, 209, 264
voice hearers, 174
  voice hearers association, 175
vulnerability, 148, 316

W

war, 342, 361–62
wear and tear on the body (allostatic load), 379
whales, 46, 410
Willi, Jürg, 103
willpower, 219, 237
Woodruff, Guy, 19–20
workplace bullying, 338
workplaces, 6, 124, 397
workshops, 3, 59, 171, 180–81
written mental dialogs, 266-71, 427–430. *See also Crucible® Neurobiological Therapy methods.*

Y

yelling, 38, 200, 240, 245, 257, 332
Yerkes-Dodson law, 445
yoga, 381, 417

Z

Zephyr™ BioModule™ system, 118

About the Author

David Schnarch, Ph.D. is a licensed clinical psychologist, Board Certified in Couple and Family Psychology, and recipient of the 2013 Award for Distinguished Contributions to Independent Practice from the American Psychological Association.

He is a long-time Clinical Member of the American Association for Marriage and Family Therapy and the recipient of the 2011 AAMFT Award for Distinguished Contributions to Marriage and Family Therapy.

David is also certified as a Sex Therapist (Diplomat status) by the American Association for Sex Educators, Counselors, and Therapists. He served on the AASECT Board of Directors and chaired professional education for eight years. He received the first AASECT Professional Standard of Excellence Award in 1994.

Over the last 30 years, David has a long-standing history of innovating novel approaches to psychotherapy. He is the founder of Crucible® Therapy, the first integrated marital and sexual therapy based on interpersonal differentiation. In 1991, he wrote *Constructing The Sexual Crucible: An Integration of Sexual and Marital Therapy*, still used as a primary textbook in graduate training programs. His subsequent books for couples–*Passionate Marriage* (1997), *Resurrecting Sex* (2002), and *Intimacy & Desire* (2009)–have been published in seven languages and become international best-sellers.

David co-directs the Crucible Institute of Evergreen, Colorado with his wife, psychologist Ruth Morehouse. David and Ruth conduct training workshops for therapists and programs for the public all around the world. For information, visit www.Crucible4Points.com or contact them at Service@CrucibleInstitute.com.

Printed in Great Britain
by Amazon